T0212526

Frontiers in Mathematics

More information about this series at http://www.springer.com/series/5388

Mohammad Masjed-Jamei

Special Functions and Generalized Sturm-Liouville Problems

 Birkhäuser

Mohammad Masjed-Jamei
Department of Mathematics
K.N. Toosi University of Technology
Tehran, Iran

ISSN 1660-8046   ISSN 1660-8054   (electronic)
Frontiers in Mathematics
ISBN 978-3-030-32819-1   ISBN 978-3-030-32820-7   (eBook)
https://doi.org/10.1007/978-3-030-32820-7

Mathematics Subject Classification (2010): 34B24, 33C45, 33C47

This book is published under the imprint Birkhäuser, www.birkhauser-science.com, by the registered company Springer Nature Switzerland AG.
The registered company address is: Gewerbestrasse 11, 6330 Cham, Switzerland

Preface

In recent years, Sturm–Liouville theory as an attractive field of research has received considerable attention by authors, since it appears in solving many problems of engineering, physics, biology, and the social sciences in a natural manner.

Such problems generally lead to some eigenvalue problems for ordinary and partial differential equations. For instance, the associated Legendre functions, Bessel functions, ultraspherical functions, Hermite functions, and trigonometric sequences related to Fourier analysis are particular solutions of some Sturm–Liouville problems. Most of these functions are symmetric and have found various applications in physics and engineering.

An extensive class of special functions includes orthogonal polynomials, whose history goes back to the eighteenth century and are closely linked with solutions of some eigenvalue problems. One of those problems was related to Newton's theory of gravity. While solving a problem in this area, A. M. Legendre introduced a sequence of orthogonal polynomials, which are known today as Legendre polynomials. Since then, many other families of continuous orthogonal polynomials have been introduced from their direct applications in various problems. Clearly, all these families can be introduced by diverse approaches. For instance, in 1929, S. Bochner found all families of polynomials satisfying a second-order differential equation, which led to the classical continuous orthogonal polynomials. In this sense, some famous classical orthogonal polynomials associated with the names of Jacobi, Laguerre, Hermite, and Bessel have been studied in the literature in detail, and their weight functions are known nowadays as Pearson distributions. More precisely, the boundary value problems corresponding to the above-mentioned cases have been considered as examples of a singular Sturm–Liouville problem to introduce classical orthogonal polynomials as solutions of a second-order linear hypergeometric differential equation.

On the other hand, in modern physics we encounter mathematical operators, eigenvalue equations, and properties of orthogonal functions. The quantum harmonic oscillator necessitates constructing an eigenvalue problem and the appearance of the discrete nature of physical quantities in quantum physics, which is considered a special case of orthogonal functions arising in the context of Sturm–Liouville theory. In other words, many important special functions are solutions of a regular or singular Sturm–Liouville problem that satisfy an orthogonality relation, which can be considered in different continuous and discrete spaces and also in one particular case of them, i.e., q-spaces.

The present book deals mainly with introducing and classifying some generalized Sturm–Liouville problems in three different continuous, discrete, and q-discrete spaces.

In this direction, some conditions under which the usual Sturm–Liouville problems with symmetric solutions can be extended to a larger class are presented.

Of course, the main motivation to write this monograph comes not only from various applications of Sturm–Liouville problems in mathematical physics, but also from the necessity of gathering and collecting the theoretical aspects of the subject and its generalization in a unified structure in one place, with special attention to the solutions of the generalized problems as new orthogonal sequences of continuous or discrete functions.

This book includes three chapters. The first chapter is devoted to introducing a principal class of symmetric orthogonal functions and their generalizations by presenting a generalized Sturm–Liouville theorem for symmetric functions. Some sequences of orthogonal functions of a discrete variable are introduced in the second chapter by presenting a similar theorem, which is now expressed in discrete spaces. New classes of q-orthogonal polynomials that are generated by some q-Sturm–Liouville problems are considered in Chap. 3.

In Chap. 1, to study six sequences of hypergeometric orthogonal polynomials, we first show that the classical Pearson distributions and Gauss's hypergeometric function satisfy a unique differential equation of hypergeometric type. Three polynomials, namely Jacobi, Laguerre, and Hermite, are known as infinite classical orthogonal polynomials, whereas three other sequences are called finite hypergeometric orthogonal polynomials, which are orthogonal with respect to the generalized T, inverse gamma, and F distributions respectively. Besides reviewing these classical cases in detail, a class of symmetric orthogonal polynomials with four free parameters is introduced in Chap. 1 and all its standard properties, such as a generic second-order differential equation together with its explicit polynomial solution, a generic orthogonality relation, an analogue of Pearson distributions as their weight function, and a generic three-term recurrence relation, are obtained. Essentially, four sequences of symmetric orthogonal polynomials can be extracted from the introduced class. They are respectively the generalized ultraspherical polynomials, generalized Hermite polynomials, and two other sequences of symmetric polynomials, which are finitely orthogonal on the real line. For instance Chebyshev polynomials of the fifth and sixth kinds can be defined as special cases of these generalizations. We know that four kinds of trigonometric orthogonal polynomials, i.e., the first, second, third, and fourth kinds of Chebyshev polynomials have been investigated in the literature up to now. In Chap. 1, we introduce two further kinds of half-trigonometric orthogonal polynomials and call them the fifth and sixth kinds of Chebyshev polynomials, since they are generated by employing the Chebyshev polynomials of the first and second kinds.

In connection to the aforementioned symmetric orthogonal class with four free parameters and according to the fact that some orthogonal polynomial systems are mapped onto each other by the Fourier transform, we compute the Fourier transforms of four introduced sequences of symmetric orthogonal polynomials and obtain their orthogonality relations via Parseval's identity.

In the sequel, using the generalized Sturm–Liouville theorem for symmetric functions, a basic class of symmetric orthogonal functions and also its generalization with six parameters are introduced. Then four orthogonal special cases of these classes are represented, and their properties are studied in detail. It must be mentioned that further important cases of these classes can still be found, which becomes a motivation for further investigation into the orthogonal sequences. Just as an example, two cases that generalize Fourier trigonometric sequences and are orthogonal with respect to the constant weight function are introduced. They are useful in finding some new trigonometric series.

As another application of the key theorem related to the extension of Sturm–Liouville problems with symmetric solutions, two specific differential equations are considered, one of whose basis solutions yields two classes of incomplete symmetric orthogonal polynomials. Such systems do not contain polynomials of every degree and therefore do not have all properties as in the classical cases. However, they can be directly applied to approximation theory of functions, since their norm square values are explicitly computed.

In Chap. 2, orthogonal functions of a discrete variable are considered as solutions of a discrete Sturm–Liouville problem that satisfies a set of discrete boundary conditions. Interest in discrete Sturm–Liouville operators comes from a number of problems in mathematical physics, scattering theory, Toda and Langmuir chains, spectral properties of operators and so on.

Similar to the continuous case, special functions of a discrete variable have found various applications in mathematics and physics. For instance, Gram polynomials and symmetric Hahn polynomials play an important role in some branches of numerical analysis, while symmetric Kravchuk polynomials appear in the so-called Fourier–Kravchuk transform used in optics and quantum mechanics as approximations of harmonic oscillator wave functions. Also, some applications of classical orthogonal polynomials of a discrete variable and their zeros can be observed in the least-squares method of approximation, queueing theory, and cross-directional control on paper machines and codes.

It is proved in Chap. 2 that the previous extensions for the continuous case also hold for discrete variables and particularly for a homogeneous second-order difference equation. Hence, by using a generalization of Sturm–Liouville problems in discrete spaces, a symmetric class of orthogonal polynomials of a discrete variable, which generalizes all classical discrete symmetric orthogonal polynomials, is introduced, and its standard properties, such as a second-order difference equation, an explicit form for polynomials, a three-term recurrence relation, and an orthogonality relation, are obtained.

As mentioned earlier, one of the important classes of special functions is q-orthogonal polynomials, which can be considered solutions of a q-Sturm–Liouville problem of regular or singular type. This type of polynomial has found various applications in quantum mechanics, the q-Schrödinger equation, q-harmonic oscillators, and algebraic combinatorics, including coding theory and theories of group representation. Therefore, it seems worthwhile to investigate new developments of the main ideas applied in Chaps. 1

and 2 in this field, too. To reach this goal, in Chap. 3, some new classes of q-orthogonal polynomials are defined through new q-Sturm–Liouville problems, and then their basic properties are studied.

It should be mentioned that each chapter provides valuable references in the section "Further Reading" for those who need some additional sources and more detailed materials.

It is hoped that this work will be useful as a coherent scientific reference for mathematicians, physicists, and engineers to carry out their research projects and to find new applications that effectively develop previous results. In this sense, any comments, suggestions, and corrections would be much appreciated.

Finally, I would like to present my special thanks to Zahra Moalemi for carefully reading this book and for her valuable assistance in typesetting. As a Humboldtian fellow for 18 months, I would also like to thank my scientific host Wolfram Koepf for his warm hospitality. The finalization of this project was made possible by the financial support of the Alexander von Humboldt foundation under the grant No. Ref 3.4-IRN-1128637-GF-E.

Tehran, Iran
December 2018

Mohammad Masjed-Jamei

Contents

Special Functions Generated by Generalized Sturm–Liouville Problems in Continuous Spaces

1.1 Introduction

Consider the second-order differential equation

$$\frac{d}{dx}\left(k(x)\frac{dy}{dx}\right) + (\lambda\rho(x) - q(x))y = 0, \quad \text{where} \quad k(x) > 0 \text{ and } \rho(x) > 0, \quad (1.1)$$

on an open interval, say (a, b), with the following boundary conditions,

$$\alpha_1 y(a) + \beta_1 y'(a) = 0,$$
$$\alpha_2 y(b) + \beta_2 y'(b) = 0, \quad (1.2)$$

in which α_1, α_2 and β_1, β_2 are given constants and $k(x)$, $k'(x)$, $q(x)$, and $\rho(x)$ in (1.1) are to be assumed continuous for $x \in [a, b]$.

The boundary value problem (1.1)–(1.2) is called a regular Sturm–Liouville problem, and if one of the points a and b is singular (i.e., $k(a) = 0$ or $k(b) = 0$), it is called a singular Sturm–Liouville problem. In the latter case, boundary conditions (1.2) can be ignored.

Let $y_n(x)$ and $y_m(x)$ be two solutions (eigenfunctions) of Eq. (1.1). Following Sturm–Liouville theory, these functions are orthogonal with respect to the positive weight function $\rho(x)$ on (a, b) under the given conditions (1.2), i.e.,

$$\int_a^b \rho(x)y_n(x)y_m(x)\, dx = \left(\int_a^b \rho(x)y_n^2(x)\, dx\right)\delta_{n,m}, \quad (1.3)$$

© Springer Nature Switzerland AG 2020
M. Masjed-Jamei, *Special Functions and Generalized Sturm–Liouville Problems*,
Frontiers in Mathematics, https://doi.org/10.1007/978-3-030-32820-7_1

where

$$\delta_{n,m} = \begin{cases} 0 & (n \neq m), \\ 1 & (n = m). \end{cases}$$

Many special functions in theoretical and mathematical physics are solutions of a regular or singular Sturm–Liouville problem satisfying the orthogonality condition (1.3). For instance, the associated Legendre functions, Bessel functions, trigonometric sequences related to Fourier analysis, ultraspherical functions, and Hermite functions are particular solutions of a Sturm–Liouville problem. Most of these functions are symmetric and have found interesting applications in physics and engineering. Hence if one can extend them symmetrically and preserve their orthogonality property, new applications can be derived that logically extend the previous known applications.

In this chapter, by achieving this goal, we first review some primary definitions by considering a hypergeometric-type differential equation and study a relationship between classical Pearson distributions and Gauss hypergeometric functions. Then, by introducing a generalized Sturm–Liouville problem of symmetric type in continuous spaces, we extend some classical symmetric orthogonal functions and obtain their orthogonality properties.

1.2   Pearson Distribution Families and Gauss Hypergeometric Functions

Consider the homogeneous differential equation

$$\sigma(x)\, y''(x) + \tau(x)\, y'(x) + \lambda\, y(x) = 0, \tag{1.4}$$

in which $\sigma(x) = ax^2 + bx + c$, $\tau(x) = dx + e$; a, b, c, d, e are free parameters, and λ is a constant depending on a, b, c, d, e.

This equation is called a hypergeometric-type differential equation, because its particular solutions can be indicated in terms of Gauss hypergeometric functions. In other words, if in (1.4),

$$\sigma(x) = x(1-x) \quad \text{and} \quad \tau(x) = \gamma - (\alpha + \beta + 1)\, x,$$

the Gauss hypergeometric differential equation

$$x(1-x)\, y''(x) + (\gamma - (\alpha + \beta + 1)x)\, y'(x) - \alpha\beta\, y(x) = 0 \tag{1.5}$$

appears, where α, β, and γ are real parameters. The indicial equation corresponding to Eq. (1.5) is

$$r^2 - (1 - \gamma)r = 0,$$

with two roots $r_1 = 0$ and $r_2 = 1 - \gamma$.

Using the Frobenius method, we obtain a series solution of Eq. (1.5) for $r_1 = 0$ as follows:

$$y_1(x) = 1 + \frac{\alpha\beta}{\gamma}\frac{x}{1!} + \frac{\alpha(\alpha+1)\beta(\beta+1)}{\gamma(\gamma+1)}\frac{x^2}{2!} + \frac{\alpha(\alpha+1)(\alpha+2)\beta(\beta+1)(\beta+2)}{\gamma(\gamma+1)(\gamma+2)}\frac{x^3}{3!} + \cdots,$$
(1.6)

where $\gamma \neq 0, -1, -2, -3, \ldots$ and the series is convergent for all $x \in [-1, 1]$.

The series (1.6) is known in the literature as a Gauss hypergeometric series, and its sum, denoted by ${}_2F_1\left(\begin{array}{cc} \alpha & \beta \\ & \gamma \end{array} \middle| x\right)$ or ${}_2F_1(\alpha, \beta, \gamma; x)$, is called the Gauss hypergeometric function. Hence, we have

$${}_2F_1\left(\begin{array}{cc} \alpha & \beta \\ & \gamma \end{array} \middle| x\right) = \sum_{k=0}^{\infty} \frac{(\alpha)_k(\beta)_k}{(\gamma)_k}\frac{x^k}{k!},$$
(1.7)

in which

$$(\alpha)_n = \prod_{j=0}^{n-1}(\alpha + j).$$

Notice that (1.7) is a special case of the generalized hypergeometric function

$${}_pF_q\left(\begin{array}{cccc} a_1 & a_2 & \ldots & a_p \\ b_1 & b_2 & \ldots & b_q \end{array} \middle| z\right) = \sum_{k=0}^{\infty} \frac{(a_1)_k(a_2)_k \cdots (a_p)_k}{(b_1)_k(b_2)_k \cdots (b_q)_k}\frac{z^k}{k!},$$
(1.8)

in which z may be a complex variable. The infinite sum (1.8) is indeed a Taylor series expansion of a function, say f, namely $\sum_{k=0}^{\infty} c_k^* z^k$ with $c_k^* = f^{(k)}(0)/k!$, for which the ratio of successive terms can be written as

$$\frac{c_{k+1}^*}{c_k^*} = \frac{(k + a_1)(k + a_2) \cdots (k + a_p)}{(k + b_1)(k + b_2) \cdots (k + b_q)(k + 1)}.$$

According to the ratio test, the hypergeometric series (1.8) is convergent for all $p \leq q + 1$. In a more precise expression, it converges in $|z| < 1$ for $p = q + 1$, converges everywhere for $p < q + 1$, and converges nowhere ($z \neq 0$) for $p > q + 1$. Moreover, for $p = q + 1$ it converges absolutely for $|z| = 1$ if the condition

$$A^* = \mathrm{Re}\left(\sum_{j=1}^{q} b_j - \sum_{j=1}^{q+1} a_j \right) > 0$$

holds and is conditionally convergent for $|z| = 1$ and $z \neq 1$ if $-1 < A^* \leq 0$, and is divergent for $|z| = 1$ and $z \neq 1$ if $A^* \leq -1$.

Since the Gauss hypergeometric function has an integral representation of the form

$$_2F_1\left(\begin{matrix} a , b \\ c \end{matrix} \middle| x \right) = \frac{\Gamma(c)}{\Gamma(c-b)\,\Gamma(b)} \int_0^1 t^{b-1}(1-t)^{c-b-1}(1-tx)^{-a}\,dt , \qquad (1.9)$$

according to Beta integral

$$B(\lambda_1, \lambda_2) = \int_0^1 x^{\lambda_1-1}(1-x)^{\lambda_2-1}dx = \int_{-1}^1 x^{2\lambda_1-1}(1-x^2)^{\lambda_2-1}dx = \int_0^\infty \frac{x^{\lambda_1-1}}{(1+x)^{\lambda_1+\lambda_2}}dx$$

$$= 2 \int_0^{\pi/2} \sin^{(2\lambda_1-1)} x \cos^{(2\lambda_2-1)} x \, dx = \frac{\Gamma(\lambda_1)\Gamma(\lambda_2)}{\Gamma(\lambda_1+\lambda_2)} = B(\lambda_2 , \lambda_1),$$

where $\Gamma(\lambda)$ denotes the well-known gamma function

$$\Gamma(\lambda) = \int_0^\infty x^{\lambda-1}e^{-x}dx \qquad \text{for} \quad \mathrm{Re}\,(\lambda) > 0,$$

we can conclude that

$$_2F_1\left(\begin{matrix} a , b \\ c \end{matrix} \middle| 1 \right) = \frac{\Gamma(c)\,\Gamma(c-b-a)}{\Gamma(c-b)\,\Gamma(c-a)}. \qquad (1.10)$$

One of the properties of the series (1.7) is that many elementary functions or special functions of mathematical physics can be directly expressed in terms of it. For instance, we have

$$\ln(1+x) = x\,_2F_1(1, 1, 2; -x),$$

$$\sec x = \,_2F_1\left(\frac{1}{2}, 1, 1; \sin^2 x \right),$$

$$\arcsin x = x\,_2F_1\left(\frac{1}{2}, \frac{1}{2}, \frac{3}{2}; x^2 \right),$$

$$\arctan x = x\,_2F_1\left(\frac{1}{2}, 1, \frac{3}{2}; -x^2 \right),$$

and/or the incomplete beta function defined by

$$B_x(a, b) = \int_0^x t^{a-1}(1-t)^{b-1}dt = \frac{1}{a}x^a{}_2F_1(a, 1-b, a+1; x).$$

Another important special function having a direct relationship with the Gauss hypergeometric function (1.7) is the family of Pearson distributions, which plays a key role in the theory of classical orthogonal polynomials. If Eq. (1.4) is written in a self-adjoint form, then the Pearson differential equation

$$\frac{d}{dx}(\sigma(x)W(x)) = \tau(x)W(x) \tag{1.11}$$

appears, whose solutions are indeed the weight functions of classical orthogonal polynomial solutions of Eq. (1.4). By solving the first-order Eq. (1.11), the Pearson distributions family appears as

$$W(x) = W\left(\begin{array}{cc|c} d & e & \\ a & b & c \end{array}x\right) = K\exp\left(\int\frac{(d-2a)x+e-b}{ax^2+bx+c}dx\right), \tag{1.12}$$

where K is a normalizing constant.

For convenience, if in (1.12) we put

$$d - 2a = d^* \quad \text{and} \quad e - b = e^*,$$

then

$$(ax^2 + bx + c)y' - (d^*x + e^*)y = 0$$

$$\Rightarrow y = W^*\left(\begin{array}{cc|c} d^* & e^* & \\ a & b & c \end{array}x\right) = K\exp\left(\int\frac{d^*x+e^*}{ax^2+bx+c}dx\right). \tag{1.13}$$

Now taking the derivative on both sides of Eq. (1.13) yields

$$(ax^2 + bx + c)y'' + ((2a - d^*)x + (b - e^*))y' - d^*y = 0, \tag{1.14}$$

which can be transformed to a special case of the Gauss equation (1.5) if we suppose $x = pt + q$ and replace it in Eq. (1.14) to get

$$\left(t^2 + \frac{2aq + b}{ap}t + \frac{aq^2 + bq + c}{ap^2}\right)\frac{d^2y}{dt^2}$$

$$+ \left(\frac{2a - d^*}{a}t + \frac{(2a - d^*)q + b - e^*}{ap}\right)\frac{dy}{dt} - \frac{d^*}{a}y = 0, \tag{1.15}$$

provided that $ap \neq 0$. Moreover, if in (1.15) we take

$$aq^2 + bq + c = 0 \quad \text{and} \quad \frac{2aq + b}{ap} = -1, \tag{1.16}$$

we obtain

$$p = -\frac{\sqrt{\Delta}}{a} \quad \text{and} \quad q = \frac{-b + \sqrt{\Delta}}{2a}, \tag{1.17}$$

where $\Delta = b^2 - 4ac$. Hence, the differential equation (1.15) is simplified as

$$t(t-1)\frac{d^2y}{dt^2} + \left(\left(2 - \frac{d^*}{a}\right)t + \frac{2ae^* - bd^* + (d^* - 2a)\sqrt{\Delta}}{2a\sqrt{\Delta}}\right)\frac{dy}{dt} - \frac{d^*}{a}y = 0. \tag{1.18}$$

In conclusion, according to relations (1.13)–(1.17), one of the general solutions of Eq. (1.18) must be

$$y = W^* \left(\begin{array}{cc} d^* & e^* \\ a \ b \ c \end{array} \middle| -\frac{\sqrt{\Delta}}{a}t + \frac{-b + \sqrt{\Delta}}{2a}\right). \tag{1.19}$$

On the other hand, we mentioned that Eq. (1.18) could be a special case of the equation

$$t(t-1)\frac{d^2y}{dt^2} + \left((\alpha + \beta + 1)t - \gamma\right)\frac{dy}{dt} + \alpha\beta\, y = 0, \tag{1.20}$$

with the general solution

$$y = A\,{}_2F_1\left(\begin{array}{cc} \alpha \ \beta \\ \gamma \end{array} \middle| t\right) + B\,t^{1-\gamma}{}_2F_1\left(\begin{array}{c} \alpha + 1 - \gamma,\ \beta + 1 - \gamma \\ 2 - \gamma \end{array} \middle| t\right),$$

in which γ, $\alpha - \beta$, and $\gamma - \alpha - \beta$ are all nonintegers and A, B are two constants. Therefore, to find the parameters α, β, and γ in Eq. (1.20) in terms of the parameters a, b, c and d^*, e^* in Eq. (1.18), we should just equate the two equations, which leads to the system

$$\alpha + \beta = 1 - \frac{d^*}{a} \quad \text{and} \quad \alpha\beta = -\frac{d^*}{a}. \tag{1.21}$$

By solving the above system and comparing (1.18) with (1.20), we get

$$\alpha = 1\ ,\quad \beta = -\frac{d^*}{a},\quad \text{and}\quad \gamma = 1 - \frac{d^*}{2a} - \frac{2ae^* - bd^*}{2a\sqrt{\Delta}}.$$

Note that since (1.21) is a symmetric system, the second solution reads as

$$\alpha = -\frac{d^*}{a} \ , \quad \beta = 1, \quad \text{and} \quad \gamma = 1 - \frac{d^*}{2a} - \frac{2ae^* - bd^*}{2a\sqrt{\Delta}}.$$

Consequently, the general solution of Eq. (1.18) is

$$y = A^*{}_2F_1 \left(\begin{array}{c} 1, \ -d^*/a \\ 1 - \frac{d^*}{2a} - \frac{2ae^* - bd^*}{2a\sqrt{\Delta}} \end{array} \middle| \ t \right)$$
$$+ B^* t^{(\frac{d^*\sqrt{\Delta} + 2ae^* - bd^*}{2a\sqrt{\Delta}})}{}_2F_1 \left(\begin{array}{c} 1 + \frac{d^*}{2a} + \frac{2ae^* - bd^*}{2a\sqrt{\Delta}}, \ \frac{-d^*\sqrt{\Delta} + 2ae^* - bd^*}{2a\sqrt{\Delta}} \\ 1 + \frac{d^*}{2a} + \frac{2ae^* - bd^*}{2a\sqrt{\Delta}} \end{array} \middle| \ t \right). \qquad (1.22)$$

But (1.22) can be further simplified, because in the second term of the right-hand side, we generally have

$$_2F_1 \left(\begin{array}{cc} u & v \\ & u \end{array} \middle| \ x \right) = \sum_{k=0}^{\infty} \frac{(v)_k}{k!} x^k = (1 - x)^{-v}.$$

So the final form would be

$$y = A^*{}_2F_1 \left(\begin{array}{c} 1, \ -d^*/a \\ 1 - \frac{d^*}{2a} - \frac{2ae^* - bd^*}{2a\sqrt{\Delta}} \end{array} \middle| \ t \right) + B^* t^{(\frac{d^*\sqrt{\Delta} + 2ae^* - bd^*}{2a\sqrt{\Delta}})} (1 - t)^{(\frac{d^*\sqrt{\Delta} - (2ae^* - bd^*)}{2a\sqrt{\Delta}})}.$$

$$(1.23)$$

On the other hand, we observed that the function (1.19) was also a solution of Eq. (1.18). This means that there must be a direct relationship between (1.23) and (1.19). To find it, without loss of generality, we assume in (1.13) that $ax^2 + bx + c = a(x + \theta_1)(x + \theta_2)$, in which

$$\theta_1 = \frac{b - \sqrt{\Delta}}{2a} \quad \text{and} \quad \theta_2 = \frac{b + \sqrt{\Delta}}{2a} \quad \text{with} \quad \Delta = b^2 - 4ac.$$

The latter relation implies that

$$W^* \left(\begin{array}{cc} d^* & e^* \\ a \ b & c \end{array} \middle| \ x \right) = R \, (x + \theta_1)^r (x + \theta_2)^s, \qquad (1.24)$$

where R is a constant and

$$r = \frac{d^*\sqrt{\Delta} + 2ae^* - bd^*}{2a\sqrt{\Delta}} \quad \text{and} \quad s = \frac{d^*\sqrt{\Delta} - (2ae^* - bd^*)}{2a\sqrt{\Delta}}. \qquad (1.25)$$

Equality (1.24) is valid because the logarithmic derivative of the function

$$\rho(x) = (x + \theta_1)^r (x + \theta_2)^s$$

equals the logarithmic derivative of $W^*(a, b, c, d^*, e^*; x)$ defined in (1.13), and since

$$\frac{u'(x)}{u(x)} = \frac{v'(x)}{v(x)} \Leftrightarrow u(x) = R\ v(x), \tag{1.26}$$

equality (1.24) holds.

Now, by noting (1.24)–(1.26), the function (1.19) is simplified as

$$W^*\left(\begin{array}{cc} d^* & e^* \\ a & b & c \end{array}\middle| -\frac{\sqrt{\Delta}}{a}t - \theta_1\right) = R(-\frac{\sqrt{\Delta}}{a}t)^r(-\frac{\sqrt{\Delta}}{a}t + \theta_2 - \theta_1)^s$$

$$= R(-1)^r(\frac{\sqrt{\Delta}}{a})^{r+s}\ t^r(1-t)^s,$$

which gives exactly the second term of (1.23) for $B^* = R(-1)^r(\frac{\sqrt{\Delta}}{a})^{r+s}$.

This means that

$$y_1(t) = t^{(\frac{d^*\sqrt{\Delta}+2ae^*-bd^*}{2a\sqrt{\Delta}})}(1-t)^{(\frac{d^*\sqrt{\Delta}-(2ae^*-bd^*)}{2a\sqrt{\Delta}})},$$

$$y_2(t) = {}_2F_1\left(\begin{array}{c} 1, -d^*/a \\ 1 - \frac{d^*}{2a} - \frac{2ae^*-bd^*}{2a\sqrt{\Delta}} \end{array}\middle|\ t\right), \tag{1.27}$$

constitute a solution basis for Eq. (1.18).

The final point is that according to the theory of ordinary differential equations, if $Y(x)$ satisfies a homogeneous second-order differential equation of the form

$$a(x)\ Y''(x) + b(x)\ Y'(x) + c(x)\ Y(x) = 0,$$

such that it has the basis solutions $y_1(x)$ and $y_2(x)$, then

$$Y(x) = c_1 y_1(x) + c_2 y_2(x),$$

where

$$y_2(x) = N^* y_1(x) \int_\lambda^x (y_1(t))^{-2} \exp\left(\int -\frac{b(t)}{a(t)}\,dt\right)dt, \tag{1.28}$$

N^* is an arbitrary constant, and λ is a parameter independent of x.

In this sense, note that the values N^* and λ can be explicitly obtained. For this purpose, it is enough to put $x = \lambda$ in (1.28) to get

$$y_2(\lambda) = N^* y_1(\lambda) \int_\lambda^\lambda (y_1(t))^{-2} \exp\left(\int -\frac{b(t)}{a(t)}\, dt\right) dt = 0.$$

Therefore, λ must be one of the known roots of the second basis solution. To obtain N^*, since

$$N^* = \frac{y_2(x)}{y_1(x)\, \int_{y_2^{-1}(0)}^x (y_1(t))^{-2} \exp\left(\int -\frac{b(t)}{a(t)}\, dt\right) dt},$$

the above equality should be valid for all values of $x = \theta$ where

$$\theta \in S = \{x \mid x \in D\, y_1(x)\, \cap D\, y_2(x)\ \&\ y_1(x) \neq 0\ \&\ y_2(x) \neq 0\}. \tag{1.29}$$

All these results eventually simplify (1.28) as

$$y_2(x) = \frac{y_2(\theta)}{y_1(\theta)\, \int_{y_2^{-1}(0)}^\theta (y_1(t))^{-2} \exp\left(\int -\frac{b(t)}{a(t)}\, dt\right) dt}\, y_1(x)$$

$$\times \int_{y_2^{-1}(0)}^x (y_1(t))^{-2} \exp\left(\int -\frac{b(t)}{a(t)}\, dt\right) dt, \tag{1.30}$$

for all $\theta \in S$ defined in (1.29).

Using the latter result and (1.27), we can now determine direct relationships between the Pearson distributions family and Gauss hypergeometric functions. According to Table 1.1, there are six special cases of the Pearson distributions family that appear in the theory of classical orthogonal polynomials as the corresponding weight functions.

For example, consider the generalized T weight function

$$W_1(x; u, v) = (1 + x^2)^{-u} \exp(2v \arctan x),$$

Table 1.1 Special cases of the Pearson distributions family

Distribution name	Definition	Interval
1. Shifted beta	$(1 - x)^u (1 + x)^v$	$[-1, 1]$
2. Gamma	$x^u \exp(-x)$	$[0, \infty)$
3. Normal	$\exp(-x^2)$	$(-\infty, \infty)$
4. Fisher F	$x^v / (1 + x)^{u+v}$	$[0, \infty)$
5. Student's-t	$(1 + x^2)^{-u}$	$(-\infty, \infty)$
5*. Generalized t	$(1 + x^2)^{-u} \exp(v \arctan x)$	$(-\infty, \infty)$
6. Inverse gamma	$x^{-u} \exp(-1/x)$	$[0, \infty)$

for $u > 0$ and $v \in \mathbb{R}$, where $x \in (-\infty, \infty)$. Taking the logarithmic derivative on both sides and then comparing with (1.13) eventually yields

$$a = 1 \, , \ b = 0 \, , \ c = 1 \, , \ d^* = -2u, \ \text{and} \ e^* = 2v.$$

Hence

$$W_1(x; u, v) = K_1 \, W^* \left(\begin{array}{cc|c} -2u & 2v & \\ 1 & 0 & 1 \end{array} \, x \right),$$

which, according to (1.14), satisfies the equation

$$(1 + x^2) \, W_1'' \, (x; u, v) + ((2u + 2)x - 2v) \, W_1'(x; u, v) + 2u \, W_1(x; u, v) = 0. \quad (1.31)$$

On the other hand, since in this example

$$\Delta = b^2 - 4ac = -4 \, , \quad p = -\sqrt{-4} = -2i, \quad \text{and} \quad q = \sqrt{-4}/2 = i \, ,$$

by applying the change of variable $x = -2\,i\,t + i$ in (1.31) (or equivalently $t = \frac{i}{2}x + \frac{1}{2}$ in the Gauss hypergeometric equation (1.20)), it turns out from (1.23) that

$$W_1(x; u, v) = (1 + x^2)^{-u} \exp(2v \arctan x)$$

$$= A \, {}_2F_1 \left(\begin{array}{c|c} 1, \, 2u & \\ 1 + u + iv & \end{array} \frac{i}{2}x + \frac{1}{2} \right) + B \, (\frac{i}{2}x + \frac{1}{2})^{-u-iv}(-\frac{i}{2}x + \frac{1}{2})^{-u+iv}. \quad (1.32)$$

Since the second term on the right-hand side of (1.32) can be simplified as

$$\left(\frac{i}{2}x + \frac{1}{2}\right)^{-u-iv} \left(-\frac{i}{2}x + \frac{1}{2}\right)^{-u+iv} = 2^{2u}(1 + x^2)^{-u} \exp(2v \arctan x), \quad (1.33)$$

it eventually follows that

$$y_1(x) = (1 + x^2)^{-u} \exp(2v \arctan x) \quad \text{and} \quad y_2(x) = {}_2F_1 \left(\begin{array}{c|c} 1, \, 2u & \\ 1 + u + iv & \end{array} \frac{i}{2}x + \frac{1}{2} \right)$$

$$(1.34)$$

are two basis solutions of Eq. (1.31).

It is interesting to observe that the Taylor expansion of the generalized T distribution function is explicitly computable, because if we reconsider (1.33) in the form

$$(1 + ix)^{-u-iv}(1 - ix)^{-u+iv} = (1 + x^2)^{-u} \exp(2v \arctan x), \quad (1.35)$$

then according to the Leibniz rule

$$\frac{d^n(f(x)g(x))}{dx^n} = \sum_{k=0}^{n} \binom{n}{k} f^{(k)}(x) g^{(n-k)}(x),$$

the following identity is valid:

$$\left.\frac{d^n((1+ix)^{-u-iv}(1-ix)^{-u+iv})}{dx^n}\right|_{x=0} = i^n \sum_{k=0}^{n} (-1)^k \binom{n}{k} (u+iv)_k (u-iv)_{n-k}$$

$$= i^n (u-iv)_n \, {}_2F_1 \left(\begin{array}{c} -n \; u+iv \\ 1-n-u+iv \end{array}\middle| -1\right).$$

Therefore

$$(1+x^2)^{-u} \exp(2v \arctan x) = \sum_{n=0}^{\infty} \left(\sum_{k=0}^{n} (-1)^k \binom{n}{k} (u+iv)_k (u-iv)_{n-k}\right) \frac{(ix)^n}{n!}$$

$$= \sum_{n=0}^{\infty} {}_2F_1 \left(\begin{array}{c} -n \; u+iv \\ 1-n-u+iv \end{array}\middle| -1\right) (u-iv)_n \frac{(ix)^n}{n!}.$$

$$(1.36)$$

Moreover, if in (1.35), $ix = z$, $iv \to v$, $-u+v = \alpha$, and $-u-v = \beta$, then (1.36) changes to

$$(1-z)^{\alpha}(1+z)^{\beta} = \sum_{n=0}^{\infty} \left(\sum_{k=0}^{n} (-1)^k \binom{n}{k} (-\beta)_k (-\alpha)_{n-k}\right) \frac{z^n}{n!}$$

$$= \sum_{n=0}^{\infty} {}_2F_1 \left(\begin{array}{c} -n \; -\beta \\ 1-n+\alpha \end{array}\middle| -1\right) (-\alpha)_n \frac{z^n}{n!}. \qquad (1.37)$$

Hence, integrating both sides of (1.37) on $[-1, 1]$ gives

$$\int_{-1}^{1} (1-z)^{\alpha}(1+z)^{\beta} \, dz = \sum_{n=0}^{\infty} {}_2F_1 \left(\begin{array}{c} -n \; -\beta \\ 1-n+\alpha \end{array}\middle| -1\right) (-\alpha)_n \frac{1+(-1)^n}{(n+1)!},$$

which is simplified as

$$\sum_{n=0}^{\infty} {}_2F_1 \left(\begin{array}{c} -2n \; -\beta \\ 1-2n+\alpha \end{array}\middle| -1\right) \frac{(-\alpha)_{2n}}{(2n+1)!} = \frac{1}{2} \int_{-1}^{1} (1-z)^{\alpha}(1+z)^{\beta} \, dz$$

$$= 2^{\alpha+\beta} \frac{\Gamma(\alpha+1)\Gamma(\beta+1)}{\Gamma(\alpha+\beta+2)}.$$

Although (1.37) shows the explicit form of the Taylor expansion of the shifted beta distribution at $z = 0$, the real beta distribution function is given as

$$W_2(x; u, v) = x^u(1 - x)^v \qquad (0 \le x \le 1),$$

which is a particular case of the general distribution (1.13) for

$$a = 1, \ b = -1, \ c = 0, \ d^* = u + v, \ \text{and} \ e^* = -u. \tag{1.38}$$

The relation (1.38) implies that

$$W_2(x; u, v) = K_2 \, W^* \left(\begin{matrix} u + v & -u \\ 1 & -1 \end{matrix} \ \begin{matrix} \\ 0 \end{matrix} \ \middle| \ x \right).$$

Consequently, if the aforementioned parameters are substituted into (1.14), then the equation

$$x(x - 1) \, y'' + \left((2 - u - v)x + v - 1 \right) y' - (u + v) \, y = 0$$

has the following two basis solutions:

$$y_1(x) = W_2(1 - x; u, v) = (1 - x)^u x^v \ \text{and} \ y_2(x) = {}_2F_1 \left(\begin{matrix} 1, -u - v \\ 1 - v \end{matrix} \ \middle| \ x \right).$$

1.3   Six Sequences of Hypergeometric Orthogonal Polynomials

Let us begin this section with a special case of Eq. (1.4) in the form

$$\sigma(x) y_n''(x) + \tau(x) y_n'(x) - \lambda_n y_n(x) = 0, \tag{1.39}$$

where

$$\sigma(x) = ax^2 + bx + c \ \text{and} \ \tau(x) = dx + e$$

are independent of n and

$$\lambda_n = n\big(d + (n - 1)a\big)$$

is the eigenvalue depending on $n = 0, 1, 2, \ldots$.

It is well known in the literature that the classical hypergeometric polynomials of Jacobi, Laguerre, and Hermite are infinite types of polynomial solutions of Eq. (1.39) that are infinitely orthogonal. However, there are three further sequences of hypergeometric polynomials that are solutions of the above equation but finitely orthogonal with respect to three specific weight functions. This means that there are altogether six sequences of hypergeometric orthogonal polynomials as follows.

1.3.1   Jacobi Polynomials

If $\sigma(x) = 1 - x^2$ and $\tau(x) = -(\alpha + \beta + 2)x + (\beta - \alpha)$ are substituted into (1.39), we obtains

- Differential equation

$$(1 - x^2)y_n''(x) - \left((\alpha + \beta + 2)x + \alpha - \beta\right)y_n'(x) + n(\alpha + \beta + n + 1)y_n(x) = 0,$$

with the polynomial solution

$$P_n^{(\alpha,\beta)}(x) = \frac{1}{2^n}\sum_{k=0}^{n}\binom{n+\alpha}{k}\binom{n+\beta}{n-k}(x-1)^{n-k}(x+1)^k.$$

- Weight function:

$$w(x) = (1-x)^\alpha(1+x)^\beta \quad \text{for } x \in [-1, 1] \quad \text{and } \alpha, \beta > -1.$$

- Rodrigues representation:

$$P_n^{(\alpha,\beta)}(x) = \frac{(-1)^n}{2^n(1-x)^\alpha(1+x)^\beta}\frac{d^n}{dx^n}\left((1-x)^{n+\alpha}(1+x)^{n+\beta}\right).$$

- Orthogonality relation:

$$\int_{-1}^{1}(1-x)^\alpha(1+x)^\beta P_n^{(\alpha,\beta)}(x)\,P_m^{(\alpha,\beta)}(x)\,dx = \frac{2^{\alpha+\beta+1}\,\Gamma(n+\alpha+1)\,\Gamma(n+\beta+1)}{(2n+\alpha+\beta+1)\,n!\,\Gamma(2n+\alpha+\beta+1)}\delta_{n,m}.$$

1.3.2   Laguerre Polynomials

If $\sigma(x) = x$ and $\tau(x) = \alpha + 1 - x$ are substituted into (1.39), we obtain:

- Differential equation:

$$x\,y_n''(x) + (-x + \alpha + 1)y_n'(x) + n\,y_n(x) = 0, \tag{1.40}$$

with the polynomial solution

$$L_n^{(\alpha)}(x) = \sum_{k=0}^{n} (-1)^k \binom{n+\alpha}{n-k} \frac{1}{k!} x^k.$$

- Weight function:

$$w(x) = x^\alpha e^{-x} \quad \text{for } x \in [0, \infty) \quad \text{and } \alpha > -1.$$

- Rodrigues representation:

$$L_n^{(\alpha)}(x) = \frac{1}{n! x^\alpha e^{-x}} \frac{d^n}{dx^n} \left(x^{n+\alpha} e^{-x} \right).$$

- Orthogonality relation:

$$\int_0^\infty x^\alpha e^{-x} L_m^{(\alpha)}(x) L_n^{(\alpha)}(x)\, dx = \frac{\Gamma(n+\alpha+1)}{n!} \delta_{n,m}. \tag{1.41}$$

1.3.3 Hermite Polynomials

If $\sigma(x) = 1$ and $\tau(x) = -2x$ are substituted into (1.39), we obtain:

- Differential equation:

$$y_n''(x) - 2x y_n'(x) + 2n y_n(x) = 0,$$

with the polynomial solution

$$H_n(x) = n! \sum_{k=0}^{[n/2]} (-1)^k \frac{(2x)^{n-2k}}{k!\,(n-2k)!}.$$

- Weight function:

$$w(x) = e^{-x^2} \quad \text{for } x \in (-\infty, \infty).$$

- Rodrigues representation:

$$H_n(x) = (-1)^n e^{x^2} \frac{d^n}{dx^n} \left(e^{-x^2} \right).$$

- Orthogonality relation:

$$\int_{-\infty}^{\infty} e^{-x^2} H_n(x) H_m(x)\, dx = \sqrt{\pi}\, 2^n n!\, \delta_{n,m}.$$

1.3.4 First Finite Sequence of Hypergeometric Orthogonal Polynomials

The first finite sequence is defined for

$$\sigma(x) = x^2 + x \quad \text{and} \quad \tau(x) = (2 - p)x + q + 1$$

in (1.39). Thus, the differential equation

$$(x^2 + x)y_n''(x) + \big((2 - p)x + q + 1\big)y_n'(x) - n(n + 1 - p)y_n(x) = 0 \qquad (1.42)$$

has a polynomial solution

$$M_n^{(p,q)}(x) = (-1)^n n! \sum_{k=0}^{n} \binom{p - n - 1}{k}\binom{q + n}{n - k}(-x)^k. \qquad (1.43)$$

Now we prove that the polynomials (1.43) are finitely orthogonal with respect to the weight function

$$W_1(x; p, q) = x^q (1 + x)^{-(p+q)}$$

on $[0, \infty)$ if and only if $p > 2\{\max n\} + 1$ and $q > -1$.

To prove this, we first consider the self-adjoint form of Eq. (1.42) as

$$\left(x^{1+q}(1 + x)^{1-p-q} y_n'(x)\right)' = n(n + 1 - p)x^q (1 + x)^{-(p+q)} y_n(x),$$

and for the index m as

$$\left(x^{1+q}(1 + x)^{1-p-q} y_m'(x)\right)' = m(m + 1 - p)x^q (1 + x)^{-(p+q)} y_m(x), \qquad (1.44)$$

where $y_n(x) = M_n^{(p,q)}(x)$. Then we multiply by $y_m(x)$ and $y_n(x)$ in (1.44) and subtract to get

$$\left[\frac{x^{q+1}}{(1 + x)^{p+q-1}}\big(y_n'(x)y_m(x) - y_m'(x)y_n(x)\big)\right]_0^{\infty}$$

$$= (\lambda_n^{(1)} - \lambda_m^{(1)}) \int_0^{\infty} \frac{x^q}{(1 + x)^{p+q}} M_n^{(p,q)}(x) M_m^{(p,q)}(x)\, dx, \qquad (1.45)$$

where $\lambda_n^{(1)} = n(n + 1 - p)$. Since

$$\max \deg \{y_n'(x)y_m(x) - y_m'(x)y_n(x)\} = m + n - 1,$$

if $q > -1$ and $p > 2N + 1$ for $N = \max\{m, n\}$, the left-hand side of (1.45) tends to zero, and we will have

$$\int_0^\infty \frac{x^q}{(1 + x)^{p+q}} M_n^{(p,q)}(x) M_m^{(p,q)}(x)\, dx = 0 \;\Leftrightarrow\; m \neq n, \; p > 2N + 1 \; \text{and} \; q > -1.$$

In other words, the finite set $\{M_n^{(p>2N+1,q>-1)}(x)\}_{n=0}^N = \{M_n^{(p,q)}(x)\}_{n=0}^{N<(p-1)/2}$ is orthogonal with respect to the weight function $W_1(x; p, q) = x^q(1 + x)^{-(p+q)}$ on $[0, \infty)$.

The differential equation (1.42) shows that the polynomials (1.43) have a direct relationship to Jacobi polynomials. So, by referring to the Rodrigues representation of Jacobi polynomials, we obtain

$$M_n^{(p,q)}(x) = (-1)^n \frac{(1 + x)^{p+q}}{x^q} \frac{d^n}{dx^n}\left(x^{n+q}(1 + x)^{n-p-q}\right), \quad n = 0, 1, 2, \dots . \quad (1.46)$$

One of the advantages of the above representation is that we can directly calculate the norm square value of the above-mentioned polynomials. For this purpose, if (1.46) is replaced in the norm square value, we first get

$$\int_0^\infty \frac{x^q}{(1 + x)^{p+q}}\left(M_n^{(p,q)}(x)\right)^2 dx$$

$$= (-1)^n \int_0^\infty M_n^{(p,q)}(x) \frac{d^n}{dx^n}\left(x^{n+q}(1 + x)^{n-p-q}\right) dx.$$

Then integration by parts yields

$$(-1)^n \int_0^\infty M_n^{(p,q)}(x) \frac{d^n}{dx^n}\left(x^{n+q}(1 + x)^{n-p-q}\right) dx$$

$$= \frac{n!(p - (n + 1))!}{(p - (2n + 1))!} \int_0^\infty x^{n+q}(1 + x)^{n-p-q}\, dx.$$

On the other hand, since

$$\int_0^\infty x^{n+q}(1 + x)^{n-p-q}\, dx = \frac{(p - (2n + 2))!(q + n)!}{(p + q - (n + 1))!},$$

we finally obtain

$$\int_0^\infty \frac{x^q}{(1 + x)^{p+q}}\left(M_n^{(p,q)}(x)\right)^2 dx = \frac{n!\,(p - n - 1)!\,(q + n)!}{(p - 2n - 1)\,(p + q - n - 1)!}.$$

The latter relation again shows that $p > 2n + 1$ is a necessary condition for orthogonality of the polynomials $M_n^{(p,q)}(x)$. Consequently, we have

$$\int_0^\infty \frac{x^q}{(1+x)^{p+q}} M_n^{(p,q)}(x) M_m^{(p,q)}(x)\, dx = \frac{n!\,(p-n-1)!\,(q+n)!}{(p-2n-1)\,(p+q-n-1)!}\, \delta_{n,m} \tag{1.47}$$

if and only if $m, n = 0, 1, 2, \ldots, N < \frac{p-1}{2}$ and $q > -1$.

For example, the polynomial set

$$\left\{ M_n^{(202,0)}(x) \right\}_{n=0}^{100} = \left\{ (-1)^n n! \sum_{k=0}^n \binom{201-n}{k} \binom{n}{k} (-x)^k \right\}_{n=0}^{100}$$

is finitely orthogonal with respect to the weight function $W_1(x, 202, 0) = (1+x)^{-202}$ on $[0, \infty)$, and we have

$$\int_0^\infty (1+x)^{-202} M_n^{(202,0)}(x) M_m^{(202,0)}(x)\, dx = \frac{(n!)^2}{201-2n} \delta_{n,m} \Leftrightarrow m, n \le 100.$$

Finally, we point out that due to the form of the weight function and relation (1.46), there is a limit relation between $M_n^{(p,q)}(x)$ and the Laguerre polynomials as follows:

$$\lim_{p\to\infty} M_n^{(p,q)}(\frac{t}{p}) = \lim_{p\to\infty} (-1)^n t^{-q} (1 + \frac{t}{p})^{p+q} \frac{d^n(t^{n+q}(1+t/p)^{n-p-q})}{dt^n}$$

$$= (-1)^n t^{-q} e^t \frac{d^n(t^{n+q} e^{-t})}{dt^n} = (-1)^n n!\, L_n^{(q)}(t).$$

1.3.5   Second Finite Sequence of Hypergeometric Orthogonal Polynomials

The second finite sequence is directly connected to the Bessel polynomials for $\sigma(x) = x^2$ and $\tau(x) = 2x + 2$, which were first studied by Krall and Frink in 1949. Of course, they established a complex orthogonality of these polynomials on the unit circle. Then in 1973, the generalized Bessel polynomials were studied by Grosswald, which are special solutions of Eq. (1.39) for

$$\sigma(x) = x^2 \quad \text{and} \quad \tau(x) = (2 + \alpha)x + 2; \quad \alpha \ne -2, -3, \ldots,$$

leading to

$$\bar{B}_n^{(\alpha)}(x) = 2^n \sum_{k=0}^n \binom{n}{k} \frac{\Gamma(n+k+\alpha+1)}{\Gamma(2n+\alpha+1)} \left(\frac{x}{2}\right)^k,$$

as the monic type of generalized Bessel polynomial.

The second finite sequence can now be defined for

$$\sigma(x) = x^2 \quad \text{and} \quad \tau(x) = (2 - p)x + 1$$

to obtain the differential equation

$$x^2 y_n''(x) + \left((2 - p)x + 1\right)y_n'(x) - n(n + 1 - p)y_n(x) = 0, \tag{1.48}$$

with the polynomial solution

$$N_n^{(p)}(x) = (-1)^n \sum_{k=0}^{n} k! \binom{p - n - 1}{k} \binom{n}{n - k}(-x)^k. \tag{1.49}$$

If Eq. (1.48) is written in the self-adjoint form

$$\left(x^{-p+2} \exp\left(-\frac{1}{x}\right) y_n'(x)\right)' = n(n + 1 - p)x^{-p} \exp\left(-\frac{1}{x}\right) y_n(x),$$

and for the index m as

$$\left(x^{-p+2} \exp\left(-\frac{1}{x}\right) y_m'(x)\right)' = m(m + 1 - p)x^{-p} \exp\left(-\frac{1}{x}\right) y_m(x), \tag{1.50}$$

where $y_n(x) = N_n^{(p)}(x)$, then on multiplying by $y_m(x)$ and $y_n(x)$ in (1.50) and subtracting, we get

$$\left[x^{-p+2}e^{-\frac{1}{x}}\left(y_n'(x)y_m(x) - y_m'(x)y_n(x)\right)\right]_0^\infty = (\lambda_n^{(2)} - \lambda_m^{(2)}) \int_0^\infty x^{-p}e^{-\frac{1}{x}} N_n^{(p)}(x)N_m^{(p)}(x)\, dx, \tag{1.51}$$

where $\lambda_n^{(2)} = n(n + 1 - p)$. Once again, since

$$\max \deg\{y_n'(x)y_m(x) - y_m'(x)y_n(x)\} = m + n - 1,$$

the condition $p > 2N + 1$ for $N = \max\{m, n\}$ causes the left-hand side of (1.51) to tend to zero, and therefore

$$\int_0^\infty x^{-p}e^{-\frac{1}{x}} N_n^{(p)}(x)N_m^{(p)}(x)\, dx = 0 \quad \Leftrightarrow \quad m \neq n \text{ and } p > 2N + 1.$$

This means that the finite set $\{N_n^{(p > 2N+1)}(x)\}_{n=0}^{N} = \{N_n^{(p)}(x)\}_{n=0}^{N < (p-1)/2}$ is orthogonal with respect to the weight function $W_2(x; p) = x^{-p} \exp(-1/x)$ on $[0, \infty)$.

The Rodrigues representation of polynomials (1.49) can be denoted by

$$N_n^{(p)}(x) = (-1)^n x^p e^{\frac{1}{x}} \frac{d^n}{dx^n}(x^{-p+2n}e^{-\frac{1}{x}}), \quad n = 0, 1, 2, \dots . \tag{1.52}$$

To complete the orthogonality relation, we can use the above formula to explicitly compute the norm square value. Accordingly, if (1.52) is replaced in

$$\int_0^\infty x^{-p} e^{-\frac{1}{x}} \left(N_n^{(p)}(x)\right)^2 dx = (-1)^n \int_0^\infty N_n^{(p)}(x) \frac{d^n}{dx^n}(x^{-p+2n}e^{-\frac{1}{x}}) dx,$$

then integration by parts yields

$$(-1)^n \int_0^\infty N_n^{(p)}(x) \frac{d^n}{dx^n}(x^{-p+2n}e^{-\frac{1}{x}}) dx = \frac{n!(p-(n+1))!}{(p-(2n+1))!} \int_0^\infty x^{-p+2n}e^{-\frac{1}{x}} dx.$$

Note that if $p > 1$, then

$$\int_0^\infty W_2(x; p) dx = \int_0^\infty x^{-p} e^{-\frac{1}{x}} dx = \Gamma(p-1).$$

Therefore,

$$\int_0^\infty x^{-p} e^{-\frac{1}{x}} \left(N_n^{(p)}(x)\right)^2 dx = \frac{n! \Gamma(p-n)}{p-(2n+1)},$$

which gives the final orthogonality relation as

$$\int_0^\infty x^{-p} e^{-\frac{1}{x}} N_n^{(p)}(x) N_m^{(p)}(x) dx = \frac{n! \Gamma(p-n)}{p-(2n+1)} \delta_{n,m} \tag{1.53}$$

if and only if $m, n = 0, 1, 2, \dots, N < \frac{p-1}{2}$.

For example, the polynomial set

$$\left\{N_n^{(202)}(x)\right\}_{n=0}^{100} = \left\{(-1)^n \sum_{k=0}^n k! \binom{201-n}{k}\binom{n}{n-k}(-x)^k\right\}_{n=0}^{100}$$

is finitely orthogonal with respect to the weight function $W_2(x; 202) = x^{-202} \exp(-1/x)$ on $[0, \infty)$.

There is a direct relation between recent polynomials and the generalized Bessel polynomials as

$$N_n^{(p)}(x) = \frac{n!}{2^n} \binom{p-1-n}{n} \bar{B}_n^{(-p)}(2x),$$

as well as the Laguerre polynomials as

$$N_n^{(p)}(x) = n!\, x^n\, L_n^{\left(p-(2n+1)\right)}\left(\frac{1}{x}\right).$$

(1.54)

The latter relation is useful for generating a definite integral for the Laguerre polynomials, because substituting (1.54) into (1.53) yields

$$\int_0^\infty x^{p-1} e^{-x} \left(L_n^{(p)}(x)\right)^2 dx = \frac{1}{p}\frac{(n+p)!}{n!}.$$

1.3.6 Third Finite Sequence of Hypergeometric Orthogonal Polynomials

In this part, we study the last finite sequence of hypergeometric polynomials, which is finitely orthogonal with respect to the generalized Student's-t weight function

$$W^{(p,q)}(x) = \left(1 + x^2\right)^{-p} \exp(q \arctan x),$$

on $(-\infty, \infty)$. This function can also be considered an important statistical distribution, because by means of it we can generalize the usual Student's-t distribution and prove that it tends to the normal distribution just like the Student's-t sampling distribution.

Before studying the third finite case, we consider a particular case of Eq. (1.39) for

$$\sigma(x) = 1 + x^2 \quad \text{and} \quad \tau(x) = (3 - 2p)x,$$

which corresponds to the same as the usual Student's-t weight function, i.e.,

$$(1 + x^2)y_n''(x) + (3 - 2p)x\, y_n'(x) - n(n + 2 - 2p)y_n(x) = 0,$$

(1.55)

with the symmetric polynomial solution

$$I_n^{(p)}(x) = n! \sum_{k=0}^{[n/2]} (-1)^k \binom{p-1}{n-k}\binom{n-k}{k} (2x)^{n-2k},$$

which is finitely orthogonal with respect to the weight function

$$\rho(x, p) = (1 + x^2)^{-(p-\frac{1}{2})}$$

on $(-\infty, \infty)$.

To prove this claim, it is enough to transform Eq. (1.55) to a Sturm–Liouville equation and apply the same technique that we applied for the first and second kinds of finite classical orthogonal polynomials. This proccess eventually leads to

$$\int_{-\infty}^{\infty} (1+x^2)^{-(p-\frac{1}{2})} I_n^{(p)}(x) I_m^{(p)}(x)\, dx = 0 \quad \Leftrightarrow \quad m \neq n \text{ and } p > N+1.$$

On the other hand, since

$$I_n^{(p)}(x) = \frac{(-2)^n (p-n)_n}{(2p-2n-1)_n} (1+x^2)^{p-\frac{1}{2}} \frac{d^n ((1+x^2)^{n-(p-\frac{1}{2})})}{dx^n}, \quad n = 0, 1, 2, \dots ,$$

the orthogonality relation finally takes the form

$$\int_{-\infty}^{\infty} (1+x^2)^{-(p-\frac{1}{2})} I_n^{(p)}(x) I_m^{(p)}(x)\, dx$$

$$= \frac{n!\, 2^{2n-1} \sqrt{\pi}\, \Gamma^2(p)\, \Gamma(2p-2n)}{(p-n-1)\, \Gamma(p-n)\, \Gamma(p-n+1/2)\, \Gamma(2p-n-1)} \delta_{n,m}$$

if and only if $m, n = 0, 1, 2, \dots, N < p-1$.

As we mentioned, the weight function of the above orthogonality relation corresponds to the well-known Student's-t distribution

$$T(x; n) = \frac{\Gamma((n+1)/2)}{\sqrt{n\pi}\, \Gamma(n/2)} \rho\left(\frac{x}{\sqrt{n}}; \frac{n}{2}+1\right) \quad \text{for } x \in (-\infty, \infty) \text{ and } n \in \mathbb{N}.$$

However, that is not the end of story, and there is still a more extensive polynomial sequence that is finitely orthogonal on $(-\infty, \infty)$ and generalizes the Student's-t distribution as its weight function. For this purpose, we first replace

$$\sigma(x) = 1 + x^2 \quad \text{and} \quad \tau(x) = 2(1-p)x + q$$

in Eq. (1.39) to obtain

$$(1+x^2)\, y_n''(x) + \big(2(1-p)x + q\big)\, y_n'(x) - n(n+1-2p)\, y_n(x) = 0. \tag{1.56}$$

The polynomial solution of above equation can be represented in terms of a Gauss hypergeometric function as

$$J_n^{(p,q)}(x) = (-i)^n (n+1-2p)_n \sum_{k=0}^{n} \binom{n}{k} {}_2F_1\left(\begin{array}{c} k-n,\ p-n-iq/2 \\ 2p-2n \end{array} \middle|\, 2\right) (-ix)^k. \tag{1.57}$$

Now write Eq. (1.56) in the self-adjoint form

$$\left(\left(1+x^2\right)^{1-p}\exp(q\arctan x)y_n'(x)\right)' = n(n+1-2p)\left(1+x^2\right)^{-p}\exp(q\arctan x)y_n(x)$$

and apply the Sturm–Liouville theorem on $(-\infty, \infty)$ to get

$$\left[\left(1+x^2\right)^{1-p}\exp(q\arctan x)(y_n'(x)y_m(x) - y_m'(x)\,y_n(x))\right]_{-\infty}^{\infty}$$

$$= \left(\lambda_n^{(3)} - \lambda_m^{(3)}\right)\int_{-\infty}^{\infty}\left(1+x^2\right)^{-p}\exp(q\arctan x)\,J_n^{(p,q)}(x)\,J_m^{(p,q)}(x)\,dx, \qquad (1.58)$$

where $\lambda_n^{(3)} = n(n+1-2p)$. Since

$$\max\deg\{y_n'(x)y_m(x) - y_m'(x)y_n(x)\} = n+m-1,$$

if the conditions $p > N + 1/2$ for $N = \max\{m, n\}$ and $q \in \mathbb{R}$ hold, the left-hand side of (1.58) tends to zero, and

$$\int_{-\infty}^{\infty}(1+x^2)^{-p}\exp(q\arctan x)\,J_n^{(p,q)}(x)J_m^{(p,q)}(x)\,dx = 0$$

$$\Leftrightarrow m \neq n, \ p > N + 1/2 \text{ and } q \in \mathbb{R}. \qquad (1.59)$$

To compute the norm square value, it is enough to first substitute the Rodrigues representation

$$J_n^{(p,q)}(x) = (-1)^n \left(1+x^2\right)^p \exp(-q\arctan x)\frac{d^n\left(\left(1+x^2\right)^{n-p}\exp(q\arctan x)\right)}{dx^n}$$

into

$$\int_{-\infty}^{\infty}(1+x^2)^{n-p}\exp(q\arctan x)\left(J_n^{(p,q)}(x)\right)^2 dx$$

$$= (-1)^n \int_{-\infty}^{\infty} J_n^{(p,q)}(x)\frac{d^n\left(\left(1+x^2\right)^{n-p}\exp(q\arctan x)\right)}{dx^n}\,dx.$$

Then integration by parts yields

$$(-1)^n \int_{-\infty}^{\infty} J_n^{(p,q)}(x)\frac{d^n\left(\left(1+x^2\right)^{n-p}\exp(q\arctan x)\right)}{dx^n}\,dx$$

$$= \frac{n! \, \Gamma(2p-n)}{\Gamma(2p-2n)} \int_{-\infty}^{\infty} \left(1+x^2\right)^{n-p} \exp(q \arctan x) \, dx$$

$$= \frac{n! \, \Gamma(2p-n)}{\Gamma(2p-2n)} \int_{-\frac{\pi}{2}}^{\frac{\pi}{2}} (\cos\theta)^{2p-2n-2} \, e^{q\theta} \, d\theta.$$

Therefore

$$\int_{-\infty}^{\infty} W^{(p,q)}(x) \, J_n^{(p,q)}(x) \, J_m^{(p,q)}(x) \, dx$$

$$= \left(\frac{n! \Gamma(2p-n)}{\Gamma(2p-2n)} \int_{-\pi/2}^{\pi/2} (\cos\theta)^{2p-2n-2} \, e^{q\theta} \, d\theta \right) \delta_{n,m} \qquad (1.60)$$

if and only if $m, n = 0, 1, 2, \ldots, N < p - 1/2$ and $q \in \mathbb{R}$.
Note that (1.60) can be simplified if $2p$ is a natural number, because

$$\int_{-\pi/2}^{\pi/2} \cos^m\theta \, e^{q\theta} \, d\theta = 2^{-m} \sum_{k=0}^{m} \binom{m}{k} \frac{1}{(m-2k)\, i + q} \left(e^{((m-2k)\, i + q)\frac{\pi}{2}} - e^{-((m-2k)\, i + q)\frac{\pi}{2}} \right).$$

Remark 1.1 If $\mathrm{Re}(a) > 0$, $\mathrm{Re}(b) > 0$ and $\mathrm{Re}(c+d) > 1$, then the Cauchy beta integral

$$\frac{1}{2\pi} \int_{-\infty}^{\infty} \frac{dt}{(a+it)^c \, (b-it)^d} = \frac{\Gamma(c+d-1)}{\Gamma(c)\Gamma(d)} \, (a+b)^{1-(c+d)}$$

appears. One of the direct consequences of Cauchy's formula is the definite integral

$$\int_{-\frac{\pi}{2}}^{\frac{\pi}{2}} e^{st} \cos^r t \, dt = \frac{\pi \, 2^{-r} \, \Gamma(r+1)}{\Gamma(1 + \frac{r+is}{2}) \, \Gamma(1 + \frac{r-is}{2})},$$

which can be directly applied for evaluating the integral in (1.60).

For instance, the polynomial set

$$S = \{J_n^{(4,1)}(x)\}_{n=0}^{3} = \{\, 1\, ,\, 6x - 1\, ,\, 20x^2 - 10x - 3\, ,\, 24x^3 - 36x^2 - 12x + 7\,\}$$

is finitely orthogonal with respect to the weight function $(1+x^2)^{-4} \exp(\arctan x)$ on $(-\infty, \infty)$, so that

$$\int_{-\infty}^{\infty} \frac{\exp(\arctan x)}{(1+x^2)^4} \, J_n^{(4,1)}(x) J_m^{(4,1)}(x) \, dx$$

$$= \frac{n!(7-n)!(2\sinh\frac{\pi}{2})}{(7-2n) \prod_{k=0}^{3-n} (1 + (6-2n-2k)^2)} \, \delta_{n,m} \qquad \Leftrightarrow \qquad m, n \le 3.$$

Table 1.2 Six sequences of hypergeometric orthogonal polynomials

Kind	Notation	a	b	c	d	e
Infinite	$P_n^{(\alpha,\beta)}(x)$	-1	0	1	$-\alpha-\beta-2$	$-\alpha+\beta$
Infinite	$L_n^{(\alpha)}(wx)$	0	1	0	$-w$	$\alpha+1$
Infinite	$H_n(wx+v/2w)$	0	0	1	$-2w^2$	$-v$
Finite	$M_n^{(p,q)}((w/v)x)$	w	v	0	$(-p+2)w$	$(q+1)v$
Finite	$N_n^{(p)}(wx)$	w	0	0	$(-p+2)w$	1
Finite	$J_n^{(p,q)}(x)$	1	0	1	$2-2p$	q

Table 1.3 General properties of six sequences of hypergeometric orthogonal polynomials

Shifted polynomials	Weight function	Interval	Parametric conditions
$H_n(wx+\frac{v}{2w})$	$\exp(-x(v+w^2x))$	$(-\infty,\infty)$	$\forall n,\ w\neq 0,\ v\in\mathbb{R}$
$I_n^{(p)}(wx+v)$	$(1+(wx+v)^2)^{-(p-\frac{1}{2})}$	$(-\infty,\infty)$	$\max n < p-1$ $w\neq 0,\ v\in\mathbb{R}$
$J_n^{(p,q)}(x)$	$(1+x^2)^{-p}\exp(q\arctan x)$	$(-\infty,\infty)$	$\max n < (p-1)/2$ $q\in\mathbb{R}$
$L_n^{(\alpha)}(wx)$	$x^\alpha e^{-wx}$	$[0,\infty)$	$\forall n,\ w>0,\ \alpha>-1$
$M_n^{(p,q)}(\frac{w}{v}x)$	$x^q(wx+v)^{-(p+q)}$	$[0,\infty)$	$\max n < (p-1)/2$ $q>-1,\ w>0,\ v>0$
$N_n^{(p)}(wx)$	$x^{-p}\exp\left(-\frac{1}{wx}\right)$	$[0,\infty)$	$\max n < (p-1)/2$ $w>0$
$P_n^{(\alpha,\beta)}(x)$	$(1-x)^\alpha(1+x)^\beta$	$[-1,1]$	$\forall n,\ \alpha>-1,\ \beta>-1$

By referring to the set S, some rather complicated integrals can be explicitly evaluated. For instance, we have

$$\int_{-\infty}^{\infty}\frac{\exp(\arctan x)}{(1+x^2)^4}(20x^2-10x-3)^2\,dx = 32\sinh\frac{\pi}{2}.$$

Also there is a limit relation between $J_n^{(p,q)}(x)$ and the Hermite polynomials as follows:

$$\lim_{p\to\infty}J_n^{(p,q)}\left(\frac{x}{\sqrt{p}}\right) = \lim_{p\to\infty}\frac{(-1)^n}{W^{(p,q)}\left(\frac{x}{\sqrt{p}}\right)}\frac{d^n\left(\left(1+\frac{x^2}{p}\right)^n W^{(p,q)}\left(\frac{x}{\sqrt{p}}\right)\right)}{dx^n}$$

$$= (-1)^n e^{+x^2}\frac{d^n(e^{-x^2})}{dx^n} = H_n(x).$$

At the end of this part, we conclude that there are altogether six sequences of hypergeometric orthogonal polynomials that are generated by the main Eq. (1.39). The parameters a, b, c, d, and e corresponding to the equations of each these six sequences are shown in Table 1.2.

Also, Table 1.3 shows the basic properties of the sequences.

1.4 A Generic Polynomial Solution for the Classical Hypergeometric Differential Equation

In this section, we reconsider the expanded form of Eq. (1.39) as

$$(ax^2 + bx + c)y_n''(x) + (dx + e)y_n'(x) - n(d + (n-1)a)y_n(x) = 0 \qquad (1.61)$$

and suppose that its polynomial solution is denoted by $y_n(x) = P_n \left(\begin{array}{cc} d & e \\ a & b & c \end{array} \middle| x \right)$ or in the

text as $y_n(x) = p_n(x; a, b, c, d, e)$.

In 1929, Bochner classified all polynomial-type solutions of Eq. (1.61) and showed that the only polynomial systems up to a linear change of variable arising as eigenfunctions of Eq. (1.61) are:

- Jacobi polynomials $\{P_n^{(\alpha,\beta)}(x)\}_{n=0}^\infty$ for $\alpha, \beta, \alpha + \beta + 1 \notin \{-1, -2, \ldots\}$,
- Laguerre polynomials $\{L_n^{(\alpha)}(x)\}_{n=0}^\infty$ for $\alpha \notin \{-1, -2, \ldots\}$,
- Hermite polynomials $\{H_n(x)\}_{n=0}^\infty$, and
- Bessel polynomials $\{B_n^{(\alpha)}(x)\}_{n=0}^\infty$ for $\alpha \notin \{-2, -3, \ldots\}$.

Then in 1988, Nikiforov and Uvarov obtained some general properties of $P_n \left(\begin{array}{cc} d & e \\ a & b & c \end{array} \middle| x \right)$

based only on its Rodrigues representation. We obtain here a generic representation for the polynomial solution of Eq. (1.61) and use the following algebraic identity to reach this goal.

If a, b, and C_k $(k = 0, 1, \ldots, n)$ are real numbers, then

$$\sum_{k=0}^n C_k (ax + b)^k = \sum_{k=0}^n \left(\sum_{i=0}^{n-k} \binom{n-i}{k} b^{n-i} C_{n-i} \right) \left(\frac{ax}{b} \right)^k$$

$$= \sum_{k=0}^n \left(\sum_{i=0}^k \binom{n-i}{k-i} b^{k-i} C_{n-i} \right) (ax)^{n-k}. \qquad (1.62)$$

Theorem 1.1 *The monic polynomial solution of Eq. (1.61) is given by*

$$y_n(x) = \bar{P}_n \left(\begin{array}{cc} d & e \\ a & b & c \end{array} \middle| x \right) = \sum_{k=0}^n \binom{n}{k} G_k^{(n)}(a, b, c, d, e) x^k, \qquad (1.63)$$

where

$$G_k^{(n)} = \left(\frac{2a}{b + \sqrt{b^2 - 4ac}}\right)^{k-n} {}_2F_1\left(\begin{matrix} k - n, \ \frac{2ae - bd}{2a\sqrt{b^2 - 4ac}} + 1 - \frac{d}{2a} - n \\ 2 - d/a - 2n \end{matrix}\ \middle|\ \frac{2\sqrt{b^2 - 4ac}}{b + \sqrt{b^2 - 4ac}}\right).$$

For the special case $a = 0$, the equality can be adapted by limit considerations and gives (1.63) *in the form*

$$G_k^{(n)}(0, b, c, d, e) = \lim_{a \to 0} G_k^{(n)}(a, b, c, d, e) = (\frac{b}{c})^{k-n} {}_2F_0\left(\begin{matrix} k - n, \ \frac{cd - be}{b^2} + 1 - n \\ - \end{matrix}\ \middle|\ \frac{b^2}{cd}\right),$$

which is valid for $c, d \neq 0$, leading to

$$\bar{P}_n\left(\begin{matrix} d & e \\ 0 & b & c \end{matrix}\ \middle|\ x\right) = \left(\frac{b}{d}\right)^n \left(\frac{eb - cd}{b^2}\right)_n {}_1F_1\left(\begin{matrix} -n \\ \frac{eb - cd}{b^2} \end{matrix}\ \middle|\ -\frac{d}{b}x - \frac{cd}{b^2}\right).$$

Also for $a = b = 0$ and $d \neq 0$, (1.63) *is transformed to*

$$\bar{P}_n\left(\begin{matrix} d & e \\ 0 & 0 & c \end{matrix}\ \middle|\ x\right) = \lim_{\substack{a \to 0 \\ b \to 0}} \bar{P}_n\left(\begin{matrix} d & e \\ a & b & c \end{matrix}\ \middle|\ x\right) = \left(x + \frac{e}{d}\right)^n {}_2F_0\left(\begin{matrix} -\frac{n}{2}, \ -\frac{n-1}{2} \\ - \end{matrix}\ \middle|\ \frac{2cd}{(dx + e)^2}\right).$$

Proof First change Eq. (1.61) by means of $x = pt + q$ to

$$\left(t^2 + \frac{2aq + b}{ap}t + \frac{aq^2 + bq + c}{ap^2}\right)\frac{d^2y}{dt^2} + \left(\frac{d}{a}t + \frac{e + dq}{pa}\right)\frac{dy}{dt} - \frac{n}{a}(d + (n-1)a)y = 0.$$

$$(1.64)$$

If $aq^2 + bq + c = 0$ and $(2aq + b)/ap = -1$ in (1.64), then

$$p = \mp\frac{\sqrt{b^2 - 4ac}}{a} \quad \text{and} \quad q = \frac{-b \pm \sqrt{b^2 - 4ac}}{2a}.$$

Therefore, we have

$$t(t - 1)\frac{d^2y}{dt^2} + \left(\frac{d}{a}t + \frac{2ae - bd \pm d\sqrt{b^2 - 4ac}}{\mp 2a\sqrt{b^2 - 4ac}}\right)\frac{dy}{dt} - \frac{n}{a}(d + (n-1)a)y = 0. \quad (1.65)$$

Equation (1.65) is a special case of the Gauss equation

$$t(t - 1)\frac{d^2y}{dt^2} + ((\alpha + \beta + 1)t - \gamma)\frac{dy}{dt} + \alpha\beta y = 0, \quad (1.66)$$

for $\alpha = -n$, $\beta = n - 1 + d/a$, and $\gamma = \frac{2ae - bd \pm d\sqrt{b^2 - 4ac}}{\mp 2a\sqrt{b^2 - 4ac}}$.

Hence, by considering $P_n \left(\begin{array}{cc|c} d & e \\ a & b & c \end{array} x \right)$ as a preassigned solution of Eq. (1.61) and comparing the two relations (1.65) and (1.66), we obtain

$$P_n \left(\begin{array}{cc|c} d & e \\ a & b & c \end{array} \mp \frac{\sqrt{b^2 - 4ac}}{a} t + \frac{-b \pm \sqrt{b^2 - 4ac}}{2a} \right) = K \, {}_2F_1 \left(\begin{array}{c} -n, \, n - 1 + d/a \\ \frac{2ae - bd \pm d\sqrt{b^2 - 4ac}}{\pm 2a\sqrt{b^2 - 4ac}} \end{array} \middle| \, t \right).$$

(1.67)

Relation (1.67) can also be written in terms of the variable x as

$$P_n \left(\begin{array}{cc|c} d & e \\ a & b & c \end{array} x \right) = K \, {}_2F_1 \left(\begin{array}{c} -n, \, n - 1 + d/a \\ \frac{2ae - bd \pm d\sqrt{b^2 - 4ac}}{\pm 2a\sqrt{b^2 - 4ac}} \end{array} \middle| \mp \left(\frac{ax}{\sqrt{b^2 - 4ac}} + \frac{b \mp \sqrt{b^2 - 4ac}}{2\sqrt{b^2 - 4ac}} \right) \right).$$

(1.68)

From relation (1.68) the following two subcases are derived:

$$(i) \quad \bar{P}_n \left(\begin{array}{cc|c} d & e \\ a & b & c \end{array} x \right) = K^* \, {}_2F_1 \left(\begin{array}{c} -n, \, n - 1 + d/a \\ \frac{2ae - bd + d\sqrt{b^2 - 4ac}}{2a\sqrt{b^2 - 4ac}} \end{array} \middle| \frac{-ax}{\sqrt{b^2 - 4ac}} - \frac{b - \sqrt{b^2 - 4ac}}{2\sqrt{b^2 - 4ac}} \right),$$

(1.69)

$$(ii) \quad \bar{P}_n \left(\begin{array}{cc|c} d & e \\ a & b & c \end{array} x \right) = K^{**} \, {}_2F_1 \left(\begin{array}{c} -n, \, n - 1 + d/a \\ \frac{-(2ae - bd) + d\sqrt{b^2 - 4ac}}{2a\sqrt{b^2 - 4ac}} \end{array} \middle| \frac{ax}{\sqrt{b^2 - 4ac}} + \frac{b + \sqrt{b^2 - 4ac}}{2\sqrt{b^2 - 4ac}} \right).$$

Note that the latter relations differ only by a minus sign in the argument of the second formula (ii), which does not affect Eq. (1.61). In other words, if in (i) we consider the case $\bar{P}_n \left(\begin{array}{cc|c} -d & -e \\ -a & -b & -c \end{array} x \right)$, we will reach the second formula of (1.69). This means that only formula (ii) must be considered as the main solution. To compute K^{**}, it is sufficient to obtain the leading coefficient of ${}_2F_1 (\ldots | \ldots)$ in (ii), which is given by

$$K^{**} = \frac{n! \, (\sqrt{b^2 - 4ac})^n \, ((bd - 2ae + d\sqrt{b^2 - 4ac})/(2a\sqrt{b^2 - 4ac}))_n}{a^n (-n)_n (n - 1 + d/a)_n}.$$

On the other hand, according to the algebraic identity (1.62), we have

$$\, {}_2F_1 \left(\begin{array}{c} -n, \, p \\ q \end{array} \middle| \, rx + s \right) = \frac{(-1)^n (p)_n}{(q)_n} \sum_{k=0}^{n} \binom{n}{k} s^{n-k} \, {}_2F_1 \left(\begin{array}{c} k - n, \, 1 - q - n \\ 1 - p - n \end{array} \middle| \frac{1}{s} \right) (rx)^k.$$

(1.70)

Hence, by considering the main solution (1.69) and assuming

$$p = n - 1 + \frac{d}{a},$$

$$q = \frac{bd - 2ae + d\sqrt{b^2 - 4ac}}{2a\sqrt{b^2 - 4ac}},$$

$$r = \frac{a}{\sqrt{b^2 - 4ac}},$$

$$s = \frac{b + \sqrt{b^2 - 4ac}}{2\sqrt{b^2 - 4ac}},$$

relation (1.70) changes to

$$
{}_2F_1\left(
\begin{array}{c}
-n,\ n - 1 + d/a \\
\frac{bd - 2ae + d\sqrt{b^2-4ac}}{2a\sqrt{b^2-4ac}}
\end{array}
\left|
\frac{ax}{\sqrt{b^2 - 4ac}} + \frac{b + \sqrt{b^2 - 4ac}}{2\sqrt{b^2 - 4ac}}
\right.
\right)
$$

$$
= \frac{(-1)^n (\frac{d}{a} + n - 1)_n}{(\frac{bd - 2ae + d\sqrt{b^2-4ac}}{2a\sqrt{b^2-4ac}})_n}
\sum_{k=0}^{n} \binom{n}{k} \left(\frac{b + \sqrt{b^2 - 4ac}}{2\sqrt{b^2 - 4ac}}\right)^{n-k} \left(\frac{a}{\sqrt{b^2 - 4ac}}\right)^k
$$

$$
\times\ {}_2F_1\left(
\begin{array}{c}
k - n,\ \frac{2ae - bd - (d + (2n-2)a)\sqrt{b^2-4ac}}{2a\sqrt{b^2-4ac}} \\
-d/a - 2n + 2
\end{array}
\left|
\frac{2\sqrt{b^2 - 4ac}}{b + \sqrt{b^2 - 4ac}}
\right.
\right) x^k.
$$

Simplifying the above relation and substituting K^{**} finally gives the monic polynomial solution of Eq. (1.61) in the form (1.63). To deduce the limiting case when $a \to 0$, we can use the limit relation

$$\lim_{a \to 0} a^r \left(-\frac{d}{a} + 2 - 2n\right)_r = (-d)^r$$

and the identity

$$
{}_1F_1\left(
\begin{array}{c}
-n \\
p
\end{array}
\left|
rx + s
\right.
\right)
= \frac{\Gamma(1 - p - n)}{\Gamma(1 - p)}
\sum_{k=0}^{n} \binom{n}{k}\ {}_2F_0\left(
\begin{array}{c}
k - n,\ 1 - p - n \\
-
\end{array}
\left|
-\frac{1}{s}
\right.
\right)\left(\frac{r}{s}x\right)^k,
$$

which is a special case of identity (1.62) for $C_k = \frac{(-n)_k}{(p)_k k!}$. □

The general formula $G_k^{(n)}(a, b, c, d, e)$ in (1.63) is a suitable tool to compute the coefficients of x^k for any fixed degree k and arbitrary a. For example, if the coefficient

x^{n-1} is needed in $\bar{P}_n \begin{pmatrix} d & e \\ a & b & c \end{pmatrix} x$, it is enough to calculate the term

$$G_{n-1}^{(n)}(a, b, c, d, e) = \left(\frac{2a}{b+\Delta}\right)^{-1} {}_2F_1\left(\begin{array}{c} -1, \frac{2ae-bd-(d+(2n-2)a)\Delta}{2a\Delta} \\ -\left(\frac{d+(2n-2)a}{a}\right) \end{array} \middle| \frac{2\Delta}{b+\Delta}\right)$$

$$= \left(\frac{b+\Delta}{2a}\right)\left(1 + \frac{2ae - bd - (d+(2n-2)a)\Delta}{2a\Delta}\frac{2\Delta}{b+\Delta}\frac{a}{d+(2n-2)a}\right)$$

$$= \frac{e+(n-1)b}{d+(2n-2)a}, \qquad (1.71)$$

in which $\Delta = \sqrt{b^2 - 4ac}$. Note in (1.71) that all parameters are free and can assume any value including zero, because it is easy to see that neither both values a and d nor both values b and e can vanish together in (1.63). After simplifying $G_k^{(n)}(a, b, c, d, e)$ for $k = n-1, n-2, \ldots$, we eventually obtain

$$\bar{P}_n \begin{pmatrix} d & e \\ a & b & c \end{pmatrix} x = x^n + \binom{n}{1}\frac{e+(n-1)b}{d+(2n-2)a}x^{n-1}$$

$$+ \binom{n}{2}\frac{(e+(n-1)b)(e+(n-2)b) + c(d+(2n-2)a)}{(d+(2n-2)a)(d+(2n-3)a)}x^{n-2} + \cdots$$

$$+ \binom{n}{n}\left(\frac{b+\sqrt{b^2-4ac}}{2a}\right)^n {}_2F_1\left(\begin{array}{c} -n, \frac{2ae-bd-(d+(2n-2)a)\sqrt{b^2-4ac}}{2a\sqrt{b^2-4ac}} \\ -\left(\frac{d+(2n-2)a}{a}\right) \end{array} \middle| \frac{2\sqrt{b^2-4ac}}{b+\sqrt{b^2-4ac}}\right).$$

In particular, the above relation gives

$$\bar{P}_n \begin{pmatrix} d & e \\ a & b & c \end{pmatrix} 0 = \left(\frac{b+\sqrt{b^2-4ac}}{2a}\right)^n$$

$$\times {}_2F_1\left(\begin{array}{c} -n, \frac{2ae-bd-(d+(2n-2)a)\sqrt{b^2-4ac}}{2a\sqrt{b^2-4ac}} \\ -\left(\frac{d+(2n-2)a}{a}\right) \end{array} \middle| \frac{2\sqrt{b^2-4ac}}{b+\sqrt{b^2-4ac}}\right).$$

For instance, we have

$$\bar{P}_0 \left(\left. \begin{matrix} d & e \\ a & b & c \end{matrix} \right| x \right) = 1 \,,$$

$$\bar{P}_1 \left(\left. \begin{matrix} d & e \\ a & b & c \end{matrix} \right| x \right) = x + \frac{e}{d} \,,$$

$$\bar{P}_2 \left(\left. \begin{matrix} d & e \\ a & b & c \end{matrix} \right| x \right) = x^2 + 2\frac{e+b}{d+2a}x + \frac{c(d+2a)+e(e+b)}{(d+2a)(d+a)} \,,$$

$$\bar{P}_3 \left(\left. \begin{matrix} d & e \\ a & b & c \end{matrix} \right| x \right) = x^3 + 3\frac{e+2b}{d+4a}x^2 + 3\frac{c(d+4a)+(e+b)(e+2b)}{(d+4a)(d+3a)}x$$
$$+ \frac{2c(d+3a)(e+2b)+ce(d+4a)+e(e+b)(e+2b)}{(d+4a)(d+3a)(d+2a)} \,.$$

Remark 1.2 We can apply the Gauss summation formula and take $\frac{2\sqrt{b^2-4ac}}{b+\sqrt{b^2-4ac}} = 1$ in (1.63) to conclude that $ac = 0$. In this case, if $c = 0$, the following special case is derived:

$$\bar{P}_n \left(\left. \begin{matrix} d & e \\ a & b & 0 \end{matrix} \right| x \right) = \sum_{k=0}^{n} \binom{n}{k} (\frac{a}{b})^{k-n} \, {}_2F_1 \left(\left. \begin{matrix} k-n, \ \frac{2ae-bd}{2ab}+1-\frac{d}{2a}-n \\ 2-d/a-2n \end{matrix} \right| 1 \right) x^k$$

$$= \sum_{k=0}^{n} \binom{n}{k} (\frac{a}{b})^{k-n} \frac{\Gamma(2-2n-d/a)\Gamma(1-k-e/b)}{\Gamma(2-n-k-d/a)\Gamma(1-n-e/b)} x^k$$

$$= \frac{b^n \, \Gamma(2-2n-d/a)\Gamma(1-e/b)}{a^n \, \Gamma(2-n-d/a)\Gamma(1-n-e/b)} \, {}_2F_1 \left(\left. \begin{matrix} -n, \ n-1+d/a \\ e/b \end{matrix} \right| -\frac{a}{b}x \right). \qquad (1.72)$$

Moreover, if we refer to the Nikiforov and Uvarov approach and consider Eq. (1.61) as a self-adjoint form, the Rodrigues representation of the monic polynomials is derived as

$$\bar{P}_n \left(\left. \begin{matrix} d & e \\ a & b & c \end{matrix} \right| x \right) = \frac{1}{\left(\prod\limits_{k=1}^{n} d + (n+k-2)a \right) W \left(\left. \begin{matrix} d & e \\ a & b & c \end{matrix} \right| x \right)}$$

$$\times \frac{d^n \left((ax^2+bx+c)^n W \left(\left. \begin{matrix} d & e \\ a & b & c \end{matrix} \right| x \right) \right)}{dx^n}, \qquad (1.73)$$

where

$$W \begin{pmatrix} d & e \\ a & b & c \end{pmatrix} x = \exp \left(\int \frac{(d - 2a)x + e - b}{ax^2 + bx + c} dx \right). \qquad (1.74)$$

Now if once again $c = 0$ in (1.73), by referring to (1.72), we obtain

$$\frac{\exp \left(- \int \frac{(d-2a)x+e-b}{ax^2+bx} dx \right) d^n \left((ax^2 + bx)^n \exp \left(\int \frac{(d-2a)x+e-b}{ax^2+bx} dx \right) \right)}{\left(\prod_{k=1}^{n} d + (n + k - 2)a \right) \qquad dx^n}$$

$$= \frac{\Gamma(2 - 2n - d/a)}{\Gamma(1 - n - e/b)} \sum_{k=0}^{n} \binom{n}{k} \left(\frac{a}{b} \right)^{k-n} \frac{\Gamma(1 - k - e/b)}{\Gamma(2 - n - k - d/a)} x^k.$$

Remark 1.3 There is another representation for $\bar{P}_n(x; a, b, c, d, e)$ in (1.63). If the weight function (1.74) is written in the same form as (1.24), i.e.,

$$W \begin{pmatrix} d & e \\ a & b & c \end{pmatrix} x = K(x + \theta_1)^A (x + \theta_2)^B,$$

where K is a constant,

$$A = \frac{d}{2a} - 1 + \frac{2ae - bd}{2a\sqrt{b^2 - 4ac}}, \quad B = \frac{d}{2a} - 1 - \frac{2ae - bd}{2a\sqrt{b^2 - 4ac}},$$

and

$$\theta_1 = \frac{b - \sqrt{b^2 - 4ac}}{2a}, \quad \theta_2 = \frac{b + \sqrt{b^2 - 4ac}}{2a}, \qquad (1.75)$$

then replacing it in the Rodrigues representation (1.73) yields

$$\bar{P}_n \begin{pmatrix} d & e \\ a & b & c \end{pmatrix} x = \frac{\frac{d^n}{dx^n} (Ka^n(x + \theta_1)^n (x + \theta_2)^n (x + \theta_1)^A (x + \theta_2)^B)}{K \left(\prod_{k=1}^{n} d + (n + k - 2)a \right) (x + \theta_1)^A (x + \theta_2)^B}$$

$$= \frac{1}{(n - 1 + d/a)_n (x + \theta_1)^A (x + \theta_2)^B} \frac{d^n}{dx^n} \left((x + \theta_1)^{n+A} (x + \theta_2)^{n+B} \right). \qquad (1.76)$$

But the Leibniz rule

$$\frac{d^n\left((x+\theta_1)^{n+A}(x+\theta_2)^{n+B}\right)}{dx^n}$$

$$= (-1)^n \sum_{k=0}^{n} \binom{n}{k} (-n-A)_k(-n-B)_{n-k}(x+\theta_1)^{n+A-k}(x+\theta_2)^{B+k},$$

implies that relation (1.76) is simplified as

$$\bar{P}_n\left(\begin{array}{cc} d & e \\ a & b & c \end{array}\bigg| x\right) = \frac{1}{(2-2n-d/a)_n}$$

$$\times \sum_{k=0}^{n} \binom{n}{k} (-n - \frac{d}{2a} + 1 - \frac{2ae-bd}{2a\sqrt{b^2-4ac}})_k(-n - \frac{d}{2a} + 1 + \frac{2ae-bd}{2a\sqrt{b^2-4ac}})_{n-k}$$

$$\times \left(x + \frac{b-\sqrt{b^2-4ac}}{2a}\right)^{n-k}\left(x + \frac{b+\sqrt{b^2-4ac}}{2a}\right)^{k}, \qquad (1.77)$$

which is another representation for the polynomial solution of Eq. (1.61). By combining (1.63) and (1.77), we straightforwardly arrive at the identity

$$\sum_{k=0}^{n} \left(\binom{n}{k} (-n - \frac{d}{2a} + 1 - \frac{2ae-bd}{2a\Delta})_k\left(-n - \frac{d}{2a} + 1 + \frac{2ae-bd}{2a\Delta}\right)_{n-k}\right.$$

$$\left.\left(x + \frac{b-\Delta}{2a}\right)^{n-k}\left(x + \frac{b+\Delta}{2a}\right)^{k}\right)$$

$$= (2-2n-d/a)_n \sum_{k=0}^{n} \binom{n}{k} \left(\frac{2a}{b+\Delta}\right)^{k-n}$$

$$\times {}_2F_1\left(\begin{array}{c} k-n, \ \frac{2ae-bd}{2a\Delta} + 1 - \frac{d}{2a} - n \\ 2 - d/a - 2n \end{array}\bigg| \frac{2\Delta}{b+\Delta}\right) x^k,$$

where $\Delta = \sqrt{b^2-4ac}$.

Relation (1.77) can also be represented in terms of the hypergeometric form

$$\bar{P}_n\left(\begin{array}{cc} d & e \\ a & b & c \end{array}\bigg| x\right) = \frac{(-n - \frac{d}{2a} + 1 + \frac{2ae-bd}{2a\Delta})_n(x+\theta_1)^n}{(2-2n-d/a)_n}$$

$$\times {}_2F_1\left(\begin{array}{c} -n,\ -n-\frac{d}{2a}+1-\frac{2ae-bd}{2a\Delta} \\ \frac{d}{2a}-\frac{2ae-bd}{2a\Delta} \end{array}\middle|\ \begin{array}{c} x+\theta_2 \\ x+\theta_1 \end{array}\right),$$

where θ_1 and θ_2 are defined by (1.75).

Remark 1.4 Using the explicit representation (1.69), one can compute the generic values of the polynomials at their boundary points of definition, $-\theta_1$ and $-\theta_2$, so that if

$$\frac{ax}{\sqrt{b^2-4ac}}+\frac{b+\sqrt{b^2-4ac}}{2\sqrt{b^2-4ac}}=\begin{cases} 0 \\ 1 \end{cases}$$

then

$$x=\frac{-b-\sqrt{b^2-4ac}}{2a}=-\theta_2 \quad \text{and} \quad x=\frac{-b+\sqrt{b^2-4ac}}{2a}=-\theta_1.$$

Hence by substituting the above values into (1.69), we respectively obtain

$$\bar{P}_n\left(\begin{array}{ccc} d & e \\ a & b & c \end{array}\middle| -\theta_2\right)=\frac{(\sqrt{b^2-4ac})^n\ ((d/2a)-(2ae-bd)/(2a\sqrt{b^2-4ac}))_n}{(-a)^n(n-1+d/a)_n}$$

and

$$\bar{P}_n\left(\begin{array}{ccc} d & e \\ a & b & c \end{array}\middle| -\theta_1\right)=\frac{(\sqrt{b^2-4ac})^n((d/2a)+(2ae-bd)/(2a\sqrt{b^2-4ac}))_n}{a^n(n-1+d/a)_n}.$$

For instance, for the monic Jacobi orthogonal polynomials $\bar{P}_n^{(\alpha,\beta)}(x)$ with the initial vector $(a,b,c,d,e)=(-1,0,1,-\alpha-\beta-2,-\alpha+\beta)$, we have

$$\bar{P}_n^{(\alpha,\beta)}(+1)=2^n\frac{(\alpha+1)_n}{(n+1+\alpha+\beta)_n}=2^n\frac{\Gamma(n+1+\alpha)\Gamma(n+1+\alpha+\beta)}{\Gamma(\alpha+1)\Gamma(2n+1+\alpha+\beta)},$$

and

$$\bar{P}_n^{(\alpha,\beta)}(-1)=(-2)^n\frac{(\beta+1)_n}{(n+1+\alpha+\beta)_n}=(-2)^n\frac{\Gamma(n+1+\beta)\Gamma(n+1+\alpha+\beta)}{\Gamma(\beta+1)\Gamma(2n+1+\alpha+\beta)}.$$

Some Other Properties of $\bar{P}_n(x;a,b,c,d,e)$

Property 1 Using the representation (1.73), a linear change of variable can be applied for the monic polynomials so that if $x = wt + v$, then we have

$$
\bar{P}_n \left(\left. \begin{matrix} d & e \\ a & b & c \end{matrix} \right| wt + v \right)
$$

$$
= \frac{w^n \, d^n \left((aw^2 t^2 + (2awv + bw)t + (av^2 + bv + c))^n \, W \left(\left. \begin{matrix} d & e \\ a & b & c \end{matrix} \right| wt + v \right) \right)}{\left(\prod\limits_{k=1}^{n} d + (n + k - 2)a \right) W \left(\left. \begin{matrix} d & e \\ a & b & c \end{matrix} \right| wt + v \right) dt^n}.
$$

(1.78)

But since

$$
W \left(\left. \begin{matrix} d & e \\ a & b & c \end{matrix} \right| wt + v \right) = \exp \left(\int \frac{(d - 2a)(wt + v) + e - b}{aw^2 t^2 + (2av + b)wt + av^2 + bv + c} \, w \, dt \right)
$$

$$
= W \left(\left. \begin{matrix} dw^2, & (dv + e)w \\ aw^2, & (2av + b)w, & av^2 + bv + c \end{matrix} \right| t \right),
$$

relation (1.78) is eventually transformed to

$$
\bar{P}_n \left(\left. \begin{matrix} d & e \\ a & b & c \end{matrix} \right| wt + v \right) = w^n \, \bar{P}_n \left(\left. \begin{matrix} dw^2, & (dv + e)w \\ aw^2, & (2av + b)w, & av^2 + bv + c \end{matrix} \right| t \right).
$$

For instance, we can straightforwardly conclude that

$$
\bar{P}_n \left(\left. \begin{matrix} d & e \\ a & b & c \end{matrix} \right| -t \right) = (-1)^n \, \bar{P}_n \left(\left. \begin{matrix} d & -e \\ a & -b & c \end{matrix} \right| t \right).
$$

Property 2 The second formula of (1.69) is a suitable tool to compute a generic three-term recurrence equation for the polynomials (1.63), since it can be applied along with various identities of the Gauss hypergeometric function for generating a recurrence relation. For example, by noting the well-known identity

$$
(p - q) \left((p - q - 1)(p - q + 1)t + 2pq + r(1 - p - q) \right) {}_2F_1(p, q, r;\ t)
$$
$$
+ q(r - p)(p - q + 1) {}_2F_1(p - 1, q + 1, r;\ t)
$$
$$
+ p(r - q)(p - q - 1) {}_2F_1(p + 1, q - 1, r;\ t) = 0, \qquad (1.79)
$$

if we assume in (1.79) that

$$p = -n,$$

$$q = \frac{d}{a} + n - 1,$$

$$r = \frac{bd - 2ae + d\sqrt{b^2 - 4ac}}{2a\sqrt{b^2 - 4ac}},$$

and

$$t = \frac{a}{\sqrt{b^2 - 4ac}}x + \frac{b + \sqrt{b^2 - 4ac}}{2\sqrt{b^2 - 4ac}},$$

then after some computations, we obtain

$$\bar{P}_{n+1}(x) = \left(x + \frac{2n(n+1)ab + (d-2a)(e+2nb)}{(d+2na)(d+(2n-2)a)}\right) \bar{P}_n(x) + n(d+(n-2)a)$$

$$\times \frac{(c(d+(2n-2)a)^2 - nb^2(d+(n-2)a) + (e-b)(a(e+b)-bd))}{(d+(2n-3)a)(d+(2n-2)a)^2(d+(2n-1)a)} \bar{P}_{n-1}(x),$$

in which $\bar{P}_n(x)$ has the same meaning as the monic polynomials of (1.63) with values

$$\bar{P}_0(x) = 1 \quad \text{and} \quad \bar{P}_1(x) = x + \frac{e}{d}.$$

Property 3 As the last property, we can compute the norm square value of the monic polynomials. Let $[L, U]$ be a predetermined orthogonality interval that (other than for finite families) consists of the zeros of $\sigma(x) = ax^2 + bx + c$ or $\pm\infty$. By noting the Rodrigues representation (1.73), we have

$$\|\bar{P}_n\|^2 = \int_L^U \bar{P}_n^2\left(\begin{array}{cc} d & e \\ a & b & c \end{array}\bigg| x\right) W\left(\begin{array}{cc} d & e \\ a & b & c \end{array}\bigg| x\right) dx = \frac{1}{\prod\limits_{k=1}^{n} d + (n+k-2)a}$$

$$\times \int_L^U \bar{P}_n\left(\begin{array}{cc} d & e \\ a & b & c \end{array}\bigg| x\right) \frac{d^n}{dx^n}\left((ax^2 + bx + c)^n W\left(\begin{array}{cc} d & e \\ a & b & c \end{array}\bigg| x\right)\right) dx. \tag{1.80}$$

So integrating by parts on the right-hand side of (1.80) yields

$$\|\bar{P}_n\|^2 = \frac{n!\,(-1)^n}{\prod\limits_{k=1}^{n} d + (n+k-2)a} \int_L^U (ax^2 + bx + c)^n \left(\exp \int \frac{(d-2a)x + e - b}{ax^2 + bx + c} dx \right) dx.$$

$$(1.81)$$

1.4.1 Six Special Cases of the Generic Polynomials $\bar{P}_n(x; a, b, c, d, e)$

As we have shown in Tables 1.2 and 1.3, there are six sequences of hypergeometric orthogonal polynomials. In this section, we apply previously obtained formulas to determine the basic properties of each of the six sequences.

1. For Jacobi polynomials we obtain

- Initial vector:

$$(a, b, c, d, e) = (-1, 0, 1, -\alpha - \beta - 2, -\alpha + \beta).$$

- Generic representation of polynomials:

$$P_n^{(\alpha,\beta)}(x) = \frac{(n+\alpha+\beta+1)_n}{2^n n!} \bar{P}_n \left(\begin{array}{cc} -\alpha-\beta-2, & -\alpha+\beta \\ -1, & 0, & 1 \end{array} \middle| x \right)$$

$$= \frac{(n+\alpha+\beta+1)_n}{2^n n!} \sum_{k=0}^{n} (-1)^{k-n} \binom{n}{k} {}_2F_1 \left(\begin{array}{cc} k-n, & -n-\alpha \\ -2n-\alpha-\beta \end{array} \middle| 2 \right) x^k.$$

- Hypergeometric representation:

$$\bar{P}_n \left(\begin{array}{cc} -\alpha-\beta-2, & -\alpha+\beta \\ -1, & 0, & 1 \end{array} \middle| x \right) = (-1)^n (1-x)^n {}_2F_1 \left(\begin{array}{cc} -n, & -\alpha-n \\ -\alpha-\beta-2n \end{array} \middle| \frac{2}{1-x} \right).$$

- Weight function:

$$W \left(\begin{array}{cc} -\alpha-\beta-2, & -\alpha+\beta \\ -1, & 0, & 1 \end{array} \middle| x \right) = \exp \left(\int \frac{-(\alpha+\beta)x + \beta - \alpha}{-x^2 + 1} dx \right)$$

$$= (1-x)^\alpha (1+x)^\beta.$$

Jacobi polynomials include some special cases, such as:

- Gegenbauer polynomials:

$$
C_n^{(\lambda)}(x) = \frac{2^n (\lambda)_n}{n!} \, \bar{P}_n \left(\begin{array}{cc|c} -2\lambda - 1, \, 0 & \\ -1 \quad 0 \quad 1 & \end{array} x \right)
$$

$$
= \frac{2^n (\lambda)_n}{n!} (-1)^n (1-x)^n {}_2F_1 \left(\begin{array}{c|c} -n, \, 1/2 - \lambda - n & \dfrac{2}{1-x} \\ 1 - 2\lambda - 2n & \end{array} \right),
$$

- Legendre polynomials:

$$
P_n(x) = \frac{(2n)!}{(n!)^2 2^n} \, \bar{P}_n \left(\begin{array}{cc|c} -2, \, 0 & \\ -1, \quad 0, \quad 1 & \end{array} x \right) = \frac{(2n)!}{(n!)^2 2^n} (-1)^n (1-x)^n {}_2F_1 \left(\begin{array}{c|c} -n, \, -n & \dfrac{2}{1-x} \\ -2n & \end{array} \right),
$$

- Chebyshev polynomials of the first kind:

$$
T_n(x) = 2^{n-1} \, \bar{P}_n \left(\begin{array}{cc|c} -1, \, 0 & \\ -1, \quad 0, \quad 1 & \end{array} x \right) = 2^{n-1} (-1)^n (1-x)^n {}_2F_1 \left(\begin{array}{c|c} -n, \, 1/2 - n & \dfrac{2}{1-x} \\ 1 - 2n & \end{array} \right),
$$

- Chebyshev polynomials of the second kind:

$$
U_n(x) = 2^n \, \bar{P}_n \left(\begin{array}{cc|c} -3, \, 0 & \\ -1, \quad 0, \quad 1 & \end{array} x \right) = 2^n (-1)^n (1-x)^n {}_2F_1 \left(\begin{array}{c|c} -n, \, -1/2 - n & \dfrac{2}{1-x} \\ -1 - 2n & \end{array} \right).
$$

2. For the Laguerre polynomials we obtain

- Initial vector:

$$
(a, b, c, d, e) = (0, 1, 0, -1, \alpha + 1).
$$

- Generic and hypergeometric representations:

$$
L_n^{(\alpha)}(x) = \frac{(-1)^n}{n!} \, \bar{P}_n \left(\begin{array}{cc|c} -1, \, \alpha + 1 & \\ 0, \quad 1, \quad 0 & \end{array} x \right) = \frac{(\alpha + 1)_n}{n!} \, {}_1F_1 \left(\begin{array}{c|c} -n & x \\ \alpha + 1 & \end{array} \right).
$$

- Weight function:

$$
W \left(\begin{array}{cc|c} -1, \, \alpha + 1 & \\ 0, \quad 1, \quad 0 & \end{array} x \right) = \exp \left(\int \frac{-x + \alpha}{x} \, dx \right) = x^\alpha e^{-x}.
$$

3. For Hermite polynomials we obtain

- Initial vector:

$$(a, b, c, d, e) = (0, 0, 1, -2, 0).$$

- Generic and hypergeometric representations:

$$H_n(x) = 2^n \, \bar{P}_n \left(\begin{matrix} -2, \ 0 \\ 0, \ \ 0, \ \ 1 \end{matrix} \, \middle| \, x \right) = (2x)^n \, {}_2F_0 \left(\begin{matrix} -\frac{n}{2}, \ -\frac{n-1}{2} \\ - \end{matrix} \, \middle| \, -\frac{1}{x^2} \right).$$

- Weight function:

$$W \left(\begin{matrix} -2, \ 0 \\ 0, \ \ 0, \ \ 1 \end{matrix} \, \middle| \, x \right) = \exp \left(\int -2x \, dx \right) = \exp(-x^2).$$

4. For the first finite sequence of hypergeometric orthogonal polynomials we respectively find

- Initial vector:

$$(a, b, c, d, e) = (1, 1, 0, -p + 2, q + 1).$$

In [4], this type of polynomials are called Romanovski-Jacobi polynomials, see also [5, 8], while in [6] they are indicated by $M_n^{(p,q)}(x)$.
- Generic and hypergeometric representations:

$$M_n^{(p,q)}(x) = (-1)^n (n + 1 - p)_n \, \bar{P}_n \left(\begin{matrix} -p + 2, \ q + 1 \\ 1, \ \ 1, \ \ 0 \end{matrix} \, \middle| \, x \right)$$

$$= (-1)^n (q + 1)_n \, {}_2F_1 \left(\begin{matrix} -n, \ n + 1 - p \\ q + 1 \end{matrix} \, \middle| \, -x \right).$$

By referring to the general formula (1.81), we can recompute their norm square value as follows

$$\int_0^\infty \frac{x^q}{(1 + x)^{p+q}} \, \bar{P}_n^2 \left(\begin{matrix} -p + 2, \ q + 1 \\ 1, \ \ 1, \ \ 0 \end{matrix} \, \middle| \, x \right) dx$$

$$= \frac{n! \, (-1)^n}{\prod\limits_{k=1}^n (-p + n + k)} \int_0^\infty (x^2 + x)^n \frac{x^q}{(1 + x)^{p+q}} \, dx,$$

which yields

$$\int_0^\infty \frac{x^q}{(1+x)^{p+q}} \left(M_n^{(p,q)}(x) \right)^2 dx = \frac{n!\,(p-n-1)!\,(q+n)!}{(p-2n-1)\,(p+q-n-1)!}.$$

5. For the second finite sequence of hypergeometric orthogonal polynomials we obtain

- Initial vector:

$$(a, b, c, d, e) = (1, 0, 0, -p+2, 1).$$

In [4], such polynomials are called Romanovski–Bessel polynomials, while in [6] they are indicated by $N_n^{(p)}(x)$.

- Generic and hypergeometric representations:

$$N_n^{(p)}(x) = (-1)^n (n+1-p)_n\, \bar{P}_n \left(\begin{array}{ccc} -p+2, & 1 & \\ 1, & 0, & 0 \end{array} \middle| x \right) = (-1)^n {}_2 F_0 \left(\begin{array}{cc} -n, & 1-p+n \\ & - \end{array} \middle| -x \right).$$

Once again, we can refer to (1.81) and recompute their norm square value as

$$\int_0^\infty x^{-p} e^{-\frac{1}{x}} \bar{P}_n^2 \left(\begin{array}{ccc} -p+2, & 1 & \\ 1, & 0, & 0 \end{array} \middle| x \right) dx = \frac{n!\,(-1)^n}{\prod_{k=1}^{n} (-p+n+k)} \int_0^\infty x^{2n} x^{-p} e^{-\frac{1}{x}} dx,$$

which yields

$$\int_0^\infty x^{-p} e^{\frac{-1}{x}} \left(N_n^{(p)}(x) \right)^2 dx = \frac{n!\,(p-(n+1))!}{p-(2n+1)}.$$

6. For the third finite sequence of hypergeometric orthogonal polynomials we obtain

- Initial vector:

$$(a, b, c, d, e) = \left(1, 0, 1, 2(1-p), q\right).$$

In [4], such polynomials are called Romanovski–Pseudo–Jacobi polynomials, while in [7] they are indicated by $J_n^{(p,q)}(x)$.

- Generic and hypergeometric representations:

$$J_n^{(p,q)}(x) = (-1)^n (n+1-2p)_n\, \bar{P}_n \left(\begin{array}{ccc} 2(1-p), & q & \\ 1, & 0, & 1 \end{array} \middle| x \right)$$

$$= (-1)^n (n + 1 - 2p)_n (x + i)^n \, {}_2F_1 \left(\begin{array}{c} -n, \, -n + p - \frac{q}{2}i \\ 2p - 2n \end{array} \middle| \, \frac{2i}{x+i} \right).$$

$$(1.82)$$

Taking into account the identity

$$\sum_{k=0}^{n} \binom{n}{k} {}_2F_1 \left(\begin{array}{c} k - n, \, 1 - q^* - n \\ 1 - p^* - n \end{array} \middle| \, \frac{1}{s^*} \right) \left(\frac{r^*}{s^*} x \right)^k$$

$$= \frac{(q^*)_n}{(-s^*)^n (p^*)_n} \, {}_2F_1 \left(\begin{array}{c} -n, \, p^* \\ q^* \end{array} \middle| \, r^* x + s^* \right),$$

we can simplify relation (1.82) as

$$J_n^{(p,q)}(x) = (2i)^n (1 - p + iq/2)_n \, {}_2F_1 \left(\begin{array}{c} -n, \, n + 1 - 2p \\ 1 - p + iq/2 \end{array} \middle| \, \frac{1 - ix}{2} \right)$$

$$= n!(2i)^n \, P_n^{(-p+iq/2, \, -p-iq/2)}(ix),$$

where $P_n^{(\alpha, \beta)}(x)$ are the same as the Jacobi orthogonal polynomials.

- Weight function:

$$W \left(\begin{array}{cc} 2(1 - p), \, q \\ 1, \quad 0, \quad 1 \end{array} \middle| \, x \right) = (1 + x^2)^{-p} \exp(q \arctan x).$$

1.4.2 How to Find the Initial Vector If a Special Case of the Main Weight Function Is Given?

The easiest way to find initial parameters a, b, c, d, e is to first compute the logarithmic derivative of the given weight function and then match the pattern with

$$\frac{W'(x)}{W(x)} = \frac{(d - 2a)x + e - b}{ax^2 + bx + c}.$$

We clarify this technique with some particular examples.

Example 1.1 Consider the weight function $W(x) = (-x^2 + 3x - 2)^{10}$ on $x \in [1, 2]$. By computing the logarithmic derivative of it, we get

$$\frac{W'(x)}{W(x)} = \frac{-20x + 30}{-x^2 + 3x - 2} = \frac{(d - 2a)x + e - b}{ax^2 + bx + c} \Rightarrow (a, b, c, d, e) = (-1, 3, -2, -22, 33).$$

Therefore, the related monic polynomials are $\bar{P}_n \begin{pmatrix} -22, & 33 \\ -1, & 3, & -2 \end{pmatrix} x$, which are orthogonal with respect to the given weight function on $[1, 2]$ for every value n. This means that it is no longer necessary to know that they are shifted Jacobi polynomials on the interval $[1, 2]$, because they can be explicitly expressed in terms of the generic polynomials (1.63).

Example 1.2 The weight function $W(x) = (2x^2 + 2x + 1)^{-10}$ is given on $x \in (-\infty, \infty)$. Thus

$$\frac{W'(x)}{W(x)} = \frac{-40x - 20}{2x^2 + 2x + 1} \Rightarrow (a, b, c, d, e) = (2, 2, 1, -38, -18),$$

and the related monic polynomials are $\bar{P}_n \begin{pmatrix} -38, & -18 \\ 2, & 2, & 1 \end{pmatrix} x$. Note that they are finitely orthogonal for $n \leq 9$, because according to (1.59), $N = \max\{n\} < 10 - (1/2)$. In other words, the finite set

$$\{ \bar{P}_n(x \, ; \, 2, 2, 1, -38, -18) \}_{n=0}^{9}$$

is orthogonal with respect to the weight function $(2x^2 + 2x + 1)^{-10}$ on $(-\infty, \infty)$.

Example 1.3 Consider the weight function $W(x) = \exp\left(x(\theta - x)\right)$ on $x \in (-\infty, \infty)$, where $\theta \in \mathbb{R}$. Then

$$\frac{W'(x)}{W(x)} = -2x + \theta \Rightarrow (a, b, c, d, e) = (0, 0, 1, -2, \theta),$$

and the related monic orthogonal polynomials are $\bar{P}_n \begin{pmatrix} -2, & \theta \\ 0, & 0, & 1 \end{pmatrix} x$ for every value n.

Example 1.4 For the weight function $W(x) = \frac{\sqrt{x}}{(x+2)^8}$ on $x \in [0, \infty)$, which is a special case of the second kind of the beta distribution, we have

$$\frac{W'(x)}{W(x)} = \frac{-15x + 2}{2x^2 + 4x} \Rightarrow (a, b, c, d, e) = (2, 4, 0, -11, 6).$$

Therefore, the related monic orthogonal polynomials are $\bar{P}_n \begin{pmatrix} -11, & 6 \\ 2, & 4, & 0 \end{pmatrix} x$, valid for $n \leq 3$. This means that the finite set $\{ \bar{P}_n(x \, ; \, 2, 4, 0, -11, 6) \}_{n=0}^{3}$ is orthogonal with respect to the weight function $\sqrt{x}/(x + 2)^8$ on $[0, \infty)$.

1.5   Fourier Transforms of Finite Sequences of Hypergeometric Orthogonal Polynomials

It is known that some orthogonal polynomial systems are mapped onto each other by the Fourier transform. The best-known examples of this type are Hermite functions, i.e., Hermite polynomials $H_n(x)$ multiplied by $\exp(-x^2/2)$, which are eigenfunctions of a Fourier transform. In this section, we introduce three examples of this type using finite sequences of hypergeometric orthogonal polynomials and obtain their orthogonality relations in the sequel.

To derive the Fourier transform of polynomials $M_n^{(p,q)}(x)$ and $N_n^{(p)}(x)$ defined in (1.43) and (1.49), we will use the two beta integrals

$$B(a,b) = \int_0^1 x^{a-1}(1-x)^{b-1}dx = \int_0^\infty \frac{t^{a-1}}{(1+t)^{a+b}}dt = \frac{\Gamma(a)\Gamma(b)}{\Gamma(a+b)}.$$

The Fourier transform of a function, say $g(x)$, is defined as

$$\mathbf{F}(s) = \mathbf{F}(g(x)) = \int_{-\infty}^\infty e^{-isx} g(x)\, dx,$$

and for the inverse transform we have

$$g(x) = \frac{1}{2\pi} \int_{-\infty}^\infty e^{isx}\mathbf{F}(s)\, ds.$$

Also for $g, h \in L^2(\mathbb{R})$, Parseval's identity is given by

$$\int_{-\infty}^\infty g(x)h(x)\, dx = \frac{1}{2\pi} \int_{-\infty}^\infty \mathbf{F}(g(x))\overline{\mathbf{F}(h(x))}\, ds. \tag{1.83}$$

We now define the specific functions

$$g(x) = \frac{e^{qx}}{(1+e^x)^{p+q}} M_n^{(r,u)}(e^x) \quad \text{and} \quad h(x) = \frac{e^{ax}}{(1+e^x)^{a+b}} M_m^{(c,d)}(e^x)$$

in terms of the polynomials $M_n^{(p,q)}(x)$.

For both functions, the Fourier transform exists. For instance, for $g(x)$ we have

$$\mathbf{F}(g(x)) = \int_{-\infty}^\infty e^{-isx} \frac{e^{qx}}{(1+e^x)^{p+q}} M_n^{(r,u)}(e^x)\, dx = \int_0^\infty t^{-is-1} \frac{t^q}{(1+t)^{p+q}} M_n^{(r,u)}(t)\, dt$$

$$= (-1)^n n! \binom{u+n}{n} \sum_{k=0}^{n} \frac{(-1)^k (-n)_k (n+1-r)_k}{(u+1)_k k!} \left(\int_0^\infty \frac{t^{q-is-1+k}}{(1+t)^{p+q}} dt \right)$$

$$= (-1)^n \frac{(u+n)!}{u!} \sum_{k=0}^{n} \frac{(-1)^k (-n)_k (n+1-r)_k}{(u+1)_k k!} \frac{\Gamma(q-is+k)\Gamma(p+is-k)}{\Gamma(p+q)}$$

$$= \frac{(-1)^n \Gamma(u+n+1)\Gamma(q-is)\Gamma(p+is)}{\Gamma(u+1)\Gamma(p+q)} \, {}_3F_2 \left(\begin{array}{c} -n,\, n+1-r,\, q-is \\ u+1,\, -p+1-is \end{array} \middle| 1 \right),$$

$$\tag{1.84}$$

in which the following identities have been used:

$$\Gamma(a+k) = \Gamma(a)(a)_k \quad \text{and} \quad \Gamma(a-k) = \frac{(-1)^k \Gamma(a)}{(1-a)_k}.$$

By substituting (1.84) into Parseval's identity (1.83), we obtain

$$2\pi \int_{-\infty}^{\infty} \frac{e^{(q+a)x}}{(1+e^x)^{(p+q+a+b)}} M_n^{(r,u)}(e^x) M_m^{(c,d)}(e^x) dx$$

$$= 2\pi \int_0^\infty \frac{t^{q+a-1}}{(1+t)^{p+q+a+b}} M_n^{(r,u)}(t) M_m^{(c,d)}(t) dt$$

$$= \frac{(-1)^{n+m} \Gamma(u+n+1)\Gamma(d+m+1)}{\Gamma(u+1)\Gamma(p+q)\Gamma(d+1)\Gamma(a+b)}$$

$$\times \int_{-\infty}^{\infty} \Gamma(q-is)\Gamma(p+is)\overline{\Gamma(a-is)\Gamma(b+is)} \, {}_3F_2 \left(\begin{array}{c} -n,\, n+1-r,\, q-is \\ u+1,\, -p+1-is \end{array} \middle| 1 \right)$$

$$\times {}_3F_2 \left(\begin{array}{c} -m,\, m+1-c,\, a-is \\ d+1,\, -b+1-is \end{array} \middle| 1 \right) ds. \tag{1.85}$$

On the other hand, if on the left-hand side of (1.85) we take

$$u = d = q+a-1 \quad \text{and} \quad r = c = p+b+1,$$

then according to the orthogonality relation (1.47), Eq. (1.85) reads

$$\frac{(2\pi)n!(p+b-n)!(q+a-1+n)!}{(p+b-2n)(p+q+a+b-n-1)!} \frac{\Gamma^2(q+a)\Gamma(p+q)\Gamma(a+b)}{(-1)^{n+m}\Gamma(q+a+n)\Gamma(q+a+m)} \delta_{n,m}$$

$$= \int_{-\infty}^{\infty} \Gamma(q-is)\Gamma(p+is)\overline{\Gamma(a-is)\Gamma(b+is)} \, {}_3F_2\left(\begin{array}{cc} -n, \ n-p-b, \ q-is \\ q+a, \ -p+1-is \end{array} \middle| 1\right)$$

$$\times \, {}_3F_2\left(\begin{array}{cc} -m, \ m-p-b, \ a-is \\ q+a, \ -b+1-is \end{array} \middle| 1\right) ds.$$

Hence, the special function

$$A_n(x; a, b, c, d) = \frac{\Gamma(a+d+n)}{\Gamma(a+d)} \, {}_3F_2\left(\begin{array}{cc} -n, \ n-b-c, \ d-x \\ a+d, \ -c+1-x \end{array} \middle| 1\right)$$

satisfies a finite orthogonality relation of the form

$$\frac{1}{2\pi} \int_{-\infty}^{\infty} \Gamma(a+ix)\Gamma(b-ix)\Gamma(c+ix)\Gamma(d-ix) A_n(ix; a, b, c, d) A_m(-ix; d, c, b, a) dx$$

$$= \frac{n!}{(b+c-2n)} \frac{\Gamma(a+d+n)\Gamma(b+c+1-n)\Gamma(c+d)\Gamma(a+b)}{\Gamma(a+b+c+d-n)} \delta_{n,m},$$

where $a+d > -n$, $b+c > 2n$, $a+b > 0$, and $c+d > 0$.

This approach can be similarly applied to the finite orthogonal polynomials $N_n^{(p)}(x)$. For this purpose, we first define the specific functions

$$u(x) = \exp(-px - \frac{1}{2}e^{-x}) N_n^{(q)}(e^x) \quad \text{and} \quad v(x) = \exp(-rx - \frac{1}{2}e^{-x}) N_m^{(u)}(e^x),$$

and take the Fourier transform of, e.g., $u(x)$ to get

$$\mathbf{F}(u(x)) = \int_{-\infty}^{\infty} e^{-isx} e^{-(px+\frac{1}{2}e^{-x})} N_n^{(q)}(e^x) dx = \int_0^{\infty} t^{-is-1-p} e^{\frac{-1}{2t}} N_n^{(q)}(t) dt$$

$$= (-1)^n n!(q-n-1)! \sum_{k=0}^{n} \frac{(-1)^k}{(q-n-1-k)!k!(n-k)!} \left(\int_0^{\infty} t^{-is-1-p+k} e^{\frac{-1}{2t}} dt\right)$$

$$= (-1)^n 2^{p+is} \Gamma(p+is) \sum_{k=0}^{n} \frac{(-n)_k (n+1-q)_k}{(1-p-is)_k} \frac{(2^{-1})^k}{k!}$$

$$= (-1)^n 2^{p+is} \Gamma(p+is) \, {}_2F_1\left(\begin{array}{cc} -n, \ n+1-q \\ 1-p-is \end{array} \middle| \frac{1}{2}\right). \tag{1.86}$$

In (1.86), we have used the definite integral

$$\int_0^{\infty} t^{-is-1-p+k} e^{\frac{-1}{2t}} dt = 2^{p+is-k} \Gamma(p+is-k).$$

Now apply Parseval's identity once again to obtain

$$2\pi \int_{-\infty}^{\infty} e^{-(p+r)x} e^{-e^{-x}} N_n^{(q)}(e^x) N_m^{(u)}(e^x)\,dx = 2\pi \int_0^{\infty} t^{-(p+r+1)} e^{\frac{-1}{t}} N_n^{(q)}(t) N_m^{(u)}(t)\,dt$$

$$= (-1)^{n+m} 2^{p+r}$$

$$\times \int_{-\infty}^{\infty} \Gamma(p+is)\overline{\Gamma(r+is)}\, {}_2F_1\left(\begin{matrix} -n,\, n+1-q \\ 1-p-is \end{matrix}\,\middle|\, \frac{1}{2}\right) \overline{{}_2F_1\left(\begin{matrix} -m,\, m+1-u \\ 1-r-is \end{matrix}\,\middle|\, \frac{1}{2}\right)}\,ds.$$

$$(1.87)$$

If $q = u = p + r + 1$ in (1.87), on noting the orthogonality relation (1.53), it can be concluded that the function

$$B_n(x; a, b) = {}_2F_1\left(\begin{matrix} -n,\, n-a-b \\ -a+1-x \end{matrix}\,\middle|\, \frac{1}{2}\right)$$

satisfies a finite orthogonality relation of the form

$$\frac{1}{2\pi} \int_{-\infty}^{\infty} \Gamma(a+ix)\Gamma(b-ix)\, B_n(ix; a, b) B_m(-ix; b, a)\,dx = \frac{n!\,\Gamma(a+b+1-n)}{(a+b-2n)\,2^{a+b}}\,\delta_{n,m},$$

provided that $a + b > 2n$.

Finally, the aforementioned approach can be applied to the third finite sequence $J_n^{(p,q)}(x)$ defined in (1.57). We now consider the specific functions

$$g(x) = (1+x^2)^{-\beta} \exp(\alpha \arctan x) J_n^{(c,d)}(x)$$

and

$$h(x) = (1+x^2)^{-u} \exp(l \arctan x) J_m^{(v,w)}(x),$$

in which $\alpha, \beta, c, d,$ and l, u, v, w are all real parameters.

For instance, for the function g, we get

$$\mathbf{F}(g(x)) = \int_{-\infty}^{\infty} e^{-isx}(1+x^2)^{-\beta} \exp(\alpha \arctan x) J_n^{(c,d)}(x)\,dx$$

$$= (2i)^n (1-c+id/2)_n$$

$$\times \int_{-\infty}^{\infty} e^{-isx}(1-ix)^{-\beta+i\frac{\alpha}{2}}(1+ix)^{-\beta-i\frac{\alpha}{2}} \left(\sum_{k=0}^{n} \frac{(-n)_k(n+1-2c)_k}{(1-c+id/2)_k k! 2^k}(1-ix)^k \right)dx$$

$$= (2i)^n (1-c+id/2)_n$$

$$\times \sum_{k=0}^{n} \frac{(-n)_k(n+1-2c)_k}{(1-c+id/2)_k k! 2^k} \int_{-\infty}^{\infty} e^{-isx}(1-ix)^{-\beta+k+i\frac{\alpha}{2}}(1+ix)^{-\beta-i\frac{\alpha}{2}}dx. \tag{1.88}$$

So it remains to evaluate in (1.88) the integral

$$A_k^*(s;\alpha,\beta) = \int_{-\infty}^{\infty} e^{-isx}(1-ix)^{-\beta+k+i\frac{\alpha}{2}}(1+ix)^{-\beta-i\frac{\alpha}{2}}dx. \tag{1.89}$$

In general, if $\mathrm{Re}(p+q) > 1$, then according to [3] we have

$$\int_{-\infty}^{\infty} e^{-isx}(1-ix)^{-p}(1+ix)^{-q}dx = \begin{cases} \frac{\pi}{\Gamma(p)}(s/2)^{\frac{p+q}{2}-1} W_{\frac{p-q}{2},\frac{1-p-q}{2}}(2s) & (s>0), \\[2mm] \frac{\pi}{\Gamma(q)}(-s/2)^{\frac{p+q}{2}-1} W_{\frac{q-p}{2},\frac{1-p-q}{2}}(-2s) & (s<0), \end{cases}$$

where $W_{a,b}(s)$ denotes Whittaker functions of the second kind, defined by

$$W_{a,b}(s) = s^{1/2}e^{-s/2}\left(\frac{\Gamma(-2b)}{\Gamma(1/2-b-a)} s^b\, {}_1F_1\left(\begin{array}{c} 1/2+b-a \\ 2b+1 \end{array} \middle| s \right) \right.$$

$$\left. + \frac{\Gamma(2b)}{\Gamma(1/2+b-a)} s^{-b}\, {}_1F_1\left(\begin{array}{c} 1/2-b-a \\ -2b+1 \end{array} \middle| s \right) \right), \quad 2b \notin \mathbb{Z}.$$

Hence, for $\mathrm{Re}(2\beta-k) < 1$ and $k+1-2\beta \notin \mathbb{Z}$, (1.89) would be equal to

$$A_k^*(s;\alpha,\beta) = \begin{cases} \frac{\pi}{\Gamma(\beta-k-i\alpha/2)}(s/2)^{\beta-1-k/2} W_{-\frac{k+i\alpha}{2},\frac{k+1}{2}-\beta}(2s) & (s>0), \\[2mm] \frac{\pi}{\Gamma(\beta+i\alpha/2)}(-s/2)^{\beta-1-k/2} W_{\frac{k+i\alpha}{2},\frac{k+1}{2}-\beta}(-2s) & (s<0), \end{cases}$$

and relation (1.88) can be simplified as

$$\mathbf{F}(g(x)) = 2^n i^n \frac{\Gamma(1-c+n+id/2)}{\Gamma(1-c+id/2)} C_n(s;\alpha,\beta,c,d), \tag{1.90}$$

where we have defined

$$C_n(s; p_1, p_2, p_3, p_4) = \sum_{k=0}^{n} \frac{(-n)_k(n+1-2p_3)_k}{(1-p_3+ip_4/2)_k k! 2^k} A_k^*(s; p_1, p_2). \tag{1.91}$$

By referring to (1.83), another consequence is that

$$\overline{\mathbf{F}(h(x))} = \frac{(-i)^m 2^m \Gamma(1-v+m-iw/2)}{\Gamma(1-v-iw/2)} C_m(s; -l, u, v, -w). \tag{1.92}$$

Therefore, on substituting (1.90) and (1.92) into Parseval's identity (1.83), we obtain

$$\int_{-\infty}^{\infty} (1+x^2)^{-(\beta+u)} \exp((\alpha+l)\arctan x)\, J_n^{(c,d)}(x)\, J_m^{(v,w)}(x)\, dx$$

$$= \frac{(-1)^m i^{n+m} 2^{m+n} \Gamma(1-c+n+id/2)\,\Gamma(1-v+m-iw/2)}{\Gamma(1-c+id/2)\,\Gamma(1-v-iw/2)}$$

$$\times \int_{-\infty}^{\infty} C_n(s;\alpha,\beta,c,d)\, C_m(s;-l,u,v,-w)\, ds. \qquad (1.93)$$

If $c = v = \beta + u$ and $d = w = \alpha + l$ in (1.93), then according to the orthogonality relation (1.60), we conclude that the sequence of functions

$$\{C_n(s;p_1,p_2,p_3,p_4)\}_{n\geq 0},$$

defined in (1.91) satisfies the following finite orthogonality relation:

$$\int_{-\infty}^{\infty} C_n(s;\alpha,\beta,v,\omega)\, C_m(s;\alpha-\omega,v-\beta,v,-\omega)\, ds$$

$$= \frac{\pi\, 2^{2-2v} n!}{(2v-2n-1)} \frac{\Gamma(2v-n)\,\Gamma^2(1-v-i\omega/2)}{\Gamma(v-n+iw/2)\Gamma(v-n-iw/2)\Gamma^2(1-v+n+i\omega/2)}\, \delta_{m,n},$$

for $m, n = 0, 1, \ldots, N = \max\{m,n\} < 2\beta - 1 \leq v - 1/2$, $\mathrm{Re}(2\beta - n) < 1$, $n+1-2\beta \notin \mathbb{Z}$, and $\alpha, \omega \in \mathbb{R}$.

1.6   A Symmetric Generalization of Sturm–Liouville Problems

In this section, we present some conditions under which ordinary Sturm–Liouville problems with symmetric solutions are extended to a larger class. The main advantage of this extension is that the corresponding solutions preserve the orthogonality property.

Let $\Phi_n(x)$ be a sequence of symmetric functions that satisfies the second-order differential equation

$$A(x)\Phi_n''(x)+B(x)\Phi_n'(x)+\left(\lambda_n C(x)+D(x)+\frac{1-(-1)^n}{2}E(x)\right)\Phi_n(x) = 0, \quad (1.94)$$

where $A(x)$, $B(x)$, $C(x)$, $D(x)$, and $E(x)$ are independent functions and $\{\lambda_n\}$ is a sequence of constants. Clearly, choosing $E(x) = 0$ in (1.94) leads to an ordinary Sturm–Liouville equation. Since $\Phi_n(x)$ is symmetric, by substituting the symmetry property

$$\Phi_n(-x) = (-1)^n \Phi_n(x),$$

into (1.94), we can immediately conclude that $A(x)$, $C(x)$, $D(x)$, and $E(x)$ are even functions, while $B(x)$ must be odd. Now we write Eq. (1.94) in a self-adjoint form to get

$$\frac{d}{dx}\left(A(x)\,R(x)\frac{d\,\Phi_n(x)}{dx}\right)$$

$$= -\left(\lambda_n\,C(x) + D(x)\right) R(x)\,\Phi_n(x) - \frac{1-(-1)^n}{2}E(x)R(x)\,\Phi_n(x),\qquad(1.95)$$

where

$$R(x) = \exp\left(\int \frac{B(x)-A'(x)}{A(x)}\,dx\right) = \frac{1}{A(x)}\exp\left(\int \frac{B(x)}{A(x)}\,dx\right).$$

Since $A(x)R(x)$ is an even function, the orthogonality interval should be also symmetric, say $[-\theta,\theta]$. Hence if $x = \theta$ is a root of the function $A(x)R(x)$, applying the Sturm–Liouville theorem to (1.95) yields

$$\left[A(x)\,R(x)\,\left(\Phi_n'(x)\Phi_m(x) - \Phi'_m(x)\Phi_n(x)\right)\right]_{-\theta}^{\theta}$$

$$= (\lambda_m - \lambda_n)\int_{-\theta}^{\theta} C(x)\,R(x)\,\Phi_n(x)\,\Phi_m(x)\,dx$$

$$- \left(\frac{(-1)^m - (-1)^n}{2}\right)\int_{-\theta}^{\theta} E(x)\,R(x)\,\Phi_n(x)\,\Phi_m(x)\,dx.\qquad(1.96)$$

Obviously, the left-hand side of (1.96) is zero. So to prove the orthogonality property, it is enough to show that the value

$$F(n,m) = \frac{(-1)^m - (-1)^n}{2}\int_{-\theta}^{\theta} E(x)\,R(x)\,\Phi_n(x)\,\Phi_m(x)\,dx$$

is always equal to zero for every $m, n \in \mathbb{Z}^+$. For this purpose, four cases should be generally considered for m and n:

If both m and n are even (or odd), then clearly $F(n,m) = 0$.

If one of the two values is odd and the other one is even (or conversely), then

$$F(2i,\,2j+1) = -\int_{-\theta}^{\theta} E(x)\,R(x)\,\Phi_{2i}(x)\,\Phi_{2j+1}(x)\,dx.\qquad(1.97)$$

Since in the above integral $E(x)$, $R(x)$, and $\Phi_{2i}(x)$ are even functions and $\Phi_{2j+1}(x)$ is odd, the integrand of (1.97) is an odd function, and consequently $F(2i,\,2j+1) = 0$. This issue also holds for the case $n = 2i+1$ and $m = 2j$, i.e., $F(2i+1,\,2j) = 0$.

Theorem 1.2 *The symmetric sequence $\Phi_n(x) = (-1)^n \Phi_n(-x)$, as a specific solution of differential equation (1.94), satisfies the orthogonality relation*

$$\int_{-\theta}^{\theta} W^*(x)\, \Phi_n(x)\, \Phi_m(x)\, dx = \left(\int_{-\theta}^{\theta} W^*(x)\, \Phi_n^2(x)\, dx \right) \delta_{n,m},$$

where

$$W^*(x) = C(x)R(x) = C(x)\exp(\int \frac{B(x) - A'(x)}{A(x)}\, dx) \tag{1.98}$$

denotes the corresponding weight function and is a positive and even function on $[-\theta, \theta]$.

Remark 1.5 Although some special functions such as Bessel functions are symmetric and satisfy a differential equation whose coefficients are alternately even and odd, they do not satisfy the conditions of the above theorem. For instance, if in Eq. (1.94) we choose

$$A(x) = x^2, \quad B(x) = x, \quad C(x) = 1, \quad D(x) = x^2, \quad E(x) = 0, \quad \text{and } \lambda_n = -n^2,$$

we obtain the Bessel differential equation

$$x^2\, \Phi_n''(x) + x\, \Phi_n'(x) + (x^2 - n^2)\, \Phi_n(x) = 0,$$

with the symmetric solution

$$(-1)^n J_n(-x) = J_n(x) = \sum_{k=0}^{\infty} \frac{(-1)^k}{k!(n+k)!} (\frac{x}{2})^{n+2k} \qquad (n \in \mathbb{Z}^+).$$

However, since the orthogonality interval of Bessel functions is not symmetric (i.e., $[0, 1]$), the theorem cannot be applied to $J_n(x)$ unless there exists a specific even function $E(x)$ for Eq. (1.94) such that the corresponding solution has infinitely many zeros, identical to the case of the usual Bessel functions of order n.

1.7   A Basic Class of Symmetric Orthogonal Polynomials

In the previous section, we determined how to generalize ordinary Sturm–Liouville problems with symmetric solutions. In this section, using Theorem 1.2, we introduce a basic class of symmetric orthogonal polynomials with four free parameters and obtain its standard properties. Then we show that four main sequences of symmetric orthogonal polynomials can be extracted from the introduced class. They are respectively the generalized ultraspherical polynomials, generalized Hermite polynomials, and two other sequences of symmetric polynomials, which are finitely orthogonal on the real line.

So referring to differential equation (1.94), let $p, q, r, s \in \mathbb{R}$ and choose

$$
\begin{aligned}
A(x) &= x^2(px^2 + q), \\
B(x) &= x(rx^2 + s), \\
C(x) &= x^2, \\
D(x) &= 0, \\
E(x) &= -s,
\end{aligned}
\tag{1.99}
$$

leading to the second-order equation

$$
x^2(px^2 + q)\, \Phi_n''(x) + x(rx^2 + s)\, \Phi_n'(x)
$$
$$
- \left(n(r + (n-1)p)x^2 + (1 - (-1)^n)\, s/2 \right) \Phi_n(x) = 0,
\tag{1.100}
$$

whose polynomial solution is

$$
\begin{aligned}
\Phi_n(x) &= S_n \left(\left. \begin{matrix} r & s \\ p & q \end{matrix} \right| x \right) \\
&= \sum_{k=0}^{[n/2]} \binom{[n/2]}{k} \left(\prod_{i=0}^{[n/2]-(k+1)} \frac{(2i + (-1)^{n+1} + 2[n/2])\, p + r}{(2i + (-1)^{n+1} + 2)\, q + s} \right) x^{n-2k},
\end{aligned}
\tag{1.101}
$$

where neither both q and s nor p and r can vanish together.

These polynomials cover almost all known symmetric orthogonal polynomials, such as Legendre polynomials, the first and second kinds of Chebyshev polynomials, ultraspherical polynomials, generalized ultraspherical polynomials, Hermite polynomials, and generalized Hermite polynomials. Furthermore, there are two symmetric sequences of finite orthogonal polynomials that are special cases of the general representation (1.101).

We define the monic type of polynomials (1.101) as

$$
\bar{S}_n \left(\left. \begin{matrix} r & s \\ p & q \end{matrix} \right| x \right) = \prod_{i=0}^{[n/2]-1} \frac{(2i + (-1)^{n+1} + 2)\, q + s}{(2i + (-1)^{n+1} + 2[n/2])\, p + r} \, S_n \left(\left. \begin{matrix} r & s \\ p & q \end{matrix} \right| x \right).
\tag{1.102}
$$

A straightforward result from (1.101) and (1.102) is that

$$
\bar{S}_{2n+1} \left(\left. \begin{matrix} r & s \\ p & q \end{matrix} \right| x \right) = x\, \bar{S}_{2n} \left(\left. \begin{matrix} r + 2p, & s + 2q \\ p, & q \end{matrix} \right| x \right),
\tag{1.103}
$$

and accordingly

$$\Phi_{2n}(x) = \bar{S}_{2n}\left(\begin{array}{cc} r & s \\ p & q \end{array}\middle| x\right) \quad \text{and} \quad \Phi_{2n+1}(x) = x\,\bar{S}_{2n}\left(\begin{array}{cc} r+2p, & s+2q \\ p, & q \end{array}\middle| x\right). \quad (1.104)$$

For instance, we have

$$\bar{S}_0\left(\begin{array}{cc} r & s \\ p & q \end{array}\middle| x\right) = 1,$$

$$\bar{S}_1\left(\begin{array}{cc} r & s \\ p & q \end{array}\middle| x\right) = x,$$

$$\bar{S}_2\left(\begin{array}{cc} r & s \\ p & q \end{array}\middle| x\right) = x^2 + \frac{q+s}{p+r},$$

$$\bar{S}_3\left(\begin{array}{cc} r & s \\ p & q \end{array}\middle| x\right) = x^3 + \frac{3q+s}{3p+r}x,$$

$$\bar{S}_4\left(\begin{array}{cc} r & s \\ p & q \end{array}\middle| x\right) = x^4 + 2\frac{3q+s}{5p+r}x^2 + \frac{(3q+s)(q+s)}{(5p+r)(3p+r)},$$

$$\bar{S}_5\left(\begin{array}{cc} r & s \\ p & q \end{array}\middle| x\right) = x^5 + 2\frac{5q+s}{7p+r}x^3 + \frac{(5q+s)(3q+s)}{(7p+r)(5p+r)}x. \quad (1.105)$$

The hypergeometric representation of monic polynomials (1.102) is

$$\bar{S}_n\left(\begin{array}{cc} r & s \\ p & q \end{array}\middle| x\right) = x^n {}_2F_1\left(\begin{array}{c} -[n/2],\ (q-s)/2q - [(n+1)/2] \\ -(r+(2n-3)p)/2p \end{array}\middle| -\frac{q}{px^2}\right).$$

Therefore, on referring to the relation (1.10), we obtain

$$\bar{S}_n\left(\begin{array}{cc} r & s \\ p & q \end{array}\middle| \sqrt{-\frac{q}{p}}\right) = \frac{(-\frac{q}{p})^{\frac{n}{2}}\Gamma(\frac{3}{2}-\frac{r}{2p})\Gamma(1+\frac{s}{2q}-\frac{r}{2p})}{(\frac{3}{2}-\frac{r}{2p}-n)_n \Gamma(\frac{3}{2}-\frac{r}{2p}+[\frac{n}{2}]-n)\Gamma(1+\frac{s}{2q}-\frac{r}{2p}-[\frac{n}{2}])}.$$

Also, a three-term recurrence relation can be derived for them as

$$\bar{S}_{n+1}(x) = x\,\bar{S}_n(x) + C_n\left(\begin{array}{cc} r & s \\ p & q \end{array}\right)\bar{S}_{n-1}(x) \quad (n \in \mathbb{N}), \quad (1.106)$$

in which

$$C_n \begin{pmatrix} r & s \\ p & q \end{pmatrix} = \frac{pq\,n^2 + \big((r-2p)q - (-1)^n ps\big)\,n + (r-2p)\,s(1-(-1)^n)/2}{(2pn+r-p)(2pn+r-3p)}.$$

$$(1.107)$$

Since the recurrence relation (1.106) is explicitly known, to determine the norm square value of the polynomials, we can now use Favard's theorem by noting that there is orthogonality with respect to a weight function. According to this theorem, if $\{P_n(x)\}_{n=0}^{\infty}$ is defined by

$$x\,P_n(x) = A_n\,P_{n+1}(x) + B_n\,P_n(x) + C_n\,P_{n-1}(x) \qquad (n = 0, 1, 2, \ldots),$$

where $P_{-1}(x) = 0$, $P_0(x) = 1$, A_n, B_n, C_n real, and $A_n C_{n+1} > 0$ for $n = 0, 1, \ldots$, then there exists a weight function $W(x)$ such that

$$\int_{-\infty}^{\infty} W(x) P_n(x)\, P_m(x)\, dx = \left(\prod_{i=0}^{n-1} \frac{C_{i+1}}{A_i} \int_{-\infty}^{\infty} W(x)\, dx\right) \delta_{n,m}. \qquad (1.108)$$

Moreover, if the positive condition $A_n C_{n+1} > 0$ holds only for $n = 0, 1, \ldots, N$ then the orthogonality relation (1.108) holds for only a finite number of m, n. The latter note will help us in the next sections obtain two new subclasses of $\bar{S}_n(x\,;\,p, q, r, s)$ that are finitely orthogonal on $(-\infty, \infty)$.

It is clear that Favard's theorem is also valid for (1.106), in which $A_n = 1$, $B_n = 0$, and $C_n = -C_n(p, q, r, s)$. Moreover, the condition $-C_n(p, q, r, s) > 0$ must always be satisfied if we want to apply Favard's theorem to (1.106).

By noting (1.98), the weight function corresponding to symmetric polynomials (1.101) is derived as

$$W \begin{pmatrix} r & s \\ p & q \end{pmatrix} x = x^2 \exp\left(\int \frac{(r-4p)x^2 + (s-2q)}{x(px^2+q)}\, dx\right) = \exp\left(\int \frac{(r-2p)x^2 + s}{x(px^2+q)}\, dx\right).$$

$$(1.109)$$

Hence, a generic orthogonality relation can be designed for (1.101) as

$$\int_{-\theta}^{\theta} W \begin{pmatrix} r & s \\ p & q \end{pmatrix} x\, \bar{S}_n \begin{pmatrix} r & s \\ p & q \end{pmatrix} x\, \bar{S}_m \begin{pmatrix} r & s \\ p & q \end{pmatrix} x\, dx$$

$$= \left((-1)^n \prod_{i=1}^{n} C_i \begin{pmatrix} r & s \\ p & q \end{pmatrix} \int_{-\theta}^{\theta} W \begin{pmatrix} r & s \\ p & q \end{pmatrix} x\, dx\right) \delta_{n,m}, \qquad (1.110)$$

where θ might be the standard values $1, \infty$ and the function $(px^2 + q)\,W(x; p, q, r, s)$ must vanish at $x = \theta$ in order that the main orthogonality relation (1.110) be valid.

The positive function (1.109) can be investigated from a statistical point of view, too. In fact, this function is an analogue of the Pearson family of distributions

$$\rho \left(\begin{array}{cc} d\ e \\ a\ b\ c \end{array} \middle| x \right) = \exp \left(\int \frac{(d - 2a)x + e - b}{ax^2 + bx + c} dx \right), \tag{1.111}$$

satisfying the first-order differential equation

$$\frac{d}{dx}\left((ax^2 + bx + c)\,\rho(x)\right) = (dx + e)\,\rho(x). \tag{1.112}$$

Therefore, similar to Eq. (1.112), the weight function (1.109) satisfies a first-order differential equation of the form

$$x\frac{d}{dx}\left((px^2 + q)\,W(x)\right) = (rx^2 + s)\,W(x), \tag{1.113}$$

which is equivalent to

$$\frac{d}{dx}(x^2(px^2 + q)\,W(x)) = x\,(r^*x^2 + s^*)\,W(x), \tag{1.114}$$

for $r^* = r + 2p$ and $s^* = s + 2q$.

It is deduced from (1.113) or (1.114) that $W(x; p, q, r, s)$ is an analytic integrable function, and since it is positive, its probability density function is available. In general, there are four main subclasses of the family of distributions (1.109) (and consequently subsolutions of Eq. (1.113)) whose explicit probability density functions are as follows:

$$K_1\,W \left(\begin{array}{cc} -2a - 2b - 2,\ 2a \\ -1, \qquad\qquad 1 \end{array} \middle| x \right) = \frac{\Gamma(a + b + 3/2)}{\Gamma(a + 1/2)\Gamma(b + 1)}\, x^{2a}(1 - x^2)^b,$$

for $x \in [-1, 1]$ and $a + 1/2 > 0,\ \ b + 1 > 0,$

$$K_2\,W \left(\begin{array}{cc} -2,\ 2a \\ 0,\ \ 1 \end{array} \middle| x \right) = \frac{1}{\Gamma(a + 1/2)}\, x^{2a}e^{-x^2},$$

for $x \in (-\infty, \infty)$ and $a + 1/2 > 0,$

$$K_3\,W \left(\begin{array}{cc} -2a - 2b + 2,\ -2a \\ 1, \qquad\qquad\ 1 \end{array} \middle| x \right) = \frac{\Gamma(b)}{\Gamma(b + a - 1/2)\Gamma(-a + 1/2)}\frac{x^{-2a}}{(1 + x^2)^b},$$

for $x \in (-\infty, \infty)$ and $b + a > 1/2,\ a < 1/2, b > 0,$

$$K_4 \, W \left(\begin{array}{cc} -2a+2, \ 2 \\ 1, \quad \ 0 \end{array} \middle| x \right) = \frac{1}{\Gamma(a-1/2)} \, x^{-2a} e^{-\frac{1}{x^2}},$$

for $x \in (-\infty, \infty)$ and $a > 1/2$.

The values $\{K_i\}_{i=1}^4$ play a normalizing constant role in the above distributions. Moreover, it is observed that the value of the distribution vanishes at $x = 0$ in each of the four cases, i.e., $W(0; p, q, r, s) = 0$ for $s \neq 0$. Note that if $s = 0$ in (1.113), the aforementioned equation will be reduced to a special case of the Pearson differential equation (1.112). Hence, we hereinafter suppose that $s \neq 0$. Since the explicit forms of $S_n(x; p, q, r, s)$ in (1.101), $C_n(p, q, r, s)$ in (1.107), and $W(x; p, q, r, s)$ in (1.109) are all known, another probability density function can be defined by referring to the orthogonality relation (1.110) as follows:

$$D_m \left(\begin{array}{cc} r & s \\ p & q \end{array} \middle| x \right) = \frac{(-1)^m}{\prod\limits_{i=1}^{m} C_i \left(\begin{array}{cc} r & s \\ p & q \end{array} \middle| x \right) \int_{-\theta}^{\theta} W \left(\begin{array}{cc} r & s \\ p & q \end{array} \middle| x \right) dx} \, W \left(\begin{array}{cc} r & s \\ p & q \end{array} \middle| x \right)$$

$$\times \left(\bar{S}_m \left(\begin{array}{cc} r & s \\ p & q \end{array} \middle| x \right) \right)^2 .$$

1.7.1 A Direct Relationship Between $\bar{P}_n(x; a, b, c, d, e)$ and $\bar{S}_n(x; p, q, r, s)$

There is a direct relation between the first kind of hypergeometric orthogonal polynomials and $S_n(x; p, q, r, s)$. To find it, we recall the differential equation

$$(ax^2 + bx + c)y_n''(x) + (dx + e)y_n'(x) - n((n-1)a + d)y_n(x) = 0, \tag{1.115}$$

together with the Rodrigues formula

$$\bar{P}_n \left(\begin{array}{cc} d & e \\ a & b & c \end{array} \middle| x \right) = \frac{1}{(\prod\limits_{k=1}^{n} d + (n+k-2)a) \, \rho \left(\begin{array}{cc} d & e \\ a & b & c \end{array} \middle| x \right)}$$

$$\times \frac{d^n}{dx^n} \left((ax^2 + bx + c)^n \rho \left(\begin{array}{cc} d & e \\ a & b & c \end{array} \middle| x \right) \right),$$

where $\rho(x; a, b, c, d, e)$ is defined in (1.111).

Since $\bar{S}_{2n}(x \,;\, p,q,r,s)$ is an even function, taking $x = wt^2 + v$ in (1.115) gives

$$t^2(aw^2t^4 + w(2av+b)t^2 + av^2 + bv + c)y_n''(t)$$
$$+ t\,((2d-a)w^2t^4 + (2wv(d-a) + w(2e-b))t^2 - (av^2+bv+c))y_n'(t)$$
$$- 4w^2n(d+(n-1)a)t^4 y_n(t) = 0. \qquad (1.116)$$

If (1.116) is equated with (1.100), we should have

$$av^2 + bv + c = 0 \quad \text{or} \quad v = \frac{-b \pm \sqrt{b^2 - 4ac}}{2a}. \qquad (1.117)$$

The condition (1.117) simplifies Eq. (1.116) as

$$t(awt^2 \pm \sqrt{b^2 - 4ac})\,y_n''(t)$$
$$+ ((2d-a)wt^2 + (\tfrac{d}{a} - 1)(-b \pm \sqrt{b^2 - 4ac}) + 2e - b)y_n'(t)$$
$$- 4wn(d+(n-1)a)t\, y_n(t) = 0$$
$$\Leftrightarrow\; y_n(t) = \bar{P}_n \left(\begin{array}{cc} d & e \\ a\; b & c \end{array} \middle|\; wt^2 + \frac{-b \pm \sqrt{b^2 - 4ac}}{2a} \right). \qquad (1.118)$$

Equation (1.118) is clearly a special case of (1.100), so that we have

$$P_n \left(\begin{array}{cc} d & e \\ a\; b & c \end{array} \middle|\; wt^2 + \frac{-b \pm \sqrt{b^2 - 4ac}}{2a} \right)$$
$$= K\, S_{2n} \left(\begin{array}{cc} (2d-a)w, & (\tfrac{d}{a}-1)(-b \pm \sqrt{b^2 - 4ac}) + 2e - b \\ aw, & \pm\sqrt{b^2 - 4ac} \end{array} \middle|\; t \right), \qquad (1.119)$$

where K is the leading coefficient of the left-hand-side polynomial of relation (1.119) divided by the leading coefficient of the right-hand-side polynomial.

As we observe, there exist five free parameters a, b, c, d, e on the left-hand side of (1.119). So one of them must be preassigned in order to obtain the explicit form of $\bar{S}_{2n}(x \,;\, p, q, r, s)$ in terms of $\bar{P}_n(wt^2 + v \,;\, a, b, c, d, e)$. For this purpose, if, for instance, $c = 0$ in (1.119), the following two cases appear:

$$\bar{S}_{2n} \left(\begin{array}{cc} r & s \\ p & q \end{array} \middle|\; t \right) = w^{-n}\, \bar{P}_n \left(\begin{array}{cc} \tfrac{r+p}{2w}, & \tfrac{s+q}{2} \\ p/w, & q, \quad 0 \end{array} \middle|\; wt^2 \right) \qquad (w \neq 0),$$

and

$$\bar{S}_{2n}\begin{pmatrix} r & s \\ p & q \end{pmatrix} \Bigg| t \Bigg) = w^{-n} \bar{P}_n \begin{pmatrix} \frac{r+p}{2w}, & \frac{sp-rq}{2p} \\ p/w, & -q, & 0 \end{pmatrix} \Bigg| wt^2 + w\frac{q}{p} \Bigg) \qquad (p, w \neq 0).$$

Moreover, if (1.103) is applied to the latter two relations, we respectively obtain

$$\bar{S}_{2n+1}\begin{pmatrix} r & s \\ p & q \end{pmatrix} \Bigg| t \Bigg) = w^{-n} t \, \bar{P}_n \begin{pmatrix} \frac{r+3p}{2w}, & \frac{s+3q}{2} \\ p/w, & q, & 0 \end{pmatrix} \Bigg| wt^2 \Bigg) \qquad (w \neq 0)$$

and

$$\bar{S}_{2n+1}\begin{pmatrix} r & s \\ p & q \end{pmatrix} \Bigg| t \Bigg) = w^{-n} t \, \bar{P}_n \begin{pmatrix} \frac{r+3p}{2w}, & \frac{sp-rq}{2p} \\ p/w, & -q, & 0 \end{pmatrix} \Bigg| wt^2 + w\frac{q}{p} \Bigg) \qquad (p, w \neq 0).$$

1.7.2　Four Special Cases of the Symmetric Polynomials $S_n(x\,;\,p,q,r,s)$

In this section, we study four special cases of $S_n(x\,;\,p,q,r,s)$ that are known as the generalized ultraspherical polynomials, generalized Hermite polynomials, and two other sequences of finite symmetric polynomials orthogonal on the real line.

1. For the generalized ultraspherical polynomials (GUP) we have

- Initial vector:

$$(p, q, r, s) = (-1,\, 1,\, -2a - 2b - 2,\, 2a).$$

- Differential equation:

$$x^2(-x^2 + 1)\,\Phi_n''(x) - 2x\big((a + b + 1)x^2 - a\big)\Phi_n'(x)$$
$$+ \Big(n(2a + 2b + n + 1)x^2 + ((-1)^n - 1)a\Big)\Phi_n(x) = 0.$$

- Explicit form of monic GUP:

$$\bar{S}_n\begin{pmatrix} -2a - 2b - 2, & 2a \\ -1, & 1 \end{pmatrix} \Bigg| x \Bigg) = \prod_{i=0}^{[n/2]-1} \frac{2i + 2a + 2 - (-1)^n}{-2i - (2b + 2a + 2 - (-1)^n + 2\,[n/2])}$$

$$\times \sum_{k=0}^{[n/2]} \binom{[n/2]}{k} \left(\prod_{i=0}^{[\frac{n}{2}]-(k+1)} \frac{-2i - (2b + 2a + 2 - (-1)^n + 2[n/2])}{2i + 2a + 2 - (-1)^n} \right) x^{n-2k}.$$

- Recurrence relation of monic polynomials:

$$\bar{S}_{n+1}(x) = x\,\bar{S}_n(x) + C_n \begin{pmatrix} -2a - 2b - 2, & 2a \\ -1, & 1 \end{pmatrix} \bar{S}_{n-1}(x),$$

where

$$C_n \begin{pmatrix} -2a - 2b - 2, & 2a \\ -1, & 1 \end{pmatrix} = \frac{-n^2 - (2b + 2(1 - (-1)^n)a)n - 2a(a+b)(1 - (-1)^n)}{(2n + 2a + 2b - 1)(2n + 2a + 2b + 1)}$$

$$= \frac{-(n + (1 - (-1)^n)a)(n + (1 - (-1)^n)a + 2b)}{(2n + 2a + 2b - 1)(2n + 2a + 2b + 1)}.$$

- Weight function:

$$W \begin{pmatrix} -2a - 2b - 2, & 2a \\ -1, & 1 \end{pmatrix} x \end{pmatrix} = x^{2a}(1 - x^2)^b.$$

- Orthogonality relation:

$$\int_{-1}^{1} x^{2a}(1 - x^2)^b \bar{S}_n \begin{pmatrix} -2a - 2b - 2, & 2a \\ -1, & 1 \end{pmatrix} x \end{pmatrix} \bar{S}_m \begin{pmatrix} -2a - 2b - 2, & 2a \\ -1, & 1 \end{pmatrix} x \end{pmatrix} dx$$

$$= \left((-1)^n \prod_{i=1}^{n} C_i \begin{pmatrix} -2a - 2b - 2, & 2a \\ -1, & 1 \end{pmatrix} \int_{-1}^{1} x^{2a}(1 - x^2)^b dx \right) \delta_{n,m}, \qquad (1.120)$$

where

$$\int_{-1}^{1} x^{2a}(1 - x^2)^b dx = B\left(a + \frac{1}{2}, b + 1\right) = \frac{\Gamma(a + 1/2)\Gamma(b + 1)}{\Gamma(a + b + 3/2)}. \qquad (1.121)$$

Relation (1.121) shows that the constraints on the parameters a and b are respectively $a + 1/2 > 0$, $(-1)^{2a} = 1$, and $b + 1 > 0$.

Since the ultraspherical, Legendre, and Chebyshev polynomials of the first and second kinds are all special cases of GUP, they can be expressed in terms of

$\bar{S}_n(x; p, q, r, s)$ directly, so that we have

- Ultraspherical polynomials:

$$C_n^a(x) = \frac{2^n (a)_n}{n!} \bar{S}_n \left(\begin{array}{cc} -2a - 1, & 0 \\ -1, & 1 \end{array} \middle| x \right).$$

- Legendre polynomials:

$$P_n(x) = \frac{(2n)!}{(n!)^2 2^n} \bar{S}_n \left(\begin{array}{cc} -2 & 0 \\ -1 & 1 \end{array} \middle| x \right).$$

- Chebyshev polynomials of the first kind:

$$T_n(x) = 2^{n-1} \bar{S}_n \left(\begin{array}{cc} -1 & 0 \\ -1 & 1 \end{array} \middle| x \right). \tag{1.122}$$

- Chebyshev polynomials of the second kind:

$$U_n(x) = 2^n \bar{S}_n \left(\begin{array}{cc} -3 & 0 \\ -1 & 1 \end{array} \middle| x \right).$$

Remark 1.6 Up to now, four kinds of trigonometric orthogonal polynomials, i.e., first, second, third, and fourth kinds of Chebyshev polynomials, have been well known in the literature. For $x = \cos \theta$, the first and second kinds are defined as

$$T_n(x) = 2^{n-1} \prod_{k=1}^{n} \left(x - \cos \frac{(2k-1)\pi}{2n} \right) = \cos(n\theta)$$

and

$$U_n(x) = 2^n \prod_{k=1}^{n} \left(x - \cos \frac{k\pi}{n+1} \right) = \frac{\sin((n+1)\theta)}{\sin \theta},$$

while for $x = \cos 2\theta$, the third and fourth kinds are defined as

$$V_n(x) = 2^n \prod_{k=1}^{n} \left(x - \cos \frac{(2k-1)\pi}{2n+1} \right) = \frac{\cos((2n+1)\theta)}{\cos(\theta)}$$

and

$$W_n(x) = 2^n \prod_{k=1}^{n} \left(x - \cos \frac{2k\pi}{2n+1} \right) = \frac{\sin((2n+1)\theta)}{\sin(\theta)}.$$

Here we would like to add that there exist two further kinds of half-trigonometric orthogonal polynomials, which are particular cases of $\bar{S}_n(x\,;\,p,q,r,s)$ and are generated by the first and second kinds of Chebyshev polynomials.

To generate them, we first refer to relation (1.103). According to (1.122), the initial vector of the first kind of Chebyshev polynomials is $(p,q,r,s) = (-1,1,-1,0)$. If this vector is replaced in (1.103), then we obtain

$$\bar{S}_{2n+1} \left(\begin{matrix} -1 & 0 \\ -1 & 1 \end{matrix} \,\middle|\, x \right) = \bar{T}_{2n+1}(x) = x\,\bar{S}_{2n} \left(\begin{matrix} -3 & 2 \\ -1 & 1 \end{matrix} \,\middle|\, x \right).$$

Hence, the secondary vector $(p,q,r,s) = (-1,1,-3,2)$ appears, and we can define the half-trigonometric polynomials

$$X_n(x) = S_n \left(\begin{matrix} -3 & 2 \\ -1 & 1 \end{matrix} \,\middle|\, x \right) = \begin{cases} \dfrac{(-1)^{n/2}}{n+1} \dfrac{\cos((n+1)\theta)}{\cos\theta} & (n=2m), \\[2mm] S_n(x\,;\,-1,1,-3,2) & (n=2m+1), \end{cases} \qquad (1.123)$$

in which $x = \cos\theta$. According to (1.123) and (1.106), $\bar{X}_n(x)$ satisfies the recurrence relation

$$\bar{X}_{n+1}(x) = x\,\bar{X}_n(x) + C_n \left(\begin{matrix} -3 & 2 \\ -1 & 1 \end{matrix} \right) \bar{X}_{n-1}(x),$$

where

$$C_n \left(\begin{matrix} -3 & 2 \\ -1 & 1 \end{matrix} \right) = \frac{-(n-(-1)^n)(n+1-(-1)^n)}{4n(n+1)}.$$

By substituting the initial vector $(-1,1,-3,2)$ into the generic relation (1.110), we obtain

$$\int_{-1}^{1} W \left(\begin{matrix} -3 & 2 \\ -1 & 1 \end{matrix} \,\middle|\, x \right) \bar{X}_n(x)\bar{X}_m(x)\,dx$$

$$= \left((-1)^n \prod_{i=1}^{n} C_i \left(\begin{matrix} -3 & 2 \\ -1 & 1 \end{matrix} \right) \int_{-1}^{1} W \left(\begin{matrix} -3 & 2 \\ -1 & 1 \end{matrix} \,\middle|\, x \right) dx \right) \delta_{n,m}. \qquad (1.124)$$

On the other hand, since

$$\int_{-1}^{1} W\left(\begin{array}{cc} -3 & 2 \\ -1 & 1 \end{array}\middle| x\right) dx = \int_{-1}^{1} \frac{x^2}{\sqrt{1-x^2}} dx = B\left(\frac{3}{2}, \frac{1}{2}\right) = \frac{\pi}{2},$$

(1.124) is simplified as

$$\int_{-1}^{1} \frac{x^2}{\sqrt{1-x^2}} \bar{X}_n(x) \bar{X}_m(x) dx = \left((-1)^n \prod_{i=1}^{n} C_i\left(\begin{array}{cc} -3 & 2 \\ -1 & 1 \end{array}\right)\right) \frac{\pi}{2} \delta_{n,m}.$$

Clearly $\bar{X}_{2n}(x)$ is decomposable as

$$\bar{X}_{2n}(x) = \prod_{k=1}^{2n} \left(x - \cos \frac{(2k-1)\pi}{2(2n+1)}\right).$$

We can similarly follow the above-mentioned approach for the second kind Chebyshev polynomials with $(p, q, r, s) = (-1, 1, -3, 0)$. If this vector is substituted into (1.103), then

$$\bar{S}_{2n+1}\left(\begin{array}{cc} -3 & 0 \\ -1 & 1 \end{array}\middle| x\right) = \bar{U}_{2n+1}(x) = x \, \bar{S}_{2n}\left(\begin{array}{cc} -5 & 2 \\ -1 & 1 \end{array}\middle| x\right),$$

which gives the secondary vector as $(p, q, r, s) = (-1, 1, -5, 2)$. By assuming $x = \cos\theta$, we can now define the polynomials

$$Y_n(x) = S_n\left(\begin{array}{cc} -5 & 2 \\ -1 & 1 \end{array}\middle| x\right) = \begin{cases} \dfrac{(-1)^{n/2}}{n+2} \dfrac{\sin((n+2)\theta)}{\cos\theta \, \sin\theta} & (n = 2m), \\[2mm] S_n(x \, ; \, -1, 1, -5, 2) & (n = 2m+1), \end{cases}$$

satisfying the recurrence relation

$$\bar{Y}_{n+1}(x) = x \, \bar{Y}_n(x) + C_n\left(\begin{array}{cc} -5 & 2 \\ -1 & 1 \end{array}\right) \bar{Y}_{n-1}(x),$$

where

$$C_n\left(\begin{array}{cc} -5 & 2 \\ -1 & 1 \end{array}\right) = \frac{-(n+1-(-1)^n)(n+2-(-1)^n)}{4(n+1)(n+2)},$$

and having the orthogonality relation

$$\int_{-1}^{1} W\left(\begin{matrix} -5 & 2 \\ -1 & 1 \end{matrix} \middle| x\right) \bar{Y}_n(x)\bar{Y}_m(x)\, dx$$

$$= \left((-1)^n \prod_{i=1}^{n} C_i \left(\begin{matrix} -5 & 2 \\ -1 & 1 \end{matrix}\right) \int_{-1}^{1} W\left(\begin{matrix} -5 & 2 \\ -1 & 1 \end{matrix} \middle| x\right) dx\right) \delta_{n,m}, \qquad (1.125)$$

where

$$\int_{-1}^{1} W\left(\begin{matrix} -5 & 2 \\ -1 & 1 \end{matrix} \middle| x\right) dx = \int_{-1}^{1} x^2\sqrt{1-x^2}\, dx = B\left(\frac{3}{2}, \frac{3}{2}\right) = \frac{\pi}{8}. \qquad (1.126)$$

The relation (1.126) simplifies (1.125) as

$$\int_{-1}^{1} x^2\sqrt{1-x^2}\,\bar{Y}_n(x)\bar{Y}_m(x)\, dx = \left((-1)^n \prod_{i=1}^{n} C_i \left(\begin{matrix} -5 & 2 \\ -1 & 1 \end{matrix}\right)\right) \frac{\pi}{8}\, \delta_{n,m}.$$

Similar to the previous case, $\bar{Y}_{2n}(x)$ is decomposable as

$$\bar{Y}_{2n}(x) = \prod_{k=1}^{2n} \left(x - \cos\frac{k\pi}{2n+2}\right).$$

Let us add that there are still two other sequences of half-trigonometric polynomials that are not orthogonal but can be expressed in terms of $\bar{S}_n(x; p, q, r, s)$. These sequences are defined as

$$\bar{S}_n\left(\begin{matrix} 1 & -2 \\ -1 & 1 \end{matrix} \middle| x\right) = \begin{cases} \bar{S}_n(x; -1, 1, 1, -2) & (n = 2m), \\ x\,\bar{T}_{n-1}(x) & (n = 2m+1), \end{cases}$$

and

$$\bar{S}_n\left(\begin{matrix} -1 & -2 \\ -1 & 1 \end{matrix} \middle| x\right) = \begin{cases} \bar{S}_n(x; -1, 1, -1, -2) & (n = 2m), \\ x\,\bar{U}_{n-1}(x) & (n = 2m+1). \end{cases}$$

However, since

$$\int_{-1}^{1} W\left(\begin{matrix} 1 & -2 \\ -1 & 1 \end{matrix} \middle| x\right) dx = \int_{-1}^{1} \frac{dx}{x^2\sqrt{1-x^2}} = +\infty$$

Table 1.4 Chebyshev polynomials in terms of $\bar{S}_n(x\,;\,p,q,r,s)$

Type	Notation	Definition	Weight
First kind	$\bar{T}_n(x)$	$\bar{S}_n\left(\begin{matrix}-1 & 0 \\ -1 & 1\end{matrix}\;\middle\vert\;x\right)$	$\dfrac{1}{\sqrt{1-x^2}}$
Second kind	$\bar{U}_n(x)$	$\bar{S}_n\left(\begin{matrix}-3 & 0 \\ -1 & 1\end{matrix}\;\middle\vert\;x\right)$	$\sqrt{1-x^2}$
Third kind	$\bar{V}_n(x)$	$2^n\,\bar{S}_{2n}\left(\begin{matrix}-3 & 2 \\ -1 & 1\end{matrix}\;\middle\vert\;\sqrt{\dfrac{1+x}{2}}\right)$	$\sqrt{\dfrac{1+x}{1-x}}$
Fourth kind	$\bar{W}_n(x)$	$2^n\,\bar{S}_{2n}\left(\begin{matrix}-3 & 2 \\ -1 & 1\end{matrix}\;\middle\vert\;\sqrt{\dfrac{1-x}{2}}\right)$	$\sqrt{\dfrac{1-x}{1+x}}$
Fifth kind	$\bar{X}_n(x)$	$\bar{S}_n\left(\begin{matrix}-3 & 2 \\ -1 & 1\end{matrix}\;\middle\vert\;x\right)$	$\dfrac{x^2}{\sqrt{1-x^2}}$
Sixth kind	$\bar{Y}_n(x)$	$\bar{S}_n\left(\begin{matrix}-5 & 2 \\ -1 & 1\end{matrix}\;\middle\vert\;x\right)$	$x^2\sqrt{1-x^2}$

and

$$\int_{-1}^{1} W\left(\begin{matrix}-1 & -2 \\ -1 & 1\end{matrix}\;\middle\vert\;x\right) dx = \int_{-1}^{1} \frac{\sqrt{1-x^2}}{x^2}dx = +\infty,$$

they cannot fall into the half-trigonometric orthogonal polynomials category. In this sense, Table 1.4 shows some properties of the first kind to the sixth kind of monic Chebyshev polynomials orthogonal on $[-1, 1]$.

2. For generalized Hermite polynomials (GHP) we have

- Initial vector:

$$(p,q,r,s) = (0,\,1,\,-2,\,2a).$$

- Differential equation:

$$x^2\varPhi_n''(x) - 2x(x^2 - a)\,\varPhi_n'(x) + \left(2n\,x^2 + ((-1)^n - 1)a\right)\varPhi_n(x) = 0.$$

- Explicit form of monic GHP:

$$\bar{S}_n\left(\begin{matrix}-2 & 2a \\ 0 & 1\end{matrix}\;\middle\vert\;x\right) = (-1)^{[\frac{n}{2}]}\left(a+1-\frac{(-1)^n}{2}\right)_{[\frac{n}{2}]}$$

$$\times \sum_{k=0}^{[n/2]}\binom{[n/2]}{k}\left(\prod_{i=0}^{[n/2]-(k+1)}\frac{-2}{2i + (-1)^{n+1} + 2 + 2a}\right)x^{n-2k}.$$

- Recurrence relation of monic polynomials:

$$\bar{S}_{n+1}(x) = x\,\bar{S}_n(x) + C_n \begin{pmatrix} -2 & 2a \\ 0 & 1 \end{pmatrix} \bar{S}_{n-1}(x),$$

where

$$C_n \begin{pmatrix} -2 & 2a \\ 0 & 1 \end{pmatrix} = -\frac{1}{2}n - \frac{1-(-1)^n}{2}a.$$

- Weight function:

$$W \begin{pmatrix} -2, & 2a \\ 0, & 1 \end{pmatrix} x = x^{2a}\exp(-x^2).$$

- Orthogonality relation:

$$\int_{-\infty}^{\infty} x^{2a} e^{-x^2} \bar{S}_n \begin{pmatrix} -2 & 2a \\ 0 & 1 \end{pmatrix} x \, \bar{S}_m \begin{pmatrix} -2 & 2a \\ 0 & 1 \end{pmatrix} x \, dx$$

$$= \left(\frac{1}{2^n} \prod_{i=1}^{n} (1 - (-1)^i)a + i \right) \Gamma(a + \frac{1}{2})\, \delta_{n,m},$$

which is valid for $a + 1/2 > 0$ and $(-1)^{2a} = 1$.

3. For the first finite sequence of symmetric orthogonal polynomials we have

- Initial vector:

$$(p, q, r, s) = (1,\ 1,\ -2a - 2b + 2,\ -2a).$$

- Differential equation:

$$x^2(x^2 + 1)\,\Phi_n''(x) - 2x\big((a + b - 1)x^2 + a\big)\Phi_n'(x)$$
$$+ \Big(n(2a + 2b - (n + 1))\,x^2 + (1 - (-1)^n)a\Big)\Phi_n(x) = 0. \tag{1.127}$$

- Explicit form of polynomials:

$$S_n \begin{pmatrix} -2a - 2b + 2, & -2a \\ 1, & 1 \end{pmatrix} x$$

$$= \sum_{k=0}^{[n/2]} \binom{[\frac{n}{2}]}{k} \left(\prod_{i=0}^{[\frac{n}{2}]-(k+1)} \frac{2i + 2[n/2] + (-1)^{n+1} + 2 - 2a - 2b}{2i + (-1)^{n+1} + 2 - 2a} \right) x^{n-2k}.$$

- Recurrence relation of monic polynomials:

$$\bar{S}_{n+1}(x) = x\,\bar{S}_n(x) + C_n \left(\begin{matrix} -2a - 2b + 2, & -2a \\ 1, & 1 \end{matrix} \right) \bar{S}_{n-1}(x),$$

where

$$C_n \left(\begin{matrix} -2a - 2b + 2, & -2a \\ 1, & 1 \end{matrix} \right) = \frac{(n - (1 - (-1)^n)\,a)\,(n - (1 - (-1)^n)\,a - 2b)}{(2n - 2a - 2b + 1)(2n - 2a - 2b - 1)}.$$

- Weight function:

$$W \left(\begin{matrix} -2a - 2b + 2, & -2a \\ 1, & 1 \end{matrix} \middle| x \right) = \frac{x^{-2a}}{(1 + x^2)^b}.$$

- Orthogonality relation:

$$\int_{-\infty}^{\infty} \frac{x^{-2a}}{(1 + x^2)^b}\, \bar{S}_n \left(\begin{matrix} -2a - 2b + 2, & -2a \\ 1, & 1 \end{matrix} \middle| x \right) \bar{S}_m \left(\begin{matrix} -2a - 2b + 2, & -2a \\ 1, & 1 \end{matrix} \middle| x \right) dx$$

$$= \left((-1)^n \prod_{i=1}^{n} C_i \left(\begin{matrix} -2a - 2b + 2, & -2a \\ 1, & 1 \end{matrix} \right) \right) \frac{\Gamma(b + a - 1/2)\Gamma(-a + 1/2)}{\Gamma(b + a)}\, \delta_{n,m},$$

$$(1.128)$$

which is valid if

$$- C_{n+1} \left(\begin{matrix} -2a - 2b + 2, & -2a \\ 1, & 1 \end{matrix} \right) > 0 \text{ for all } n \in \mathbb{Z}^+,\ b + a > 1/2,\ a < 1/2,\ \text{and } b > 0.$$

Here we explain how to determine these conditions in order to establish the orthogonality property (1.128). For this purpose, we reconsider Eq. (1.127) and write it in a self-adjoint form to get

$$\left[x^{-2a}(1 + x^2)^{-b+1}(\Phi_n'(x)\Phi_m(x) - \Phi_m'(x)\Phi_n(x)) \right]_{-\infty}^{\infty} = 0. \qquad (1.129)$$

Since

$$\max \deg\{\Phi_n'(x)\Phi_m(x) - \Phi_m'(x)\Phi_n(x)\} = n + m - 1, \qquad (1.130)$$

relations (1.129) and (1.130) yield

$$-2a + 2(-b + 1) + n + m - 1 \leq 0,$$

which gives the final result as

$$-2a - 2b + n + m + 1 \leq 0 \quad \text{for} \quad a < 1/2 \text{ and } b > 0.$$

In other words, (1.128) holds if and only if

$$m, n = 0, 1, \ldots, N \leq a + b - 1/2, \quad a < 1/2, \quad \text{and} \quad b > 0.$$

Corollary 1.1 *The symmetric polynomial set* $\{S_n(x \, ; \, 1, \, 1, \, -2a - 2b + 2, \, -2a)\}_{n=0}^{N}$ *is finitely orthogonal with respect to the weight function* $x^{-2a}(1 + x^2)^{-b}$ *on* $(-\infty, \infty)$ *if and only if* $N \leq a + b - 1/2$, $a < 1/2$, $b > 0$, *and* $(-1)^{2a} = 1$.

4. For the second finite sequence of symmetric orthogonal polynomials we have

- Initial vector:

$$(p, q, r, s) = (1, 0, -2a + 2, 2).$$

- Differential equation:

$$x^4 \, \Phi_n''(x) + 2x\big((1-a)x^2 + 1\big)\Phi_n'(x) - \Big(n(n + 1 - 2a)\,x^2 + 1 - (-1)^n\Big)\Phi_n(x) = 0.$$

- Explicit form of polynomials:

$$S_n \left(\begin{array}{cc} -2a + 2, \, 2 \\ 1, \quad 0 \end{array} \middle| \, x \right)$$

$$= \sum_{k=0}^{[n/2]} \binom{[n/2]}{k} \left(\prod_{i=0}^{[\frac{n}{2}]-(k+1)} \frac{2i + 2[n/2] + (-1)^{n+1} + 2 - 2a}{2} \right) x^{n-2k}.$$

- Recurrence relation of monic polynomials:

$$\bar{S}_{n+1}(x) = x \, \bar{S}_n(x) + C_n \left(\begin{array}{cc} -2a + 2, \, 2 \\ 1, \quad 0 \end{array} \right) \bar{S}_{n-1}(x),$$

where

$$C_n \begin{pmatrix} -2a+2, \ 2 \\ 1, \quad 0 \end{pmatrix} = \frac{-2\,(-1)^n\,(n-a)-2a}{(2n-2a+1)(2n-2a-1)}.$$

- Weight function:

$$W \left(\begin{matrix} -2a+2, \ 2 \\ 1, \quad 0 \end{matrix} \ \middle| \ x \right) = x^{-2a}\exp\left(-\frac{1}{x^2}\right).$$

- Orthogonality relation:

$$\int_{-\infty}^{\infty} x^{-2a} e^{-\frac{1}{x^2}} \bar{S}_n\left(\begin{matrix} -2a+2, \ 2 \\ 1, \quad 0 \end{matrix} \ \middle| \ x \right) \bar{S}_m\left(\begin{matrix} -2a+2, \ 2 \\ 1, \quad 0 \end{matrix} \ \middle| \ x \right) dx$$

$$= \left((-1)^n \prod_{i=1}^{n} C_i \begin{pmatrix} -2a+2, \ 2 \\ 1, \quad 0 \end{pmatrix} \right) \Gamma\left(a-\frac{1}{2}\right) \delta_{n,m} . \tag{1.131}$$

Relation (1.131) is valid only if

$$\left[x^{2-2a} \exp(-\frac{1}{x^2}) \left(\Phi_n'(x)\Phi_m(x) - \Phi_m'(x)\Phi_n(x) \right) \right]_{-\infty}^{\infty} = 0,$$

which is equivalent to

$$2-2a+n+m-1 \le 0 \quad \Leftrightarrow \quad N \le a-\frac{1}{2} \quad \text{for} \quad N = \max\{m,n\}.$$

Corollary 1.2 *The symmetric polynomial set* $\{S_n(x\,;\ 1,\ 0,\ -2a+2,\ 2)\}_{n=0}^{N}$ *is finitely orthogonal with respect to the weight function* $x^{-2a}e^{-1/x^2}$ *on* $(-\infty,\infty)$ *if and only if* $N \le a-1/2$ *and* $(-1)^{2a}=1$.

1.7.3   A Unified Approach for the Classification of $\bar{S}_n(x\,;\ p,q,r,s)$

We observed that each of the four introduced classes of symmetric orthogonal polynomials was directly determined by $\bar{S}_n(x\,;\ p,q,r,s)$, and it was sufficient only to obtain the initial vector corresponding to them. On the other hand, since the orthogonality interval of all classes, except the first one (GUP), is $(-\infty,\infty)$, applying a linear transformation preserves the orthogonality interval. This means that by having the initial vector, we can have access to all other standard properties and design a unified approach for some cases that may occur. Here we consider two special cases of this approach.

First Case: How to Find Initial Parameters If a Special Case of the Weight Function is Given

An easy way to derive p, q, r, s is to first compute the logarithmic derivative of the weight function as before and then equate the pattern with (1.109), as the following general example shows.

Example 1.5 The weight functions

$$
\begin{array}{llll}
(i) & W_1(x) = -x^6 + 4x^4 & & x \in [-2, 2], \\
(ii) & W_2(x) = (16x^2 - 8x + 1)\exp(2x(1 - 2x)) & & x \in (-\infty, \infty), \\
(iii) & W_3(x) = (2x + 1)^{-2}(2x^2 + 2x + 1)^{-5} & & x \in (-\infty, \infty),
\end{array}
$$

with their orthogonality intervals are given. To find initial vectors corresponding to each given weight function, we first compute the logarithmic derivative of the first weight function as

$$
\frac{W'_1(x)}{W_1(x)} = \frac{6x^2 - 16}{x(x^2 - 4)} = \frac{(r - 2p)x^2 + s}{x(px^2 + q)} \Rightarrow (p, q, r, s) = (1, -4, 8, -16).
$$

Hence the related monic polynomials are $\left\{ \bar{S}_n(x; \ 1, -4, 8, -16) \right\}_{n=0}^{\infty}$, which are orthogonal with respect to $x^4(4 - x^2)$ on $[-2, 2]$ for every n. Note that it is not necessary to know that they are the same as shifted GUP on $[-2, 2]$, because they can be explicitly and independently expressed by $\bar{S}_n(x; \ p, q, r, s)$.

For the weight function $W_2(x)$, it differs somewhat, so that we have

$$
W_2(x) = 4e^{\frac{1}{4}}(2x - \tfrac{1}{2})^2 \exp(-(2x - \tfrac{1}{2})^2) \Rightarrow \tfrac{1}{4}e^{-\frac{1}{4}} W_2(\tfrac{2t+1}{4}) = t^2 e^{-t^2} = W_2^*(t)
$$

$$
\rightarrow \quad \frac{W_2^{*'}(t)}{W_2^*(t)} = \frac{-2t^2 + 2}{t} \Rightarrow (p, q, r, s) = (0, 1, -2, 2).
$$

Hence the related orthogonal polynomials are $\left\{ S_n(2x - \tfrac{1}{2}; \ 0, 1, -2, 2) \right\}_{n=0}^{\infty}$.

Finally, for the third one we have

$$
W_3(x) = 2^5 \frac{(2x+1)^{-2}}{(1+(2x+1)^2)^5} \Rightarrow 2^{-5} W_3(\tfrac{t-1}{2}) = \frac{t^{-2}}{(1+t^2)^5} = W_3^*(t)
$$

$$
\rightarrow \quad \frac{W_3^{*'}(t)}{W_3^*(t)} = \frac{-12t^2 - 2}{t(t^2 + 1)} \Rightarrow (p, q, r, s) = (1, 1, -10, -2).
$$

Consequently, the finite set $\{S_n(2x + 1; \ 1, 1, -10, -2)\}_{n=0}^{5}$ is orthogonal with respect to $W_3(x)$ on $(-\infty, \infty)$, and the upper bound of this set is determined based on the condition $N \le b + 1/2$ for $b = 5$.

Second Case: How to Find Initial Parameters If a Special Case of the Main Three-Term Recurrence Equation Is Given?

There are two ways to determine a special case of $\bar{S}_n(x; p, q, r, s)$ when a three-term recurrence equation is given. The first way is to directly compare the given recurrence equation with (1.106). This leads to a system of polynomial equations in terms of the parameters p, q, r, s. The second way is to equate the first four terms of the two recurrence equations, which leads to a polynomial system with four equations the four unknowns p, q, r, and s. The following example illustrates these methods.

Example 1.6 The recurrence equation

$$\bar{S}_{n+1}(x) = x\,\bar{S}_n(x) - 2\,\frac{6 + (-1)^n(n-6)}{(2n-11)(2n-13)}\,\bar{S}_{n-1}(x)$$

with $\bar{S}_0(x) = 1$ and $\bar{S}_1(x) = x$ is given. Find its explicit polynomial solution, differential equation of polynomials, the related weight function, and orthogonality relation of polynomials.

First Method If the above recurrence equation is directly compared with the main Eq. (1.106), we can directly obtain the values $(p, q, r, s) = (1, 0, -10, 2)$. Hence, the explicit solution of the above recurrence equation is $\bar{S}_n(x; 1, 0, -10, 2)$, satisfying the differential equation

$$x^4\,\Phi_n''(x) + x\,(-10\,x^2 + 2)\,\Phi_n'(x) - \left(n\,(n-11)\,x^2 + 1 - (-1)^n\right)\Phi_n(x) = 0.$$

Moreover, by substituting the initial vector into the main weight function (1.109) as

$$W\left(\begin{array}{cc} -10, & 2 \\ 1, & 0 \end{array}\middle|\, x\right) = \exp\left(\int \frac{-12x^2 + 2}{x^3}\,dx\right) = x^{-12}e^{-\frac{1}{x^2}},$$

we find out that the related polynomials are a particular case of the fourth introduced class, so that

$$\int_{-\infty}^{\infty} x^{-12}e^{-\frac{1}{x^2}}\,\bar{S}_n\left(\begin{array}{cc} -10, & 2 \\ 1, & 0 \end{array}\middle|\, x\right)\bar{S}_m\left(\begin{array}{cc} -10, & 2 \\ 1, & 0 \end{array}\middle|\, x\right)dx$$

$$= \left((-1)^n \prod_{i=1}^{n} C_i\left(\begin{array}{cc} -10, & 2 \\ 1, & 0 \end{array}\right)\right)\Gamma(\frac{11}{2})\,\delta_{n,m} \Leftrightarrow m, n \le 5.$$

Table 1.5 Four special cases of $S_n(x; p, q, r, s)$

Definition	Weight function	Interval and kind	Parameter constraint
$S_n\left(\begin{matrix} -2a - 2b - 2,\ 2a \\ -1, \qquad\quad 1 \end{matrix}\ \middle|\ x\right)$	$x^{2a}(1 - x^2)^b$	$[-1, 1]$, Infinite	$a > -1/2$ $b > -1$
$S_n\left(\begin{matrix} -2,\ 2a \\ 0,\ 1 \end{matrix}\ \middle|\ x\right)$	$x^{2a}\exp(-x^2)$	$(-\infty, \infty)$, Infinite	$a > -\frac{1}{2}$
$S_n\left(\begin{matrix} -2a - 2b + 2,\ -2a \\ 1, \qquad\qquad 1 \end{matrix}\ \middle|\ x\right)$	$\dfrac{x^{-2a}}{(1 + x^2)^b}$	$(-\infty, \infty)$, Finite	$N \le a + b - 1/2$ $a < 1/2,\ b > 0$
$S_n\left(\begin{matrix} -2a + 2,\ 2 \\ 1, \qquad\ 0 \end{matrix}\ \middle|\ x\right)$	$x^{-2a}\exp(-1/x^2)$	$(-\infty, \infty)$, Finite	$N \le a - \frac{1}{2}$

Second Method If the given recurrence relation is expanded only for $n = 2, 3, 4, 5$ and then equated with (1.105), the following system will be derived:

$$\bar{S}_2(x) = x^2 - \frac{2}{9} = x^2 + \frac{q + s}{p + r} \Rightarrow \frac{q + s}{p + r} = -\frac{2}{9},$$

$$\bar{S}_3(x) = x\bar{S}_2(x) - \frac{4}{63}\bar{S}_1(x) = x^3 + \frac{3q + s}{3p + r}x,$$

$$\bar{S}_4(x) = x\bar{S}_3(x) - \frac{18}{35}\bar{S}_2(x) = x^4 + 2\frac{3q + s}{5p + r}x^2 + \frac{(3q + s)(q + s)}{(5p + r)(3p + r)},$$

$$2\frac{5q + s}{7p + r} = -\frac{4}{3}.$$

Solving this system again results in $(p, q, r, s) = (1, 0, -10, 2)$.

Finally, Table 1.5 shows the four polynomials in terms of $S_n(x\ ;\ p, q, r, s)$ as well as their weight functions, kind of polynomials (finite or infinite), orthogonality interval, and constraint on the parameters.

Note that since all weights are even functions in this table, the condition $(-1)^{2a} = 1$ must be always satisfied. Therefore they can be considered in the forms $|x|^{2a}(1 - x^2)^b$, $|x|^{2a}e^{-x^2}$, $|x|^{-2a}(1 + x^2)^{-b}$ and $|x|^{-2a}e^{-1/x^2}$ respectively.

1.8 Fourier Transforms of Symmetric Orthogonal Polynomials

Similar to Sect. 1.5, in this section we introduce four classes of orthogonal functions as the Fourier transforms of the introduced cases of $S_n(x; p, q, r, s)$ and then obtain their orthogonality relations using Parseval's identity.

1.8.1 Fourier Transform of Generalized Ultraspherical Polynomials

For the generalized ultraspherical polynomials

$$U_n^{(a,b)}(x) = x^n {}_2F_1\left(\begin{array}{c} -[n/2],\ -a + 1/2 - [(n+1)/2] \\ -a - b - n + 1/2 \end{array} \middle|\ \frac{1}{x^2} \right),$$

let us define the specific functions

$$g(x) = (\tanh x)^{2\alpha}(1 - \tanh^2 x)^{\beta}\, U_n^{(c,d)}(\tanh x) \quad \text{with } (-1)^{2\alpha} = 1$$

and   (1.132)

$$h(x) = (\tanh x)^{2l}(1 - \tanh^2 x)^{u}\, U_m^{(v,w)}(\tanh x) \quad \text{with } (-1)^{2l} = 1.$$

The Fourier transform of the function, e.g., $g(x)$ is computed as

$$\mathbf{F}(g(x)) = \int_{-\infty}^{\infty} e^{-isx} (\tanh x)^{2\alpha}(1 - \tanh^2 x)^{\beta}\, U_n^{(c,d)}(\tanh x)\, dx$$

$$= \int_{-1}^{1} (1+t)^{-\frac{is}{2}}(1-t)^{\frac{is}{2}} t^{2\alpha}(1-t^2)^{\beta-1}\, U_n^{(c,d)}(t)\, dt$$

$$= 2^{2\beta-1} \int_0^1 (1-z)^{-\frac{is}{2}} z^{\frac{is}{2}} (1-2z)^{2\alpha} z^{\beta-1}(1-z)^{\beta-1}\, U_n^{(c,d)}(1-2z)\, dz$$

$$= 2^{2\beta-1} \int_0^1 (1-z)^{\beta-1-\frac{is}{2}} z^{\beta-1+\frac{is}{2}} (1-2z)^{2\alpha+n} \left(\sum_{k=0}^{[n/2]} \frac{(-[\frac{n}{2}])_k (\frac{1}{2} - c - [\frac{n+1}{2}])_k}{(-c-d-n+1/2)_k k!} \frac{1}{(1-2z)^{2k}} \right) dz$$

$$= 2^{2\beta-1} \sum_{k=0}^{[n/2]} \frac{(-[\frac{n}{2}])_k (\frac{1}{2} - c - [\frac{n+1}{2}])_k}{(-c-d-n+1/2)_k k!} \left(\int_0^1 (1-z)^{\beta-1-\frac{is}{2}} z^{\beta-1+\frac{is}{2}} (1-2z)^{2\alpha+n-2k} dz \right).$$

(1.133)

According to the integral representation of Gauss's hypergeometric function (1.9), for Re $\beta > 0$ and Re $(2\alpha + n - 2k) > -1$, relation (1.133) is simplified as

$$\mathbf{F}(g(x)) = 2^{2\beta-1} B\left(\beta + \frac{is}{2}, \beta - \frac{is}{2} \right)$$

$$\times \sum_{k=0}^{[n/2]} \frac{(-[\frac{n}{2}])_k(\frac{1}{2}-c-[\frac{n+1}{2}])_k}{(-c-d-n+1/2)_k k!} {}_2F_1\left(\begin{array}{c} 2k-n-2\alpha \ \beta+\frac{is}{2} \\ 2\beta \end{array}\middle| 2\right). \tag{1.134}$$

If for simplicity we define

$$K_n(x; p_1, p_2, p_3, p_4) = \sum_{k=0}^{[n/2]} \frac{(-[\frac{n}{2}])_k(\frac{1}{2}-p_3-[\frac{n+1}{2}])_k}{(-p_3-p_4-n+1/2)_k k!} {}_2F_1\left(\begin{array}{c} 2k-n-2p_1 \ p_2+\frac{x}{2} \\ 2p_2 \end{array}\middle| 2\right),$$
$$\tag{1.135}$$

then (1.134) can be written as

$$\mathbf{F}(g(x)) = \frac{2^{2\beta-1}}{\Gamma(2\beta)} \Gamma\left(\beta+\frac{is}{2}\right) \Gamma\left(\beta-\frac{is}{2}\right) K_n(is; \alpha, \beta, c, d). \tag{1.136}$$

Now, by substituting (1.136) in Parseval's identity (1.83) and noting (1.132), we get

$$2\pi \int_{-\infty}^{\infty} (\tanh x)^{2(\alpha+l)}(1-\tanh^2 x)^{\beta+u} U_n^{(c,d)}(\tanh x) U_m^{(v,w)}(\tanh x)\, dx$$

$$= 2\pi \int_{-1}^{1} t^{2(\alpha+l)}(1-t^2)^{\beta+u-1} U_n^{(c,d)}(t) U_m^{(v,w)}(t)\, dt = \frac{2^{2\beta+2u-2}}{\Gamma(2\beta)\Gamma(2u)}$$

$$\times \int_{-\infty}^{\infty} \Gamma(\beta+\frac{is}{2})\Gamma(\beta-\frac{is}{2})\,\overline{\Gamma(u+\frac{is}{2})\Gamma(u-\frac{is}{2})}\, K_n(is; a, b, c, d)\,\overline{K_m(is; l, u, v, w)}\, ds\,.$$
$$\tag{1.137}$$

If on the left-hand side of (1.137) we take

$$c = v = \alpha + l \quad \text{and} \quad d = w = \beta + u - 1,$$

then according to the orthogonality relation (1.120), Eq. (1.137) finally reads as

$$\frac{1}{2\pi} \int_{-\infty}^{\infty} \Gamma(\beta+\frac{is}{2})\Gamma(\beta-\frac{is}{2})\Gamma(u+\frac{is}{2})\Gamma(u-\frac{is}{2})$$

$$K_n(is; \alpha, \beta, \alpha+l, \beta+u-1)K_m(-is; l, u, \alpha+l, \beta+u-1)ds$$

$$= \frac{\Gamma(2\beta)\Gamma(2u)\Gamma(\alpha+l+1/2)\Gamma(\beta+u)}{2^{2\beta+2u+2n-2}\,\Gamma(\alpha+\beta+l+u+1/2)}$$

$$\times \prod_{j=1}^{n} \frac{\left(j+(1-(-1)^j)(\alpha+l)\right)\left(j+(1-(-1)^j)(\alpha+l)+2\beta+2u-2\right)}{(j+\alpha+\beta+l+u-3/2)(j+\alpha+\beta+l+u-1/2)}\, \delta_{n,m},$$
$$\tag{1.138}$$

provided that $\beta, u > 0$ and $\alpha + l > -1/2$.

Remark 1.7 The weight function corresponding to (1.138) can be simplified using the Cauchy beta integral

$$\frac{1}{2\pi} \int_{-\infty}^{\infty} \frac{dt}{(a+it)^c (b-it)^d} = \frac{\Gamma(c+d-1)}{\Gamma(c)\Gamma(d)} (a+b)^{1-(c+d)}$$

and one of its consequences, i.e.,

$$\Gamma(p+iq)\,\Gamma(p-iq) = \frac{2^{2-2p}\pi\,\Gamma(2p-1)}{\int_{-\pi/2}^{\pi/2} e^{2qx}\cos^{2p-2}x\,dx},$$

which results in

$$\Gamma\left(\beta+\frac{is}{2}\right)\Gamma\left(\beta-\frac{is}{2}\right)\Gamma\left(u+\frac{is}{2}\right)\Gamma\left(u-\frac{is}{2}\right) = \frac{2^{4-2\beta-2u}\pi^2\Gamma(2\beta-1)\Gamma(2u-1)}{\int_{-\pi/2}^{\pi/2} e^{sx}\cos^{2\beta-2}x\,dx \int_{-\pi/2}^{\pi/2} e^{sx}\cos^{2u-2}x\,dx}.$$

Therefore, if we define

$$W^*(x;\lambda) = \int_{-\pi/2}^{\pi/2} e^{x\theta}\cos^{2\lambda-2}\theta\,d\theta,$$

then the function $K_n(x; p_1, p_2, p_3, p_4)$ defined in (1.135) satisfies the orthogonality relation

$$\int_{-\infty}^{\infty} \frac{K_n\,(ix; \alpha, \beta, \alpha+l, \beta+u-1)\,K_m\,(-ix; l, u, \alpha+l, \beta+u-1)}{W^*(x;\beta)\,W^*(x;u)}\,dx$$

$$= \frac{(2u-1)(2\beta-1)}{\pi\,2^{2n+1}} \frac{\Gamma(\alpha+l+1/2)\,\Gamma(\beta+u)}{\Gamma(\alpha+\beta+l+u+1/2)}$$

$$\times \prod_{j=1}^{n} \frac{\left(j + (1-(-1)^j)(\alpha+l)\right)\left(j + (1-(-1)^j)(\alpha+l) + 2\beta + 2u - 2\right)}{(j+\alpha+\beta+l+u-3/2)\,(j+\alpha+\beta+l+u-1/2)}\,\delta_{n,m},$$

where $\beta, u > 1/2, \alpha+l > -1/2$.

1.8.2 Fourier Transform of Generalized Hermite Polynomials

The above-mentioned technique can be similarly applied to the generalized Hermite polynomials

$$H_n^{(a)}(x) = x^n\,{}_2F_0\left(\begin{array}{c} -[n/2],\ -[n/2]-a+(-1)^n/2 \\ - \end{array}\middle|\ -\frac{1}{x^2}\right).$$

First define the specific functions

$$u(x) = x^{2a} e^{-\frac{1}{2}x^2} H_n^{(b)}(x) \qquad \text{with } (-1)^{2a} = 1$$

and

$$v(x) = x^{2c} e^{-\frac{1}{2}x^2} H_m^{(d)}(x) \qquad \text{with } (-1)^{2c} = 1.$$

(1.139)

The Fourier transform of the function, e.g., $u(x)$ is computed as

$$\mathbf{F}\big(u(x)\big) = \int_{-\infty}^{\infty} e^{-isx} x^{2a} e^{-\frac{1}{2}x^2} H_n^{(b)}(x)\, dx$$

$$= \int_{-\infty}^{\infty} e^{-isx} e^{-\frac{1}{2}x^2} x^{2a+n} \left(\sum_{k=0}^{[n/2]} \frac{(-[n/2])_k (-[n/2] - b + (-1)^n/2)_k}{k!} (-x^{-2})^k \right) dx$$

$$= \sum_{k=0}^{[n/2]} \frac{(-[n/2])_k (-[n/2] - b + (-1)^n/2)_k}{k!} (-1)^k \left(\int_{-\infty}^{\infty} e^{-isx} e^{-\frac{1}{2}x^2} x^{2a+n-2k}\, dx \right).$$

(1.140)

So it remains in (1.140) to evaluate the integral

$$I_{n,k}(s; a) = \int_{-\infty}^{\infty} e^{-isx} e^{-\frac{1}{2}x^2} x^{2a+n-2k}\, dx.$$

There are two ways to compute this integral. In the first method, by noting that $(-1)^{2a} = 1$, we can directly compute $I_{n,k}(s; a)$ for $n = 2m$ as follows:

$$I_{2m,k}(s; a) = \int_{-\infty}^{\infty} \left(\sum_{j=0}^{\infty} \frac{(-isx)^j}{j!} \right) x^{2a+2m-2k} e^{-\frac{1}{2}x^2}\, dx$$

$$= \sum_{j=0}^{\infty} \frac{(-1)^j i^j s^j}{j!} \left(\int_{-\infty}^{\infty} x^{j+2a+2m-2k} e^{-\frac{1}{2}x^2}\, dx \right)$$

$$= \sum_{r=0}^{\infty} \frac{(-1)^r s^{2r}}{(2r)!} \left(2 \int_{0}^{\infty} x^{2r+2a+2m-2k} e^{-\frac{1}{2}x^2}\, dx \right)$$

$$= \sum_{r=0}^{\infty} \frac{(-1)^r s^{2r}}{(2r)!} 2^{r+a+m-k+\frac{1}{2}} \Gamma\left(r + a + m - k + \frac{1}{2}\right)$$

$$= \sum_{r=0}^{\infty} \frac{(-1)^r s^{2r} 2^{r+a+m-k+\frac{1}{2}} \Gamma(a+m-k+\frac{1}{2}) \, (a+m-k+\frac{1}{2})_r}{(1/2)_r \, 2^{2r} r!}$$

$$= 2^{a+m-k+1/2} \Gamma(a+m-k+1/2) \, {}_1F_1 \left(\begin{array}{c} a+m-k+1/2 \\ 1/2 \end{array} \middle| -\frac{s^2}{2} \right).$$

(1.141)

This technique can be also applied to $I_{2m+1,k}(s; a)$. We should note that

$$\int_{-\infty}^{\infty} x^{j+2a+2m+1-2k} e^{-x^2/2} \, dx = 0 \quad \text{for all} \quad j = 0, 2, 4, \dots .$$

After some computations we obtain

$$I_{2m+1,k}(s; a) = (-is) \, 2^{a+m-k+3/2} \Gamma(a+m-k+3/2) \, {}_1F_1 \left(\begin{array}{c} a+m-k+3/2 \\ 3/2 \end{array} \middle| -\frac{s^2}{2} \right).$$

(1.142)

By combining relations (1.141) and (1.142) and using the well-known identity

$$[\frac{n+1}{2}] - [\frac{n}{2}] = \frac{1-(-1)^n}{2},$$

we finally obtain

$$I_{n,k}(s; a) = 2^{a-k+\frac{1}{2}+[\frac{n+1}{2}]}$$

$$\times \Gamma \left(a-k+\frac{1}{2}+[\frac{n+1}{2}] \right) (-is)^{\frac{1-(-1)^n}{2}} \, {}_1F_1 \left(\begin{array}{c} \frac{1}{2}+a-k+[\frac{n+1}{2}] \\ 1-(-1)^n/2 \end{array} \middle| -\frac{s^2}{2} \right).$$

(1.143)

The second way to compute $I_{n,k}(s; a)$ is that we consider $n = 2m$ and $n = 2m+1$ and directly apply the cosine and sine Fourier transforms of the function $e^{-x^2/2} x^{2a+n-2k}$. So, by noting that $(-1)^{2a} = 1$, we have

$$I_{2m,k}(s; a) = \int_{-\infty}^{\infty} \cos(sx) \, x^{2a+2m-2k} e^{-\frac{1}{2}x^2} \, dx - i \int_{-\infty}^{\infty} \sin(sx) \, x^{2a+2m-2k} e^{-\frac{1}{2}x^2} \, dx$$

$$= 2 \int_{0}^{\infty} \cos(sx) x^{2a+2m-2k} e^{-\frac{1}{2}x^2} \, dx$$

$$= 2^{a+m-k+\frac{1}{2}} \Gamma(a+m-k+\frac{1}{2}) \, {}_1F_1 \left(\begin{array}{c} a+m-k+\frac{1}{2} \\ 1/2 \end{array} \middle| -\frac{s^2}{2} \right)$$

and

$$I_{2m+1,k}(s;a) = \int_{-\infty}^{\infty} \cos(sx)\, x^{2a+2m+1-2k} e^{-\frac{1}{2}x^2}\, dx - i\int_{-\infty}^{\infty} \sin(sx)\, x^{2a+2m+1-2k} e^{-\frac{1}{2}x^2}\, dx$$

$$= -2i\int_0^{\infty} \sin(sx)x^{2a+2m+1-2k} e^{-\frac{1}{2}x^2}\, dx$$

$$= (-is)2^{a+m-k+\frac{3}{2}}\,\Gamma\!\left(a+m-k+\frac{3}{2}\right){}_1F_1\!\left(\begin{array}{c} a+m-k+\frac{3}{2} \\ 3/2 \end{array}\Bigg| -\frac{s^2}{2}\right).$$

Consequently, the result (1.143) simplifies (1.140) as

$$\mathbf{F}(u(x)) = \Gamma\!\left(a+\frac{1}{2}+\left[\frac{n+1}{2}\right]\right) 2^{a+\frac{1}{2}+\left[\frac{n+1}{2}\right]}$$

$$\times (-is)^{\frac{1-(-1)^n}{2}} \sum_{k=0}^{[n/2]} \frac{(-[\frac{n}{2}])_k(\frac{1}{2}-b-[\frac{n+1}{2}])_k}{(\frac{1}{2}-a-[\frac{n+1}{2}])_k 2^k k!}\,{}_1F_1\!\left(\begin{array}{c} \frac{1}{2}+a-k+[\frac{n+1}{2}] \\ 1-(-1)^n/2 \end{array}\Bigg| -\frac{s^2}{2}\right),$$

and according to Kummer's formula

$$_1F_1\!\left(\begin{array}{c} a \\ c \end{array}\Bigg| x\right) = e^x\,{}_1F_1\!\left(\begin{array}{c} c-a \\ c \end{array}\Bigg| -x\right),$$

it will be transformed to

$$\mathbf{F}(u(x)) = \Gamma\!\left(a+\frac{1}{2}+\left[\frac{n+1}{2}\right]\right) 2^{a+\frac{1}{2}+\left[\frac{n+1}{2}\right]} e^{-\frac{1}{2}s^2}$$

$$\times (-is)^{\frac{1-(-1)^n}{2}} \sum_{k=0}^{[n/2]} \frac{(-[\frac{n}{2}])_k(\frac{1}{2}-b-[\frac{n+1}{2}])_k}{(\frac{1}{2}-a-[\frac{n+1}{2}])_k 2^k k!}\,{}_1F_1\!\left(\begin{array}{c} -a+k-[\frac{n}{2}] \\ 1-(-1)^n/2 \end{array}\Bigg| \frac{1}{2}s^2\right).$$

If for simplicity we define

$$J_n(x;q_1,q_2) = x^{\frac{1-(-1)^n}{2}} \sum_{k=0}^{[n/2]} \frac{(-[\frac{n}{2}])_k(\frac{1}{2}-q_2-[\frac{n+1}{2}])_k}{(\frac{1}{2}-q_1-[\frac{n+1}{2}])_k 2^k k!}\,{}_1F_1\!\left(\begin{array}{c} -q_1+k-[\frac{n}{2}] \\ 1-(-1)^n/2 \end{array}\Bigg| \frac{1}{2}x^2\right),$$

then by referring to definitions (1.139) and applying Parseval's identity, we get

$$2\pi\int_{-\infty}^{\infty} x^{2(a+c)} e^{-x^2} H_n^{(b)}(x)\, H_m^{(d)}(x)\, dx$$

$$= i^{\frac{(-1)^n-(-1)^m}{2}} \frac{\Gamma(a+\frac{1}{2}+[\frac{n+1}{2}])\,\Gamma(c+\frac{1}{2}+[\frac{m+1}{2}])}{2^{-(a+\frac{1}{2}+[\frac{n+1}{2}])}2^{-(c+\frac{1}{2}+[\frac{m+1}{2}])}} \int_{-\infty}^{\infty} e^{-s^2} J_n(s;a,b)\, J_m(s;c,d)\, ds.$$

Finally, taking $b = d = a + c$ and noting the orthogonality relation, we obtain

$$\int_{-\infty}^{\infty} e^{-x^2} J_n(x; a, b) J_m(x; b - a, b)\, dx = \frac{\pi 2^{-b-2[\frac{n+1}{2}]} \Gamma([\frac{n}{2}] + 1) \Gamma(b + \frac{1}{2} + [\frac{n+1}{2}])}{\Gamma(a + \frac{1}{2} + [\frac{n+1}{2}]) \Gamma(b - a + \frac{1}{2} + [\frac{n+1}{2}])} \delta_{n,m},$$

where $a, b > -1/2$ and $(-1)^{2b} = 1$.

1.8.3 Fourier Transform of the First Finite Sequence of Symmetric Orthogonal Polynomials

To derive the Fourier transform of the first finite sequence of symmetric polynomials

$$A_n^{(a,b)}(x) = \bar{S}_n \left(\begin{array}{cc} -2a - 2b + 2, & -2a \\ 1, & 1 \end{array} \middle| x \right)$$

$$= x^n {}_2F_1 \left(\begin{array}{c} -[n/2],\, a + 1/2 - [(n+1)/2] \\ a + b - n + 1/2 \end{array} \middle| -\frac{1}{x^2} \right),$$

we first define

$$g(x) = x^{-2\alpha}(1 + x^2)^{-\beta} A_n^{(c,d)}(x) \quad \text{with} \quad (-1)^{2\alpha} = 1$$

and   (1.144)

$$h(x) = x^{-2l}(1 + x^2)^{-u} A_m^{(v,w)}(x) \quad \text{with} \quad (-1)^{2l} = 1.$$

Then the Fourier transform of the function, e.g., $g(x)$ is computed as

$$\mathbf{F}(g(x)) = \int_{-\infty}^{\infty} e^{-isx}(1 + x^2)^{-\beta} x^{-2\alpha} A_n^{(c,d)}(x)\, dx$$

$$= \int_{-\infty}^{\infty} e^{-isx}(1 + x^2)^{-\beta} x^{-2\alpha+n} \left(\sum_{k=0}^{[n/2]} \frac{(-[n/2])_k (c + 1/2 - [(n+1)/2])_k}{(c + d - n + 1/2)_k k!} \frac{(-1)^k}{x^{2k}} \right) dx$$

$$= \sum_{k=0}^{[n/2]} \frac{(-[n/2])_k (c + 1/2 - [(n+1)/2])_k (-1)^k}{(c + d - n + 1/2)_k k!} \left(\int_{-\infty}^{\infty} e^{-isx}(1 + x^2)^{-\beta} x^{-2\alpha+n-2k}\, dx \right).$$

(1.145)

Now it remains in (1.145) to evaluate the integral

$$I_{n,k}(s; \alpha, \beta) = \int_{-\infty}^{\infty} e^{-isx}(1 + x^2)^{-\beta} x^{-2\alpha+n-2k} dx. \tag{1.146}$$

Once again, there are two ways to compute the integral (1.146). In the first method, by noting that $(-1)^{2\alpha} = 1$, we can directly compute $I_{n,k}(s; \alpha, \beta)$ for $n = 2m$ as follows:

$$I_{2m,k}(s; \alpha, \beta) = \int_{-\infty}^{\infty} \left(\sum_{j=0}^{\infty} \frac{(-isx)^j}{j!} \right) x^{-2\alpha+2m-2k}(1 + x^2)^{-\beta} dx$$

$$= \sum_{j=0}^{\infty} \frac{(-1)^j i^j s^j}{j!} \left(\int_{-\infty}^{\infty} x^{j-2\alpha+2m-2k}(1 + x^2)^{-\beta} dx \right)$$

$$= \sum_{r=0}^{\infty} \frac{(-1)^r s^{2r}}{(2r)!} \left(2 \int_{0}^{\infty} x^{2r-2\alpha+2m-2k}(1 + x^2)^{-\beta} dx \right)$$

$$= \sum_{r=0}^{\infty} \frac{(-1)^r s^{2r}}{(2r)!} B\left(r - \alpha + m - k + \frac{1}{2}; \beta - r + \alpha - m + k - \frac{1}{2} \right), \tag{1.147}$$

where we have used the definition of the beta integral of the third kind.
Since the last sum in (1.147) can be represented in terms of a hypergeometric function, we have

$$I_{2m,k}(s; \alpha, \beta) = \frac{\Gamma(-\alpha + m - k + 1/2)\, \Gamma(\beta + \alpha - m + k - 1/2)}{\Gamma(\beta)}$$

$$\times {}_1F_2 \left(\begin{array}{c} -\alpha + m - k + 1/2 \\ 1/2, \ -\beta - \alpha + m - k + 3/2 \end{array} \Big| \frac{s^2}{4} \right). \tag{1.148}$$

By knowing that

$$\int_{-\infty}^{\infty} x^{j-2\alpha+2m+1-2k}(1 + x^2)^{-\beta} dx = 0 \qquad \text{for all} \qquad j = 0, 2, 4, \ldots,$$

this method can be similarly applied to $I_{2m+1,k}(s; \alpha, \beta)$, so that after some computations we obtain

$$
I_{2m+1,k}(s; \alpha, \beta) = (-is) \frac{\Gamma(-\alpha + m - k + 3/2) \, \Gamma(\beta + \alpha - m + k - 3/2)}{\Gamma(\beta)}
$$

$$
\times {}_1F_2 \left(\begin{array}{c} -\alpha + m - k + 3/2 \\ 3/2, \, -\beta - \alpha + m - k + 5/2 \end{array} \middle| \frac{s^2}{4} \right). \qquad (1.149)
$$

Hence, combining relations (1.148) and (1.149) gives the final form of (1.146) as

$$
I_{n,k}(s; \alpha, \beta) = \Gamma\left(-\alpha - k + \frac{1}{2} + \left[\frac{n+1}{2} \right] \right) \Gamma\left(\beta + \alpha + k - \frac{1}{2} - \left[\frac{n+1}{2} \right] \right)
$$

$$
\times \frac{(-is)^{\frac{1-(-1)^n}{2}}}{\Gamma(\beta)} {}_1F_2 \left(\begin{array}{c} 1/2 - \alpha - k + [(n+1)/2] \\ 1 - \frac{(-1)^n}{2}, \, -\beta - \alpha - k + 3/2 + [(n+1)/2] \end{array} \middle| \frac{s^2}{4} \right).
$$

$$
(1.150)
$$

The second way of computing $I_{n,k}(s; \alpha, \beta)$ is by taking $n = 2m$ and $n = 2m + 1$ as before and applying the cosine and sine Fourier transforms to the function $(1 + x^2)^{-\beta} x^{-2\alpha + n - 2k}$. In this sense, by noting that $(-1)^{2\alpha} = 1$, we have

$$
I_{2m,k}(s; \alpha, \beta)
$$

$$
= \int_{-\infty}^{\infty} \cos(sx) (1 + x^2)^{-\beta} x^{-2\alpha + 2m - 2k} dx - i \int_{-\infty}^{\infty} \sin(sx) (1 + x^2)^{-\beta} x^{-2\alpha + 2m - 2k} dx
$$

$$
= 2 \int_{0}^{\infty} \cos(sx) (1 + x^2)^{-\beta} x^{-2\alpha + 2m - 2k} dx
$$

$$
= \frac{\Gamma(-\alpha + m - k + 1/2) \, \Gamma(\beta + \alpha - m + k - 1/2)}{\Gamma(\beta)}
$$

$$
\times {}_1F_2 \left(\begin{array}{c} -\alpha + m - k + 1/2 \\ 1/2, \, -\beta - \alpha + m - k + 3/2 \end{array} \middle| \frac{s^2}{4} \right),
$$

as well as

$$
I_{2m+1,k}(s; \alpha, \beta)
$$

$$
= \int_{-\infty}^{\infty} \cos(sx) (1 + x^2)^{-\beta} x^{-2\alpha + 2m + 1 - 2k} dx - i \int_{-\infty}^{\infty} \sin(sx) (1 + x^2)^{-\beta} x^{-2\alpha + 2m + 1 - 2k} dx
$$

$$
= (-2i) \int_{0}^{\infty} \sin(sx) (1 + x^2)^{-\beta} x^{-2\alpha + 2m + 1 - 2k} dx
$$

$$= (-is)\Gamma\left(-\alpha+m-k+\frac{3}{2}\right)\Gamma\left(\beta+\alpha-m+k-\frac{3}{2}\right){}_1F_2\left(\begin{array}{c}-\alpha+m-k+3/2\\ 3/2,\ -\beta-\alpha+m-k+5/2\end{array}\middle|\frac{s^2}{4}\right).$$

Therefore, the result (1.150) simplifies (1.145) as

$$\mathbf{F}(g(x)) = \frac{1}{\Gamma(\beta)}\Gamma\left(-\alpha+\frac{1}{2}+\left[\frac{n+1}{2}\right]\right)\Gamma\left(\beta+\alpha-\frac{1}{2}-\left[\frac{n+1}{2}\right]\right)(-is)^{\frac{1-(-1)^n}{2}}$$

$$\times\sum_{k=0}^{[n/2]}\frac{(-[n/2])_k(c+1/2-[(n+1)/2])_k(\beta+\alpha-1/2-[(n+1)/2])_k}{(c+d-n+1/2)_k(1/2+\alpha-[(n+1)/2])_k k!}$$

$$\times{}_1F_2\left(\begin{array}{c}1/2-\alpha-k+[(n+1)/2]\\ 1-\frac{(-1)^n}{2},\ -\beta-\alpha-k+3/2+[(n+1)/2]\end{array}\middle|\frac{s^2}{4}\right). \qquad (1.151)$$

If in (1.151) we define the symmetric function

$$\mathscr{A}_n(x;\ p_1,\ p_2,\ p_3,\ p_4) = x^{\frac{1-(-1)^n}{2}}$$

$$\times\sum_{k=0}^{[n/2]}\frac{(-[n/2])_k(p_3+1/2-[(n+1)/2])_k(p_1+p_2-1/2-[(n+1)/2])_k}{(p_3+p_4-n+1/2)_k(1/2+p_1-[(n+1)/2])_k k!}$$

$$\times{}_1F_2\left(\begin{array}{c}1/2-p_1-k+[(n+1)/2]\\ 1-\frac{(-1)^n}{2},\ -p_1-p_2-k+3/2+[(n+1)/2]\end{array}\middle|\frac{x^2}{4}\right),$$

then clearly

$$\mathbf{F}(g(x)) =$$

$$(-i)^{\frac{1-(-1)^n}{2}}\frac{1}{\Gamma(\beta)}\Gamma\left(-\alpha+\frac{1}{2}+\left[\frac{n+1}{2}\right]\right)\Gamma\left(\beta+\alpha-\frac{1}{2}-\left[\frac{n+1}{2}\right]\right)\mathscr{A}_n(s;\alpha,\beta,c,d).$$

$$(1.152)$$

By substituting (1.152) in Parseval's identity and noting (1.144), we obtain

$$2\pi\int_{-\infty}^{\infty}x^{-2(\alpha+l)}(1+x^2)^{-(\beta+u)}A_n^{(c,d)}(x)A_m^{(v,w)}(x)\,dx$$

$$= i^{\frac{(-1)^n-(-1)^m}{2}}\Gamma\left(-\alpha+\frac{1}{2}+\left[\frac{n+1}{2}\right]\right)$$

$$\times\frac{1}{\Gamma(\beta)\Gamma(u)}\Gamma\left(\beta+\alpha-\frac{1}{2}-\left[\frac{n+1}{2}\right]\right)\Gamma\left(-l+\frac{1}{2}+\left[\frac{m+1}{2}\right]\right)\Gamma\left(u+l-\frac{1}{2}-\left[\frac{m+1}{2}\right]\right)$$

$$\times \int_{-\infty}^{\infty} \mathscr{A}_n(s; \alpha, \beta, c, d) \, \mathscr{A}_m(s; l, u, v, w) \, ds \,. \qquad (1.153)$$

Now, if on the left-hand side of (1.153) we take

$$c = v = \alpha + l \quad \text{and} \quad d = w = \beta + u,$$

then according to the orthogonality relation (1.128) we finally obtain

$$\frac{1}{2\pi} \int_{-\infty}^{\infty} \mathscr{A}_n(x; \alpha, \beta, p, q) \, \mathscr{A}_m(x; p - \alpha, q - \beta, p, q) \, dx$$

$$= \prod_{j=1}^{n} \frac{\left(-j + (1 - (-1)^j)p\right)\left(j - (1 - (-1)^j)p - 2q\right)}{(2j - 2p - 2q + 1)(2j - 2p - 2q - 1)}$$

$$\times \frac{\Gamma(\beta)\,\Gamma(q-\beta)\,\Gamma(p+q-1/2)\,\Gamma(-p+1/2)\delta_{n,m}}{\Gamma(q)\Gamma\left(-\alpha+\frac{1}{2}+\left[\frac{n+1}{2}\right]\right)\Gamma\left(\alpha+\beta-\frac{1}{2}-\left[\frac{n+1}{2}\right]\right)\Gamma\left(\alpha-p+\frac{1}{2}+\left[\frac{n+1}{2}\right]\right)\Gamma\left(p+q-\alpha-\beta-\frac{1}{2}-\left[\frac{n+1}{2}\right]\right)},$$

where $m, n = 0, 1, \ldots, N = \max\{m, n\} \le p+q-1/2, p < 1/2, (-1)^{2p} = 1, q > \beta > 0,$
$0 < \alpha < 1/2,$ and $\alpha + \beta > 1/2.$

1.8.4 Fourier Transform of the Second Finite Sequence of Symmetric Orthogonal Polynomials

To derive the Fourier transform of the second finite sequence of symmetric orthogonal polynomials

$$B_n^{(a)}(x) = \bar{S}_n \left(\begin{array}{cc} -2a + 2 & 2 \\ 1 & 0 \end{array} \middle| x \right) = x^n \, {}_1F_1 \left(\begin{array}{c} -[n/2] \\ a + (-1)^n/2 \end{array} \middle| \frac{1}{x^2} \right),$$

let us define the specific functions

$$u(x) = x^{-2a} e^{\frac{-1}{2x^2}} B_n^{(b)}(x) \quad \text{with} \quad (-1)^{2a} = 1$$

and

$$v(x) = x^{-2c} e^{\frac{-1}{2x^2}} B_m^{(d)}(x) \quad \text{with} \quad (-1)^{2c} = 1.$$

(1.154)

The Fourier transform of, e.g., $u(x)$ is computed as

$$F(u(x)) = \int_{-\infty}^{\infty} e^{-isx} x^{-2a} e^{\frac{-1}{2x^2}} B_n^{(b)}(x)\, dx$$

$$= \int_{-\infty}^{\infty} e^{-isx} e^{\frac{-1}{2x^2}} x^{-2a+n} \left(\sum_{k=0}^{[n/2]} \frac{(-[n/2])_k}{(b+(-1)^n/2)_k} \frac{x^{-2k}}{k!} \right) dx$$

$$= \sum_{k=0}^{[n/2]} \frac{(-[n/2])_k}{(b+(-1)^n/2)_k k!} \left(\int_{-\infty}^{\infty} e^{-isx} e^{\frac{-1}{2x^2}} x^{-2a+n-2k}\, dx \right).$$

So the following integral should be evaluated:

$$R_{n,k}(s;a) = \int_{-\infty}^{\infty} e^{-isx} e^{\frac{-1}{2x^2}} x^{-2a+n-2k}\, dx.$$

We have

$$R_{2m,k}(s;a) = \int_{-\infty}^{\infty} \left(\sum_{j=0}^{\infty} \frac{(-isx)^j}{j!} \right) e^{\frac{-1}{2x^2}} x^{-2a+2m-2k}\, dx$$

$$= \sum_{j=0}^{\infty} \frac{(-1)^j i^j s^j}{j!} \left(\int_{-\infty}^{\infty} e^{\frac{-1}{2x^2}} x^{j-2a+2m-2k}\, dx \right)$$

$$= \sum_{r=0}^{\infty} \frac{(-1)^r s^{2r}}{(2r)!} \left(2\int_{0}^{\infty} x^{2r-2a+2m-2k} e^{\frac{-1}{2x^2}}\, dx \right)$$

$$= \sum_{r=0}^{\infty} \frac{(-1)^r s^{2r}}{(2r)!} 2^{-r+a-m+k-\frac{1}{2}} \Gamma\left(-r+a-m+k-\frac{1}{2} \right)$$

$$= 2^{a-m+k-\frac{1}{2}} \Gamma\left(a-m+k-\frac{1}{2} \right) {}_0F_2 \left(\begin{array}{c} - \\ \frac{1}{2}, \frac{3}{2}-a+m-k \end{array} \Bigg| \frac{s^2}{8} \right),$$

as well as

$$R_{2m+1,k}(s;a) = \sum_{j=0}^{\infty} \frac{(-1)^j i^j s^j}{j!} \left(\int_{-\infty}^{\infty} e^{\frac{-1}{2x^2}} x^{j-2a+2m+1-2k}\, dx \right)$$

$$= (-is) \sum_{r=0}^{\infty} \frac{(-1)^r s^{2r}}{(2r+1)!} \left(2\int_{0}^{\infty} x^{2r-2a+2m-2k+2} e^{\frac{-1}{2x^2}}\, dx \right)$$

$$= (-is) 2^{a-m+k-\frac{3}{2}} \Gamma\left(a-m+k-\frac{3}{2} \right) {}_0F_2 \left(\begin{array}{c} - \\ \frac{3}{2}, \frac{5}{2}-a+m-k \end{array} \Bigg| \frac{s^2}{8} \right).$$

Consequently,

$$R_{n,k}(s;a) = 2^{a+k-\frac{1}{2}-[\frac{n+1}{2}]} \Gamma\left(a+k-\frac{1}{2}-\left[\frac{n+1}{2}\right]\right)(-is)^{\frac{1-(-1)^n}{2}}$$

$$\times {}_0 F_2\left(\begin{array}{c} - \\ 1-\frac{(-1)^n}{2},\ -a-k+\frac{3}{2}+[\frac{n+1}{2}] \end{array}\middle|\ \frac{s^2}{8}\right)$$

and

$$\mathbf{F}(u(x)) = \Gamma\left(a-\frac{1}{2}-\left[\frac{n+1}{2}\right]\right) 2^{a-\frac{1}{2}-[\frac{n+1}{2}]}(-is)^{\frac{1-(-1)^n}{2}}$$

$$\times \sum_{k=0}^{[n/2]} \frac{(-[\frac{n}{2}])_k (a-\frac{1}{2}-[\frac{n+1}{2}])_k}{(b+(-1)^n/2)_k} \frac{2^k}{k!}\ {}_0 F_2\left(\begin{array}{c} - \\ 1-\frac{(-1)^n}{2},\ -a-k+\frac{3}{2}+[\frac{n+1}{2}] \end{array}\middle|\ \frac{s^2}{8}\right).$$

$$(1.155)$$

If in (1.155) we define the symmetric function

$$\mathscr{B}_n(x;q_1,q_2) = x^{\frac{1-(-1)^n}{2}} \tag{1.156}$$

$$\times \sum_{k=0}^{[n/2]} \frac{(-[\frac{n}{2}])_k (q_1-\frac{1}{2}-[\frac{n+1}{2}])_k}{(q_2+(-1)^n/2)_k} \frac{2^k}{k!}\ {}_0 F_2\left(\begin{array}{c} - \\ 1-\frac{(-1)^n}{2},\ -q_1-k+\frac{3}{2}+[\frac{n+1}{2}] \end{array}\middle|\ \frac{x^2}{8}\right),$$

then by referring to definitions (1.154) and applying Parseval's identity, we get

$$2\pi \int_{-\infty}^{\infty} x^{-2(a+c)} e^{\frac{-1}{x^2}} B_n^{(b)}(x) B_m^{(d)}(x)\, dx$$

$$= i^{\frac{(-1)^n-(-1)^m}{2}} \frac{\Gamma(a-\frac{1}{2}-[\frac{n+1}{2}])\Gamma(c-\frac{1}{2}-[\frac{m+1}{2}])}{2^{-a+\frac{1}{2}+[\frac{n+1}{2}]} 2^{-c+\frac{1}{2}+[\frac{m+1}{2}]}} \int_{-\infty}^{\infty} \mathscr{B}_n(s;a,b) \mathscr{B}_m(s;c,d)\, ds.$$

$$(1.157)$$

Now if in (1.157) we take $b = d = a + c$ and then refer to the finite orthogonality relation (1.131), we conclude that the special function $\mathscr{B}_n(x;q_1,q_2)$ defined in (1.156) satisfies the orthogonality relation

$$\frac{1}{2\pi} \int_{-\infty}^{\infty} \mathscr{B}_n(x;a,b) \mathscr{B}_m(x;b-a,b)\, dx = 2^{-b+1+2[\frac{n+1}{2}]} \prod_{j=1}^{n} \frac{2(-1)^j(j-b)+2b}{(2j-2b+1)(2j-2b-1)}$$

$$\times \frac{\Gamma(b-1/2)}{\Gamma(a-\frac{1}{2}-[\frac{n+1}{2}])\Gamma(b-a-\frac{1}{2}-[\frac{n+1}{2}])} \delta_{n,m},$$

for $m, n = 0, 1, \ldots, N = \max\{m,n\} \le b-\frac{1}{2}$, $(-1)^{2b} = 1$, and $\frac{1}{2} < a < b-\frac{1}{2}$.

1.9   A Class of Symmetric Orthogonal Functions

In Sect. 1.7, we introduced a class of symmetric orthogonal polynomials satisfying the equation

$$x^2(px^2 + q)\, \Phi_n''(x) + x(rx^2 + s)\, \Phi_n'(x)$$
$$- \left(n(r + (n-1)p)x^2 + (1 - (-1)^n)\, s/2 \right) \Phi_n(x) = 0.$$

In this section, we introduce a class of symmetric orthogonal functions as an extension of $S_n(x; p, q, r, s)$ and obtain its standard properties. We show that this new class satisfies the equation

$$x^2(px^2 + q)\, \Phi_n''(x) + x(rx^2 + s)\, \Phi_n'(x) - \left(\alpha_n\, x^2 + (1 - (-1)^n)\, \beta/2 \right) \Phi_n(x) = 0,$$

(1.158)

in which β is a free parameter and $-\alpha_n$ denotes the corresponding eigenvalues. We then consider four cases of the introduced class and study their properties in detail.

Let us replace the following options in the generic differential equation (1.94):

$$\begin{aligned}
A(x) &= x^2(px^2 + q), \\
B(x) &= x(rx^2 + s), \\
C(x) &= x^2 > 0, \\
D(x) &= 0, \\
E(x) &= -\beta \in \mathbb{R},
\end{aligned}$$

(1.159)

to reach Eq. (1.158), where $\lambda_n = -\alpha_n$ is the corresponding eigenvalue to be derived. Although there is a small difference between the above two differential equations, we must solve Eq. (1.158) separately. Since it is independent of β for $n = 2m$ ($m \in \mathbb{Z}^+$), without loss of generality we can assume that its solution is almost similar to relation (1.104) as

$$\Phi_{2n}(x) = \bar{S}_{2n} \left(\begin{matrix} r & s \\ p & q \end{matrix} \middle| x \right) \text{ and } \Phi_{2n+1}(x) = x^\theta\, \bar{S}_{2n} \left(\begin{matrix} r^* & s^* \\ p^* & q^* \end{matrix} \middle| x \right) \text{ for } (\theta \in \mathbb{R} - \{0\}),$$

(1.160)

and try to obtain the parameters p^*, q^*, r^*, s^*, and θ in terms of the five known parameters p, q, r, s, and β to somehow get to Eq. (1.158) and determine the eigenvalues $-\alpha_n$ too. In this sense, the condition $(-1)^\theta = -1$ is necessary in the second definition of (1.160) because $\Phi_n(x)$ must be symmetric and $\Phi_{2n+1}(-x) = -\Phi_{2n+1}(x)$. To solve this problem, as Eq. (1.100) and its generic solution show, if $n = 2m$, then

$$x(px^2 + q)\, \Phi_{2m}''(x) + (rx^2 + s)\, \Phi_{2m}'(x) - 2m\, (r + (2m - 1)p)\, x\, \Phi_{2m}(x) = 0$$

has the general solution

$$\Phi_{2m}(x) = S_{2m}\left(\begin{matrix} r & s \\ p & q \end{matrix}\middle| x\right) = \sum_{k=0}^{m}\binom{m}{k}\left(\prod_{j=0}^{m-(k+1)}\frac{(2j-1+2m)p+r}{(2j+1)q+s}\right)x^{2m-2k}.$$

Hence the second equality of (1.160) must satisfy the equation

$$x(p^{*}x^{2}+q^{*})\frac{d^{2}}{dx^{2}}(x^{-\theta}\Phi_{2m+1}(x)) + (r^{*}x^{2}+s^{*})\frac{d}{dx}(x^{-\theta}\Phi_{2m+1}(x))$$

$$- 2m\,(r^{*}+(2m-1)p^{*})\,x^{1-\theta}\Phi_{2m+1}(x) = 0. \qquad (1.161)$$

After some calculations, (1.161) is simplified as

$$x^{2}(p^{*}x^{2}+q^{*})\,\Phi''_{2m+1}(x) + x((r^{*}-2\theta\,p^{*})x^{2}+s^{*}-2\theta\,q^{*})\,\Phi'_{2m+1}(x)$$

$$+ \left((-2m(r^{*}+(2m-1)p^{*})-\theta(r^{*}-(\theta+1)p^{*}))\,x^{2} + \theta((\theta+1)q^{*}-s^{*})\right)\Phi_{2m+1}(x)$$

$$= 0. \qquad (1.162)$$

If (1.162) is compared with the special case of (1.158) for $n = 2m + 1$, i.e.,

$$x^{2}(px^{2}+q)\,\Phi''_{2m+1}(x) + x(rx^{2}+s)\,\Phi'_{2m+1}(x) - \left(\alpha_{2m+1}\,x^{2}+\beta\right)\Phi_{2m+1}(x) = 0,$$

$$(1.163)$$

then equating Eqs. (1.162) and (1.163) yields

$$p^{*} = p, \quad q^{*} = q, \quad r^{*} = r + 2\theta\,p, \quad \text{and} \quad s^{*} = s + 2\theta\,q.$$

Moreover, the values $-\alpha_{2m+1}$ and $-\beta$ in (1.163) will be

$$-\alpha_{2m+1} = -2m(r + 2\theta\,p + (2m-1)p) - \theta(r + 2\theta\,p - (\theta+1)p)$$

$$= -(\theta+2m)(r + (\theta+2m-1)p)$$

and

$$-\beta = \theta((\theta+1)q - (s + 2\theta\,q)) = -\theta\,(s + (\theta-1)q).$$

All these results eventually lead to the following corollary.

Corollary 1.3 *Let* $\sigma_n = \frac{1-(-1)^n}{2}$. *Then the symmetric sequence*

$$\Phi_n(x) = S_n^{(\theta)} \left(\begin{matrix} r & s \\ p & q \end{matrix} \middle| x \right) = x^{\theta \sigma_n} \; \bar{S}_{2\left[\frac{n}{2}\right]} \left(\begin{matrix} r + 2\sigma_n \, \theta \, p, \; s + 2\sigma_n \, \theta \, q \\ p, \end{matrix} \middle| x \right),$$

having the explicit definition

$$S_n^{(\theta)} \left(\begin{matrix} r & s \\ p & q \end{matrix} \middle| x \right) = \prod_{j=0}^{[n/2]-1} \frac{(2j + 1 + 2\sigma_n \theta) \, q + s}{(2j - 1 + n + (2\theta - 1)\sigma_n) \, p + r} \, x^{(\theta - 1)\sigma_n}$$

$$\times \sum_{k=0}^{[n/2]} \binom{[n/2]}{k} \left(\prod_{j=0}^{[n/2]-(k+1)} \frac{(2j - 1 + n + (2\theta - 1)\sigma_n) \, p + r}{(2j + 1 + 2\sigma_n \theta) \, q + s} \right) x^{n-2k}, \qquad (1.164)$$

satisfies the equation

$$x^2 (px^2 + q) \, \Phi_n''(x) + x(rx^2 + s) \, \Phi_n'(x) - (\alpha_n x^2 + \beta \, \sigma_n) \, \Phi_n(x) = 0, \qquad (1.165)$$

where

$$\alpha_n = (n + (\theta - 1)\sigma_n) \, (r + (n - 1 + (\theta - 1)\sigma_n) p)$$

and

$$\beta = \theta(s + (\theta - 1)q).$$

There is a direct relationship between $S_n^{(\theta)}(x; p, q, r, s)$ and $\bar{S}_n(x; p, q, r, s)$, namely

$$S_n^{(\theta)} \left(\begin{matrix} r & s \\ p & q \end{matrix} \middle| x \right) = x^{(\theta - 1)\sigma_n} \; \bar{S}_n \left(\begin{matrix} r + 2\sigma_n(\theta - 1)p, \; s + 2\sigma_n(\theta - 1)q \\ p, \end{matrix} \middle| x \right),$$

which leads to the hypergeometric representation

$$S_n^{(\theta)} \left(\begin{matrix} r & s \\ p & q \end{matrix} \middle| x \right)$$

$$= x^{(n + \sigma_n (\theta - 1))} \, {}_2F_1 \left(\begin{matrix} -[n/2], \; \frac{2 - \theta + (-1)^n (\theta - 1)}{2} - \frac{s}{2q} - [(n + 1)/2] \\ \frac{4 - \theta - 2n + (-1)^n (\theta - 1)}{2} - \frac{r}{2p} \end{matrix} \middle| -\frac{q}{px^2} \right).$$

Also, they satisfy a three-term recurrence relation as

$$
S_{n+1}^{(\theta)}\left(\begin{array}{cc} r & s \\ p & q \end{array} \middle| x\right) = \left(x^{1+(-1)^n(\theta-1)}\right) S_n^{(\theta)}\left(\begin{array}{cc} r & s \\ p & q \end{array} \middle| x\right) + C_n^{(\theta)}\left(\begin{array}{cc} r & s \\ p & q \end{array}\right) S_{n-1}^{(\theta)}\left(\begin{array}{cc} r & s \\ p & q \end{array} \middle| x\right),
$$

where

$$
C_n^{(\theta)}\left(\begin{array}{cc} r & s \\ p & q \end{array}\right) =
$$

$$
\frac{A_n^{(\theta)}(p,q,r,s)}{\left(((-1)^n(\theta-1)+2n+\theta-2)p+r\right)\left(((-1)^n(1-\theta)+2n+\theta-4)p+r\right)},
\tag{1.166}
$$

with

$$
A_n^{(\theta)}(p,q,r,s) = \left(1+(-1)^n(\theta-1)\right)pq\,n^2
$$
$$
+ \left(\left(\theta-3+3(-1)^n(1-\theta)\right)pq + \left(1+(-1)^n(\theta-1)\right)qr - (-1)^n\theta\,ps\right)n
$$
$$
+ \left((1-\theta)(p-r)q + \left((\theta-3)p+r\right)s\right)(1-(-1)^n)/2.
$$

For instance, we have

$$
S_0^{(\theta)}\left(\begin{array}{cc} r & s \\ p & q \end{array} \middle| x\right) = 1,
$$

$$
S_1^{(\theta)}\left(\begin{array}{cc} r & s \\ p & q \end{array} \middle| x\right) = x^\theta,
$$

$$
S_2^{(\theta)}\left(\begin{array}{cc} r & s \\ p & q \end{array} \middle| x\right) = x^2 + \frac{q+s}{p+r},
$$

$$
S_3^{(\theta)}\left(\begin{array}{cc} r & s \\ p & q \end{array} \middle| x\right) = x^\theta\left(x^2 + \frac{(2\theta+1)q+s}{(2\theta+1)p+r}\right),
$$

$$
S_4^{(\theta)}\left(\begin{array}{cc} r & s \\ p & q \end{array} \middle| x\right) = x^4 + 2\frac{3q+s}{5p+r}x^2 + \frac{(3q+s)(q+s)}{(5p+r)(3p+r)},
$$

$$
S_5^{(\theta)}\left(\begin{array}{cc} r & s \\ p & q \end{array} \middle| x\right) = x^\theta\left(x^4 + 2\frac{(2\theta+3)q+s}{(2\theta+5)p+r}x^2 + \frac{((2\theta+3)q+s)((2\theta+1)q+s)}{((2\theta+5)p+r)((2\theta+3)p+r)}\right).
\tag{1.167}
$$

Relations (1.167) show that the $S_{2n}^{(\theta)}(x; p, q, r, s)$ are independent of θ. Hence the condition $(-1)^\theta = -1$ must be satisfied only for odd n. Moreover, for $\theta = 1$ in (1.164) and (1.166) we respectively obtain

$$S_n^{(1)}\left(\begin{array}{cc} r & s \\ p & q \end{array}\middle| x\right) = \bar{S}_n\left(\begin{array}{cc} r & s \\ p & q \end{array}\middle| x\right) \quad \text{and} \quad C_n^{(1)}\left(\begin{array}{cc} r & s \\ p & q \end{array}\right) = C_n\left(\begin{array}{cc} r & s \\ p & q \end{array}\right).$$

It can be verified that the weight function corresponding to $S_n^{(\theta)}(x; p, q, r, s)$ is the same $W(x; p, q, r, s)$ as defined in (1.109) (i.e., independent of θ). Therefore, we can design a generic orthogonality relation as

$$\int_{-\alpha}^{\alpha} W\left(\begin{array}{cc} r & s \\ p & q \end{array}\middle| x\right) S_n^{(\theta)}\left(\begin{array}{cc} r & s \\ p & q \end{array}\middle| x\right) S_m^{(\theta)}\left(\begin{array}{cc} r & s \\ p & q \end{array}\middle| x\right) dx = N_n\, \delta_{n,m}, \qquad (1.168)$$

where

$$N_n = \int_{-\alpha}^{\alpha} W\left(\begin{array}{cc} r & s \\ p & q \end{array}\middle| x\right) \left(S_n^{(\theta)}\left(\begin{array}{cc} r & s \\ p & q \end{array}\middle| x\right)\right)^2 dx$$

and $(px^2 + q) W(x; p, q, r, s)$ vanishes at $x = \alpha$.

To compute the norm square value, we can directly use the orthogonality relation (1.110), so that for $n = 2m$ we have

$$N_{2m} = \int_{-\alpha}^{\alpha} W\left(\begin{array}{cc} r & s \\ p & q \end{array}\middle| x\right) \left(S_{2m}^{(\theta)}\left(\begin{array}{cc} r & s \\ p & q \end{array}\middle| x\right)\right)^2 dx$$

$$= \int_{-\alpha}^{\alpha} W\left(\begin{array}{cc} r & s \\ p & q \end{array}\middle| x\right) \left(\bar{S}_{2m}\left(\begin{array}{cc} r & s \\ p & q \end{array}\middle| x\right)\right)^2 dx$$

$$= \left(\prod_{i=1}^{2m} C_i\left(\begin{array}{cc} r & s \\ p & q \end{array}\right)\right) \int_{-\alpha}^{\alpha} W\left(\begin{array}{cc} r & s \\ p & q \end{array}\middle| x\right) dx, \qquad (1.169)$$

while for $n = 2m + 1$ we get

$$N_{2m+1} = \int_{-\alpha}^{\alpha} W\left(\begin{array}{cc} r & s \\ p & q \end{array}\middle| x\right) \left(S_{2m+1}^{(\theta)}\left(\begin{array}{cc} r & s \\ p & q \end{array}\middle| x\right)\right)^2 dx$$

$$= \int_{-\alpha}^{\alpha} W\left(\begin{array}{cc} r & s \\ p & q \end{array}\middle| x\right) x^{2\theta} \left(\bar{S}_{2m}\left(\begin{array}{cc} r + 2\theta\, p, & s + 2\theta\, q \\ p, & q \end{array}\middle| x\right)\right)^2 dx.$$

On the other hand, since

$$x^{2\theta} W \left(\begin{matrix} r & s \\ p & q \end{matrix} \bigg| x \right) = \exp(\int \frac{2\theta}{x} dx) \exp(\int \frac{(r - 2p)x^2 + s}{x(px^2 + q)} dx)$$

$$= \exp \left(\int \frac{(r + 2\theta\, p - 2p)x^2 + s + 2\theta\, q}{x(px^2 + q)} dx \right) = W \left(\begin{matrix} r + 2\theta\, p, \; s + 2\theta\, q \\ p, \qquad q \end{matrix} \bigg| x \right),$$

by noting (1.169), N_{2m+1} is simplified as

$$N_{2m+1} = \int_{-\alpha}^{\alpha} W \left(\begin{matrix} r + 2\theta\, p, \; s + 2\theta\, q \\ p, \qquad q \end{matrix} \bigg| x \right) \left(\bar{S}_{2m} \left(\begin{matrix} r + 2\theta\, p, \; s + 2\theta\, q \\ p, \qquad q \end{matrix} \bigg| x \right) \right)^2 dx$$

$$= \left(\prod_{i=1}^{2m} C_i \left(\begin{matrix} r + 2\theta\, p, \; s + 2\theta\, q \\ p, \qquad q \end{matrix} \right) \right) \int_{-\alpha}^{\alpha} W \left(\begin{matrix} r + 2\theta\, p, \; s + 2\theta\, q \\ p, \qquad q \end{matrix} \bigg| x \right) dx. \qquad (1.170)$$

Combining relations (1.169) and (1.170) finally gives

$$N_n = \prod_{i=1}^{2[n/2]} C_i \left(\begin{matrix} r + 2\sigma_n\theta\, p, \; s + 2\sigma_n\theta\, q \\ p, \qquad q \end{matrix} \right)$$

$$\times \int_{-\alpha}^{\alpha} W \left(\begin{matrix} r + 2\sigma_n\theta\, p, \; s + 2\sigma_n\theta\, q \\ p, \qquad q \end{matrix} \bigg| x \right) dx. \qquad (1.171)$$

1.9.1   Four Special Cases of $S_n^{(\theta)}(x; p, q, r, s)$

Case 1 Consider the equation

$$x^2(1 - x^2)\, \Phi_n''(x) - 2x\left((a + b + 1)x^2 - a\right) \Phi_n'(x)$$

$$+ \left((n + (\theta - 1)\sigma_n)(n + 2a + 2b + 1 + (\theta - 1)\sigma_n)\, x^2 \right.$$

$$\left. - \theta\, (\theta + 2a - 1)\sigma_n \right) \Phi_n(x) = 0, \qquad (1.172)$$

together with its solution

$$\Phi_n(x) = S_n^{(\theta)} \left(\begin{matrix} -2a - 2b - 2, \; 2a \\ -1, \qquad 1 \end{matrix} \bigg| x \right). \qquad (1.173)$$

According to the generalized Sturm–Liouville theorem for symmetric functions, the sequence (1.173) satisfies the orthogonality relation

$$\int_{-1}^{1} x^{2a} (1 - x^2)^b \, S_n^{(\theta)} \left(\begin{array}{cc} -2a - 2b - 2, \; 2a \\ -1, \quad 1 \end{array} \middle| x \right) S_m^{(\theta)} \left(\begin{array}{cc} -2a - 2b - 2, \; 2a \\ -1, \quad 1 \end{array} \middle| x \right) dx$$

$$= \prod_{i=1}^{2[n/2]} \frac{(i + 2\sigma_i (a + \theta\sigma_n)) \, (i + 2\sigma_i (a + \theta\sigma_n) + 2b)}{(2i + 2(a + \theta\sigma_n) + 2b - 1) \, (2i + 2(a + \theta\sigma_n) + 2b + 1)}$$

$$\times \; B \left(a + \theta\sigma_n + \frac{1}{2}; \, b + 1 \right) \delta_{n,m},$$

provided that $a + 1/2 > 0$, $a + \theta + 1/2 > 0$, $b + 1 > 0$, $(-1)^{2a} = 1$, and $(-1)^\theta = -1$.

Case 2 Consider the equation

$$x^2 \Phi_n''(x) - 2x(x^2 - a) \, \Phi_n'(x) + \left(2 \left(n + (\theta - 1)\sigma_n \right) x^2 - \theta \, (2a + \theta - 1)\sigma_n \right) \Phi_n(x) = 0,$$

together with its solution

$$\Phi_n(x) = S_n^{(\theta)} \left(\begin{array}{cc} -2, \; 2a \\ 0, \quad 1 \end{array} \middle| x \right). \tag{1.174}$$

The solution (1.174) satisfies the orthogonality relation

$$\int_{-\infty}^{\infty} x^{2a} \exp(-x^2) \, S_n^{(\theta)} \left(\begin{array}{cc} -2 \; 2a \\ 0 \; 1 \end{array} \middle| x \right) S_m^{(\theta)} \left(\begin{array}{cc} -2 \; 2a \\ 0 \; 1 \end{array} \middle| x \right) dx$$

$$= \left(2^{-2[n/2]} \prod_{i=1}^{2[n/2]} i + 2\sigma_i (a + \theta\sigma_n) \right) \Gamma\left(a + \theta\sigma_n + \frac{1}{2} \right) \delta_{n,m},$$

provided that $a + 1/2 > 0$, $a + \theta + 1/2 > 0$, $(-1)^{2a} = 1$, and $(-1)^\theta = -1$.

Case 3 Consider the equation

$$x^2(x^2 + 1) \, \Phi_n''(x) - 2x \left((a + b - 1)x^2 + a \right) \Phi_n'(x)$$

$$- \left((n + (\theta - 1)\sigma_n)(n - 2a - 2b + 1 + (\theta - 1)\sigma_n) x^2 \right.$$

$$\left. + \theta \, (-2a + \theta - 1)\sigma_n \right) \Phi_n(x) = 0, \tag{1.175}$$

together with its solution

$$\Phi_n(x) = S_n^{(\theta)}\left(\begin{array}{cc} -2a - 2b + 2, & -2a \\ 1, & 1 \end{array}\middle| x\right),$$ (1.176)

which satisfies the orthogonality relation

$$\int_{-\infty}^{\infty} \frac{x^{-2a}}{(1+x^2)^b} S_n^{(\theta)}\left(\begin{array}{cc} -2a - 2b + 2, & -2a \\ 1, & 1 \end{array}\middle| x\right) S_m^{(\theta)}\left(\begin{array}{cc} -2a - 2b + 2, & -2a \\ 1, & 1 \end{array}\middle| x\right) dx$$

$$= \prod_{i=1}^{2[n/2]} C_i \left(\begin{array}{cc} -2a - 2b + 2 + 2\theta\sigma_n, & -2a + 2\theta\sigma_n \\ 1, & 1 \end{array}\right)$$

$$\times \frac{\Gamma(b + a - \theta\sigma_n - 1/2)\Gamma(-a + \theta\sigma_n + 1/2)}{\Gamma(b)} \delta_{n,m}.$$ (1.177)

Now the question is how to determine the parameter constraint in (1.177). For this purpose, we write Eq. (1.175) in a self-adjoint form to get

$$\left[x^{-2a}(1+x^2)^{-b+1}(\Phi_n'(x)\Phi_m(x) - \Phi_m'(x)\Phi_n(x))\right]_{-\infty}^{\infty} = 0.$$ (1.178)

Since $\Phi_n(x)$ defined in (1.176) is a function of degree at most $n + (\theta - 1)\sigma_n$, we have

$$\max \deg\{(\Phi_n'(x)\Phi_m(x) - \Phi_m'(x)\Phi_n(x))\} = n + m - 1 + (\theta - 1)(\sigma_n + \sigma_m).$$ (1.179)

Hence it is deduced from (1.178) and (1.179) that

$$- 2a + 2(-b + 1) + n + m - 1 + (\theta - 1)(\sigma_n + \sigma_m) \le 0.$$ (1.180)

Furthermore, the right-hand side of (1.177) shows that

$$b + a - 1/2 > 0, \quad b + a - 1/2 - \theta > 0, \quad -a + 1/2 > 0, \quad -a + 1/2 + \theta > 0, \quad \text{and } b > 0,$$

which are equivalent to

$$b + a - \frac{1}{2} > 0, \ a < \frac{1}{2}, \ b > 0, \ \text{and } a - \frac{1}{2} < \theta < b + a - \frac{1}{2}.$$

By considering the simplified form of (1.180) as

$$- 2a - 2b + n + m + 1 + (\theta - 1)(\sigma_n + \sigma_m) \le 0,$$ (1.181)

four cases may occur for n and m as follows:

$$(i) \begin{cases} n = 2i \\ m = 2j+1 \end{cases} \quad (ii) \begin{cases} n = 2i+1 \\ m = 2j \end{cases} \quad (iii) \begin{cases} n = 2i \\ m = 2j \end{cases} \quad (iv) \begin{cases} n = 2i+1 \\ m = 2j+1 \end{cases}$$

$$(1.182)$$

If each of the above cases is replaced in (1.181), then by taking $N = \max\{m, n\}$ we get

$$N \le b+a-\theta/2, \qquad N \le b+a-1/2, \qquad \text{and} \qquad N \le b+a-\theta+1/2. \qquad (1.183)$$

Finally, by taking

$$\min\ \{b+a-\theta/2,\ b+a-1/2,\ b+a-\theta+1/2\} = M_\theta^{(b+a)},$$

the following corollary is derived.

Corollary 1.4 *The finite set of symmetric functions*

$$\left\{ S_n^{(\theta)}(x;\ 1,\ 1,\ -2a-2b+2,\ -2a) \right\}_{n=0}^{N}$$

is orthogonal with respect to the weight function $x^{-2a}(1+x^2)^{-b}$ *on* $(-\infty, \infty)$ *if and only if* $N \le M_\theta^{(b+a)}$, $a < 1/2$, $b > 0$, $b+a-1/2 > 0$, $a-1/2 < \theta < b+a-1/2$, $(-1)^{2a} = 1$, *and* $(-1)^\theta = -1$.

For example, suppose that the even weight function $x^{-2/3}(1+x^2)^{-10}$ is given on $(-\infty, \infty)$. Therefore $a = 1/3 < 1/2$, $b = 10 > 0$, $-1/6 < \theta < 59/6$, and

$$M_\theta^{(31/3)} = \min\ \{62/6 - \theta/2,\ 61/6,\ 63/6 - \theta\},$$

for all $-1/6 < \theta < 59/6$ and $(-1)^\theta = -1$. This means that the finite set of symmetric functions

$$\left\{ S_n^{(\theta)}(x;\ 1,\ 1,\ -56/3,\ -2/3) \right\}_{n=0}^{N \le M_\theta^{(31/3)}}$$

is orthogonal with respect to the weight function $x^{-2/3}(1+x^2)^{-10}$ on $(-\infty, \infty)$. For instance, if $\theta = 1/3 \in (-1/6,\ 59/6)$, then $(-1)^{1/3} = -1$,

$$M_{1/3}^{(31/3)} = \min\ \{61/6,\ 61/6,\ 61/6\} = 61/6,$$

and $N = 10$, which yields

$$\int_{-\infty}^{\infty} \frac{1}{\sqrt[3]{x^2}(1+x^2)^{10}} S_n^{(1/3)} \left(\begin{array}{cc} -56/3, & -2/3 \\ 1, & 1 \end{array} \middle| x \right) S_m^{(1/3)} \left(\begin{array}{cc} -56/3, & -2/3 \\ 1, & 1 \end{array} \middle| x \right) dx$$

$$= \prod_{i=1}^{2[n/2]} C_i \left(\begin{array}{cc} -(55+(-1)^n)/3, & -(1+(-1)^n)/3 \\ 1, & 1 \end{array} \right)$$

$$\times \frac{\Gamma((58+(-1)^n)/6)\Gamma((2-(-1)^n)/6)}{9!} \delta_{n,m} \qquad \Leftrightarrow m, n \leq 10.$$

Case 4 Consider the equation

$$x^4 \Phi_n''(x) + 2x\big((1-a)x^2 + 1\big) \Phi_n'(x) -$$

$$\left((n+(\theta-1)\sigma_n)(n+1-2a+(\theta-1)\sigma_n)x^2 + \theta\sigma_n \right)\Phi_n(x) = 0,$$

together with its solution

$$\Phi_n(x) = S_n^{(\theta)} \left(\begin{array}{cc} -2a+2, & 2 \\ 1 & 0 \end{array} \middle| x \right),$$

which satisfies the orthogonality relation

$$\int_{-\infty}^{\infty} x^{-2a} \exp(-\frac{1}{x^2}) S_n^{(\theta)} \left(\begin{array}{cc} -2a+2, & 2 \\ 1, & 0 \end{array} \middle| x \right) S_n^{(\theta)} \left(\begin{array}{cc} -2a+2, & 2 \\ 1, & 0 \end{array} \middle| x \right) dx$$

$$= \prod_{i=1}^{2[n/2]} C_i \left(\begin{array}{cc} -2a+2+2\theta\sigma_n, & 2 \\ 1, & 0 \end{array} \right) \Gamma\left(a - \frac{1}{2} - \theta\sigma_n\right) \delta_{n,m}. \qquad (1.184)$$

Once again, to determine the parameter constraint in (1.184), we should apply the described technique for the first finite case, i.e., we must have

$$-2a + n + m + 1 + (\theta - 1)(\sigma_n + \sigma_m) \leq 0,$$

which is the same condition as (1.181) for $b = 0$. Therefore, by referring to (1.182) and (1.183), if we define

$$M_\theta^{(a)} = \min \{a - \theta/2, \ a - 1/2, \ a - \theta + 1/2\},$$

the following corollary is deduced.

Corollary 1.5 *The finite set of symmetric functions* $\left\{ S_n^{(\theta)}(x; 1, 0, -2a + 2, 2) \right\}_{n=0}^{N}$ *is orthogonal with respect to the weight function* $x^{-2a} \exp(-1/x^2)$ *on* $(-\infty, \infty)$ *if and only if* $N \leq M_\theta^{(a)}$, $a - 1/2 > 0$, $a - \theta - 1/2 > 0$, $(-1)^{2a} = 1$, *and* $(-1)^\theta = -1$.

1.10   An Extension of $S_n^{(\theta)}(x; p, q, r, s)$

As we observe in relations (1.99) and (1.159), the function $D(x)$ is equal to zero in both cases. Here one may ask what happens if $D(x) \neq 0$ in (1.99) or (1.159). In this section, we answer this question by considering the following options in the generic differential equation (1.94):

$$
\begin{aligned}
A(x) &= x^2(px^2 + q), \\
B(x) &= x(rx^2 + s), \\
C(x) &= x^2 > 0, \\
D(x) &= -(c + d) = -\alpha\,(s + (\alpha - 1)\,q), \\
E(x) &= 2c = (\alpha - \beta)s + \big(\alpha(\alpha - 1) - \beta(\beta - 1)\big)q,
\end{aligned}
$$

where p, q, r, s and α, β are free parameters.

   We show that the solution of above equation is another class of symmetric orthogonal functions that generalizes $S_n^{(\theta)}(x; p, q, r, s)$. For this purpose, we first showed in relation (1.164) that the symmetric sequence

$$
\Phi_{2n}(x) = \bar{S}_{2n} \left(\begin{array}{cc} r & s \\ p & q \end{array} \middle| x \right) \quad \text{and} \quad \Phi_{2n+1}(x) = x^\theta\, \bar{S}_{2n} \left(\begin{array}{cc} r + 2\theta\,p, & s + 2\theta\,q \\ p, & q \end{array} \middle| x \right)
$$

$$(1.185)$$

is a basis solution for the generalized Sturm–Liouville equation (1.165). Now, without loss of generality, consider the symmetric sequence

$$
\Phi_{2n}(x) = x^\alpha\, \bar{S}_{2n} \left(\begin{array}{cc} r_1 & s_1 \\ p_1 & q_1 \end{array} \middle| x \right) \quad \text{and} \quad \Phi_{2n+1}(x) = x^\beta\, \bar{S}_{2n} \left(\begin{array}{cc} r_2 & s_2 \\ p_2 & q_2 \end{array} \middle| x \right), \qquad (1.186)
$$

for $(-1)^\alpha = 1$ and $(-1)^\beta = -1$. It is clear that (1.186) is an extension of (1.185) for $\alpha = 0$. According to Eq. (1.100) and its generic solution, if $n = 2m$, then the equation

$$
x(p_1 x^2 + q_1)\, y_{2m}''(x) + (r_1 x^2 + s_1)\, y_{2m}'(x) - 2m\,(r_1 + (2m - 1)p_1)\, x\, y_{2m}(x) = 0
$$

has the general solution

$$
y_{2m}(x) = S_{2m} \left(\begin{array}{cc} r_1 & s_1 \\ p_1 & q_1 \end{array} \middle| x \right) = \sum_{k=0}^{m} \binom{m}{k} \left(\prod_{j=0}^{m-(k+1)} \frac{(2j-1+2m)p_1 + r_1}{(2j+1)\,q_1 + s_1} \right) x^{2m-2k}.
$$

Therefore, the first equality of (1.186) must satisfy the equation

$$
x(p_1 x^2 + q_1)\frac{d^2}{dx^2}(x^{-\alpha}\Phi_{2m}(x)) + (r_1 x^2 + s_1)\frac{d}{dx}(x^{-\alpha}\Phi_{2m}(x))
$$
$$
- 2m\,(r_1 + (2m-1)p_1)\,x^{1-\alpha}\Phi_{2m}(x) = 0. \tag{1.187}
$$

After some calculations, (1.187) is simplified as

$$
x^2(p_1 x^2 + q_1)\,\Phi''_{2m}(x) + x\big((r_1 - 2\alpha\,p_1)x^2 + s_1 - 2\alpha\,q_1\big)\Phi'_{2m}(x)
$$
$$
+ \big((-2m(r_1 + (2m-1)p_1) - \alpha\,(r_1 - (\alpha+1)p_1))\,x^2
$$
$$
+ \alpha\,((\alpha+1)q_1 - s_1)\big)\Phi_{2m}(x) = 0. \tag{1.188}
$$

Similarly, for the second equality of (1.186) we get

$$
x(p_2 x^2 + q_2)\frac{d^2}{dx^2}(x^{-\beta}\Phi_{2m+1}(x)) + (r_2 x^2 + s_2)\frac{d}{dx}(x^{-\beta}\Phi_{2m+1}(x))
$$
$$
- 2m\,(r_2 + (2m-1)p_2)\,x^{1-\beta}\Phi_{2m+1}(x) = 0,
$$

which is simplified as

$$
x^2(p_2 x^2 + q_2)\,\Phi''_{2m+1}(x) + x\big((r_2 - 2\beta\,p_2)x^2 + s_2 - 2\beta\,q_2\big)\Phi'_{2m+1}(x)
$$
$$
+ \big((-2m(r_2 + (2m-1)p_2) - \beta\,(r_2 - (\beta+1)p_2))\,x^2
$$
$$
+ \beta\,((\beta+1)\,q_2 - s_2)\big)\Phi_{2m+1}(x) = 0. \tag{1.189}
$$

By comparing and then equating the differential equations (1.188) and (1.189), it can be concluded that

$$
p_2 = p_1, \quad q_2 = q_1, \quad r_2 = r_1 + 2p_1(\beta - \alpha), \quad \text{and} \quad s_2 = s_1 + 2q_1(\beta - \alpha).
$$

If these results are substituted into (1.186), the following corollary is derived.

Corollary 1.6 *Let $\sigma_n = \frac{1-(-1)^n}{2}$. Then the symmetric sequence*

$$S_n^{(\alpha,\beta)} \left(\begin{matrix} r & s \\ p & q \end{matrix} \,\middle|\, x \right)$$

$$= x^{\alpha+(\beta-\alpha)\,\sigma_n} \, \bar{S}_{2\left[\frac{n}{2}\right]} \left(\begin{matrix} r + 2p(\alpha + (\beta - \alpha)\,\sigma_n), \ s + 2q\,(\alpha + (\beta - \alpha)\,\sigma_n) \\ p, \qquad\qquad\qquad q \end{matrix} \,\middle|\, x \right),$$

which is equivalent to the explicit definition

$$S_n^{(\alpha,\beta)} \left(\begin{matrix} r & s \\ p & q \end{matrix} \,\middle|\, x \right)$$

$$= \prod_{j=0}^{[n/2]-1} \frac{(2j + 1 + 2\alpha + 2\,(\beta - \alpha)\,\sigma_n)\,q + s}{(2j - 1 + n + 2\alpha + (2\beta - 2\alpha - 1)\,\sigma_n)\,p + r} \, x^{\alpha+(\beta-\alpha-1)\,\sigma_n}$$

$$\times \sum_{k=0}^{[n/2]} \binom{[n/2]}{k} \left(\prod_{j=0}^{[n/2]-(k+1)} \frac{(2j - 1 + n + 2\alpha + (2\beta - 2\alpha - 1)\,\sigma_n)\,p + r}{(2j + 1 + 2\alpha + 2\,(\beta - \alpha)\,\sigma_n)\,q + s} \right) x^{n-2k},$$

$$(1.190)$$

satisfies the equation

$$x^2(px^2 + q)\,\Phi_n''(x) + x(rx^2 + s)\,\Phi_n'(x) - \left(a_n\,x^2 + (-1)^n c + d \right) \Phi_n(x) = 0, \qquad (1.191)$$

where

$$a_n = (n + \alpha + (\beta - \alpha - 1)\,\sigma_n)\,(r + (n - 1 + \alpha + (\beta - \alpha - 1)\,\sigma_n)p),$$

$$2c = (\alpha - \beta)s + (\alpha(\alpha - 1) - \beta(\beta - 1))q,$$

and

$$2d = (\alpha + \beta)s + (\alpha(\alpha - 1) + \beta(\beta - 1))q.$$

There is a direct relationship between $S_n^{(\alpha,\beta)}(x; p, q, r, s)$ and $\bar{S}_n(x; p, q, r, s)$ as follows:

$$S_n^{(\alpha,\beta)} \left(\begin{matrix} r & s \\ p & q \end{matrix} \,\middle|\, x \right) = x^{\alpha+(\beta-\alpha-1)\,\sigma_n}$$

$$\times \bar{S}_n \left(\begin{matrix} r + 2p(\alpha + (\beta - \alpha - 1)\,\sigma_n), \ s + 2q\,(\alpha + (\beta - \alpha - 1)\,\sigma_n) \\ p, \qquad\qquad\qquad q \end{matrix} \,\middle|\, x \right), \qquad (1.192)$$

which leads to the hypergeometric representation

$$S_n^{(\alpha,\beta)}\left(\left.\begin{matrix} r & s \\ p & q \end{matrix}\right| x\right) = x^{n+\alpha+(\beta-\alpha-1)\sigma_n}$$

$$\times {}_2F_1\left(\left.\begin{matrix} -[\frac{n}{2}],\ \frac{q-s}{2q} - (\alpha + (\beta - \alpha - 1)\sigma_n) - [\frac{n+1}{2}] \\ -\frac{r}{2p} - (\alpha + (\beta - \alpha - 1)\sigma_n) - n + 3/2 \end{matrix}\right| -\frac{q}{px^2}\right).$$

Also, they satisfy a three-term recurrence relation as

$$S_{n+1}^{(\alpha,\beta)}\left(\left.\begin{matrix} r & s \\ p & q \end{matrix}\right| x\right) = \left(x^{1+(-1)^n(\beta-\alpha-1)}\right) S_n^{(\alpha,\beta)}\left(\left.\begin{matrix} r & s \\ p & q \end{matrix}\right| x\right)$$

$$+ C_n^{(\alpha,\beta)}\left(\begin{matrix} r & s \\ p & q \end{matrix}\right) S_{n-1}^{(\alpha,\beta)}\left(\left.\begin{matrix} r & s \\ p & q \end{matrix}\right| x\right), \tag{1.193}$$

where

$$C_n^{(\alpha,\beta)}\left(\begin{matrix} r & s \\ p & q \end{matrix}\right) =$$

$$\frac{A_n^{(\alpha,\beta)}(p,q,r,s)}{\left(((-1)^n(\beta-\alpha-1) + 2n+\beta+\alpha - 2)p + r\right)\left(((-1)^n(1-\beta+\alpha) + 2n + \beta+\alpha-4)p + r\right)},$$

with

$$A_n^{(\alpha,\beta)}(p,q,r,s) = \left(1 + (\beta - \alpha - 1)(-1)^n\right) pq\, n^2$$

$$+ \left((1 - \beta + \alpha)((1 - 2\alpha)p - r)q + ((\alpha + \beta - 3)p + r)(s + 2\alpha q)\right)\sigma_n$$

$$+ \left((\alpha + \beta - 3 + (-1)^n(\alpha + 3 - 3\beta))pq\right.$$

$$+ \left. (1 + (-1)^n(\beta - \alpha - 1))qr + (-1)^n(\alpha - \beta)\, ps\right)n.$$

For instance, we have

$$S_0^{(\alpha,\beta)}\left(\left.\begin{matrix} r & s \\ p & q \end{matrix}\right| x\right) = x^{\alpha},$$

$$S_1^{(\alpha,\beta)}\left(\left.\begin{matrix} r & s \\ p & q \end{matrix}\right| x\right) = x^{\beta},$$

$$S_2^{(\alpha,\beta)} \left(\begin{array}{cc} r & s \\ p & q \end{array} \middle| x \right) = x^\alpha \left(x^2 + \frac{(2\alpha+1)q+s}{(2\alpha+1)p+r} \right),$$

$$S_3^{(\alpha,\beta)} \left(\begin{array}{cc} r & s \\ p & q \end{array} \middle| x \right) = x^\beta \left(x^2 + \frac{(2\beta+1)q+s}{(2\beta+1)p+r} \right),$$

$$S_4^{(\alpha,\beta)} \left(\begin{array}{cc} r & s \\ p & q \end{array} \middle| x \right) = x^\alpha \left(x^4 + 2\frac{(2\alpha+3)q+s}{(2\alpha+5)p+r}x^2 + \frac{((2\alpha+3)q+s)((2\alpha+1)q+s)}{((2\alpha+5)p+r)((2\alpha+3)p+r)} \right),$$

$$S_5^{(\alpha,\beta)} \left(\begin{array}{cc} r & s \\ p & q \end{array} \middle| x \right) = x^\beta \left(x^4 + 2\frac{(2\beta+3)q+s}{(2\beta+5)p+r}x^2 + \frac{((2\beta+3)q+s)((2\beta+1)q+s)}{((2\beta+5)p+r)((2\beta+3)p+r)} \right).$$

Note that both conditions $(-1)^\alpha = 1$ and $(-1)^\beta = -1$ are necessary, because

$$S_n^{(\alpha,\beta)}(-x; p, q, r, s) = (-1)^n S_n^{(\alpha,\beta)}(x; p, q, r, s).$$

An interesting case for the symmetric functions (1.190) occurs when $\beta - \alpha = 1$ and $(-1)^\alpha = 1$, because in this case, (1.193) is reduced to a recurrence relation of polynomial type as

$$S_{n+1}^{(\alpha,\alpha+1)} \left(\begin{array}{cc} r & s \\ p & q \end{array} \middle| x \right) = x\, S_n^{(\alpha,\alpha+1)} \left(\begin{array}{cc} r & s \\ p & q \end{array} \middle| x \right) + C_n^{(\alpha,\alpha+1)} \left(\begin{array}{cc} r & s \\ p & q \end{array} \right) S_{n-1}^{(\alpha,\alpha+1)} \left(\begin{array}{cc} r & s \\ p & q \end{array} \middle| x \right),$$

though the corresponding initial values are $S_0^{(\alpha,\alpha+1)}(x) = x^\alpha$ and $S_1^{(\alpha,\alpha+1)}(x) = x^{\alpha+1}$ if and only if $(-1)^\alpha = 1$. On the other hand, by referring to the key identity (1.192), we have

$$S_n^{(\alpha,\alpha+1)} \left(\begin{array}{cc} r & s \\ p & q \end{array} \middle| x \right) = x^\alpha\, \bar{S}_n \left(\begin{array}{cc} r+2p\alpha, & s+2q\alpha \\ p, & q \end{array} \middle| x \right).$$

Therefore, the polynomials $\bar{S}_n(x; p, q, r+2p\alpha, s+2q\alpha)$ are orthogonal with respect to the weight function $x^{2\alpha} W(x; p, q, r, s)$.

As Eq. (1.191) and relation (1.98) show, the weight function corresponding to symmetric functions (1.190) is the same $W(x; p, q, r, s)$ as defined in (1.109) (i.e., independent of α, β). Hence we can design a generic orthogonality relation as follows:

$$\int_{-v}^{v} W \left(\begin{array}{cc} r & s \\ p & q \end{array} \middle| x \right) S_n^{(\alpha,\beta)} \left(\begin{array}{cc} r & s \\ p & q \end{array} \middle| x \right) S_m^{(\alpha,\beta)} \left(\begin{array}{cc} r & s \\ p & q \end{array} \middle| x \right) dx = N_n\, \delta_{n,m}, \qquad (1.194)$$

where

$$
N_n = \int_{-v}^{v} W\left(\begin{matrix} r & s \\ p & q \end{matrix}\,\middle|\,x\right) \left(S_n^{(\alpha,\beta)}\left(\begin{matrix} r & s \\ p & q \end{matrix}\,\middle|\,x\right)\right)^2 dx, \tag{1.195}
$$

and $(px^2+q)\,W(x;p,q,r,s)$ vanishes at $x=v$.

To compute the norm square value, we can directly use the orthogonality relation (1.110), so that for $n=2m$ in (1.195) we have

$$
N_{2m} = \int_{-v}^{v} W\left(\begin{matrix} r & s \\ p & q \end{matrix}\,\middle|\,x\right) \left(S_{2m}^{(\alpha,\beta)}\left(\begin{matrix} r & s \\ p & q \end{matrix}\,\middle|\,x\right)\right)^2 dx
$$

$$
= \int_{-v}^{v} W\left(\begin{matrix} r & s \\ p & q \end{matrix}\,\middle|\,x\right) x^{2\alpha} \left(\bar{S}_{2m}\left(\begin{matrix} r+2\alpha\,p, & s+2\alpha\,q \\ p, & q \end{matrix}\,\middle|\,x\right)\right)^2 dx.
$$

On the other hand, since

$$
x^{2\alpha}\,W\left(\begin{matrix} r & s \\ p & q \end{matrix}\,\middle|\,x\right) = \exp\left(\int \frac{2\alpha}{x}\,dx\right) \exp\left(\int \frac{(r-2p)x^2+s}{x(px^2+q)}\,dx\right)
$$

$$
= \exp\left(\int \frac{(r+2\alpha\,p-2p)x^2+s+2\alpha\,q}{x(px^2+q)}\,dx\right) = W\left(\begin{matrix} r+2\alpha\,p, & s+2\alpha\,q \\ p, & q \end{matrix}\,\middle|\,x\right),
$$

by noting (1.110), N_{2m} is simplified as

$$
N_{2m} = \int_{-v}^{v} W\left(\begin{matrix} r+2\alpha\,p, & s+2\alpha\,q \\ p, & q \end{matrix}\,\middle|\,x\right) \left(\bar{S}_{2m}\left(\begin{matrix} r+2\alpha\,p, & s+2\alpha\,q \\ p, & q \end{matrix}\,\middle|\,x\right)\right)^2 dx
$$

$$
= \left(\prod_{i=1}^{2m} C_i\left(\begin{matrix} r+2\alpha\,p, & s+2\alpha\,q \\ p, & q \end{matrix}\right)\right) \int_{-v}^{v} W\left(\begin{matrix} r+2\alpha\,p, & s+2\alpha\,q \\ p, & q \end{matrix}\,\middle|\,x\right) dx, \tag{1.196}
$$

in which $C_i(p,q,r,s) = C_i^{(0,1)}(p,q,r,s)$. Similarly, for $n=2m+1$, we get

$$
N_{2m+1} = \int_{-v}^{v} W\left(\begin{matrix} r+2\beta\,p, & s+2\beta\,q \\ p, & q \end{matrix}\,\middle|\,x\right) \left(\bar{S}_{2m}\left(\begin{matrix} r+2\beta\,p, & s+2\beta\,q \\ p, & q \end{matrix}\,\middle|\,x\right)\right)^2 dx
$$

$$
= \left(\prod_{i=1}^{2m} C_i\left(\begin{matrix} r+2\beta\,p, & s+2\beta\,q \\ p, & q \end{matrix}\right)\right) \int_{-v}^{v} W\left(\begin{matrix} r+2\beta\,p, & s+2\beta\,q \\ p, & q \end{matrix}\,\middle|\,x\right) dx. \tag{1.197}
$$

Combining relations (1.196) and (1.197) eventually gives

$$N_n = \prod_{i=1}^{2[n/2]} C_i \left(\begin{matrix} r + 2p(\alpha + (\beta - \alpha - 1)\,\sigma_n), \ s + 2q\,(\alpha + (\beta - \alpha - 1)\,\sigma_n) \\ p, \qquad\qquad\qquad\qquad q \end{matrix} \right)$$

$$\times \int_{-v}^{v} W \left(\begin{matrix} r + 2p(\alpha + (\beta - \alpha - 1)\,\sigma_n), \ s + 2q\,(\alpha + (\beta - \alpha - 1)\,\sigma_n) \\ p, \qquad\qquad\qquad\qquad q \end{matrix} \ \middle| \ x \right) dx \,.$$

1.10.1 Four Special Cases of $S_n^{(\alpha,\beta)}(x; p, q, r, s)$

Case 1 Consider the differential equation

$$x^2(1 - x^2)\,\Phi_n''(x) - 2x\big((a + b + 1)x^2 - a\big)\Phi_n'(x) + \Big((n + \alpha + (\beta - \alpha - 1)\,\sigma_n)\times$$

$$(n + 2a + 2b + 1 + \alpha + (\beta - \alpha - 1)\,\sigma_n)\,x^2 + \frac{\beta(\beta - 1 + 2a) - \alpha(\alpha - 1 + 2a)}{2}(-1)^n$$

$$- \frac{\beta(\beta - 1 + 2a) + \alpha(\alpha - 1 + 2a)}{2}\Big)\Phi_n(x) = 0,$$

with the solution

$$\Phi_n(x) = S_n^{(\alpha,\beta)} \left(\begin{matrix} -2a - 2b - 2, \ 2a \\ -1, \qquad\quad 1 \end{matrix} \ \middle| \ x \right). \tag{1.198}$$

According to the generalized Sturm–Liouville theorem for symmetric functions, the sequence (1.198) satisfies the orthogonality relation

$$\int_{-1}^{1} x^{2a}(1 - x^2)^b \, S_n^{(\alpha,\beta)} \left(\begin{matrix} -2a - 2b - 2, \ 2a \\ -1, \qquad\quad 1 \end{matrix} \ \middle| \ x \right) S_m^{(\alpha,\beta)} \left(\begin{matrix} -2a - 2b - 2, \ 2a \\ -1, \qquad\quad 1 \end{matrix} \ \middle| \ x \right) dx$$

$$= \prod_{i=1}^{2[n/2]} \frac{(i + 2\sigma_i(a + \alpha + (\beta - \alpha)\,\sigma_n))\,(i + 2\sigma_i(a + \alpha + (\beta - \alpha)\,\sigma_n) + 2b)}{(2i + 2(a + \alpha + (\beta - \alpha)\,\sigma_n) + 2b - 1)\,(2i + 2(a + \alpha + (\beta - \alpha)\,\sigma_n) + 2b + 1)}$$

$$\times B\left(a + \alpha + (\beta - \alpha)\,\sigma_n + \frac{1}{2}\,;\, b + 1 \right) \delta_{n,m},$$

provided that $a + \alpha + 1/2 > 0$, $a + \beta + 1/2 > 0$, $b + 1 > 0$, $(-1)^{2a} = 1$, $(-1)^{\alpha} = 1$, and $(-1)^{\beta} = -1$.

Case 2 Consider the differential equation

$$x^2 \Phi_n''(x) - 2x(x^2 - a)\Phi_n'(x) + \Bigg(2(n + \alpha + (\beta - \alpha - 1)\sigma_n)x^2$$

$$+ \frac{\beta(\beta - 1 + 2a) - \alpha(\alpha - 1 + 2a)}{2}(-1)^n$$

$$- \frac{\beta(\beta - 1 + 2a) + \alpha(\alpha - 1 + 2a)}{2}\Bigg)\Phi_n(x) = 0,$$

with the solution

$$\Phi_n(x) = S_n^{(\alpha,\beta)}\left(\begin{matrix} -2, & 2a \\ 0, & 1 \end{matrix}\,\Bigg|\, x\right).$$

This solution satisfies the orthogonality relation

$$\int_{-\infty}^{\infty} x^{2a}\exp(-x^2)\, S_n^{(\alpha,\beta)}\left(\begin{matrix} -2 & 2a \\ 0 & 1 \end{matrix}\,\Bigg|\, x\right) S_m^{(\alpha,\beta)}\left(\begin{matrix} -2 & 2a \\ 0 & 1 \end{matrix}\,\Bigg|\, x\right) dx$$

$$= \left(\frac{1}{2^{n-\sigma_n}} \prod_{i=1}^{2[n/2]} i + 2\sigma_i(a + \alpha + (\beta - \alpha)\sigma_n)\right) \Gamma\left(a + \alpha + (\beta - \alpha)\sigma_n + \frac{1}{2}\right)\delta_{n,m},$$

provided that $a + \alpha + 1/2 > 0$, $a + \beta + 1/2 > 0$, $(-1)^{2a} = 1$, $(-1)^{\alpha} = 1$, and $(-1)^{\beta} = -1$.

Case 3 Consider the differential equation

$$x^2(x^2 + 1)\,\Phi_n''(x) - 2x((a + b - 1)x^2 + a)\,\Phi_n'(x)$$

$$- \Bigg((n + \alpha + (\beta - \alpha - 1)\sigma_n)(n - 2a - 2b + 1 + \alpha + (\beta - \alpha - 1)\sigma_n)x^2$$

$$+ \frac{\alpha(\alpha - 1 - 2a) - \beta(\beta - 1 - 2a)}{2}(-1)^n$$

$$+ \frac{\alpha(\alpha - 1 - 2a) + \beta(\beta - 1 - 2a)}{2}\Bigg)\Phi_n(x) = 0, \qquad (1.199)$$

with the solution

$$\Phi_n(x) = S_n^{(\alpha,\beta)}\left(\begin{matrix} -2a - 2b + 2, & -2a \\ 1, & 1 \end{matrix}\,\Bigg|\, x\right), \qquad (1.200)$$

which satisfies the orthogonality relation

$$\int_{-\infty}^{\infty} \frac{x^{-2a}}{(1+x^2)^b} S_n^{(\alpha,\beta)} \left(\begin{array}{cc} -2a-2b+2, & -2a \\ 1, & 1 \end{array} \middle| x \right) S_m^{(\alpha,\beta)} \left(\begin{array}{cc} -2a-2b+2, & -2a \\ 1, & 1 \end{array} \middle| x \right) dx$$

$$= \prod_{i=1}^{2\lfloor n/2 \rfloor} C_i \left(\begin{array}{cc} -2a-2b+2+2(\alpha+(\beta-\alpha-1)\sigma_n), & -2a+2(\alpha+(\beta-\alpha-1)\sigma_n) \\ 1, & 1 \end{array} \right)$$

$$\times \frac{\Gamma(b+a-(\alpha+(\beta-\alpha-1)\sigma_n)-1/2)\Gamma(-a+\alpha+(\beta-\alpha-1)\sigma_n+1/2)}{\Gamma(b)} \delta_{n,m}.$$

$$(1.201)$$

To determine the parameter constraint in (1.201), we write Eq. (1.199) in a self-adjoint form to obtain

$$\left[x^{-2a}(1+x^2)^{-b+1} \left(\Phi_n'(x)\Phi_m(x) - \Phi_m'(x)\Phi_n(x) \right) \right]_{-\infty}^{\infty} = 0. \qquad (1.202)$$

Since $\Phi_n(x)$ defined in (1.200) is a symmetric sequence of degree at most $n + \alpha + (\beta - \alpha - 1)\sigma_n$, we have

$$\max \deg\{\Phi_n'(x)\Phi_m(x) - \Phi_m'(x)\Phi_n(x)\} = n + m - 1 + 2\alpha + (\beta - \alpha - 1)(\sigma_n + \sigma_m).$$

$$(1.203)$$

Therefore, it is deduced from (1.202) and (1.203) that

$$- 2a + 2(-b + 1) + n + m - 1 + 2\alpha + (\beta - \alpha - 1)(\sigma_n + \sigma_m) \leq 0. \qquad (1.204)$$

Moreover, the right-hand side of (1.201) shows that

$$b + a - \alpha - 1/2 > 0, \quad b + a - \beta - 1/2 > 0, \quad -a + \alpha + 1/2 > 0, \quad -a + \beta + 1/2 > 0,$$

and $b > 0$.

Once again, four cases may occur for n and m in inequality (1.204):

(i) $\begin{cases} n = 2i \\ m = 2j+1 \end{cases}$ (ii) $\begin{cases} n = 2i+1 \\ m = 2j \end{cases}$ (iii) $\begin{cases} n = 2i \\ m = 2j \end{cases}$ (iv) $\begin{cases} n = 2i+1 \\ m = 2j+1 \end{cases}$

$$(1.205)$$

If each of the above cases is replaced in (1.204), then by taking $N = \max\{m, n\}$, we obtain

$$N \leq b+a-(\alpha+\beta)/2, \quad N \leq b+a-\alpha-1/2, \quad \text{and} \quad N \leq b+a-\beta+1/2. \qquad (1.206)$$

Finally, by taking

$$\min \{b + a - (\alpha + \beta)/2, \ b + a - \alpha - 1/2, \ b + a - \beta + 1/2\} = M_{\alpha,\beta}^{(b+a)},$$

we obtain the following corollary.

Corollary 1.7 *The finite set of symmetric functions*

$$\left\{ S_n^{(\alpha,\beta)}(x; \ 1, \ 1, \ -2a - 2b + 2, \ -2a) \right\}_{n=0}^{N}$$

is orthogonal with respect to the weight function $x^{-2a}(1 + x^2)^{-b}$ *on* $(-\infty, \infty)$ *if and only if* $N \leq M_{\alpha,\beta}^{(b+a)}$, $b + a - \alpha - 1/2 > 0$, $b + a - \beta - 1/2 > 0$, $-a + \alpha + 1/2 > 0$, $-a + \beta + 1/2 > 0$, $b > 0$, $(-1)^{2a} = 1$, $(-1)^{\alpha} = 1$, *and* $(-1)^{\beta} = -1$.

Case 4 Consider the differential equation

$$x^4 \Phi_n''(x) + 2x\left((1-a)x^2 + 1\right)\Phi_n'(x) -$$

$$\left((n + \alpha + (\beta - \alpha - 1)\sigma_n)(n + 1 - 2a + \alpha + (\beta - \alpha - 1)\sigma_n) x^2 \right.$$

$$\left. + (\alpha - \beta)(-1)^n + \alpha + \beta \right) \Phi_n(x) = 0,$$

with the solution

$$\Phi_n(x) = S_n^{(\alpha,\beta)}\left(\begin{array}{cc} -2a+2, & 2 \\ 1 & 0 \end{array} \middle| x \right),$$

which satisfies the orthogonality relation

$$\int_{-\infty}^{\infty} x^{-2a} \exp(-\frac{1}{x^2}) \ S_n^{(\alpha,\beta)}\left(\begin{array}{cc} -2a+2, & 2 \\ 1, & 0 \end{array} \middle| x \right) S_m^{(\alpha,\beta)}\left(\begin{array}{cc} -2a+2, & 2 \\ 1, & 0 \end{array} \middle| x \right) dx$$

$$= \prod_{i=1}^{2[n/2]} C_i \left(\begin{array}{cc} -2a+2+2(\alpha+(\beta-\alpha)\sigma_n), & 2 \\ 1, & 0 \end{array} \right) \Gamma(a - \frac{1}{2} - (\alpha + (\beta - \alpha)\sigma_n)) \ \delta_{n,m} .$$

$$(1.207)$$

To determine the parameter constraint in (1.207), noting

$$\left[x^{2-2a} \exp(-\frac{1}{x^2})(\Phi_n'(x)\Phi_m(x) - \Phi_m'(x)\Phi_n(x)) \right]_{-\infty}^{\infty} = 0,$$

Table 1.6 Four special cases of $S_n^{(\alpha,\beta)}(x; p, q, r, s)$

Definition	Weight function	Interval and kind
$S_n^{(\alpha,\beta)}\left(\begin{array}{cc} -2a - 2b - 2, \ 2a \\ -1, \quad 1 \end{array} \Bigg| x \right)$	$x^{2a}(1 - x^2)^b$	$[-1, 1]$, Infinite
$S_n^{(\alpha,\beta)}\left(\begin{array}{cc} -2, \ 2a \\ 0, \ 1 \end{array} \Bigg| x \right)$	$x^{2a}\exp(-x^2)$	$(-\infty, \infty)$, Infinite
$S_n^{(\alpha,\beta)}\left(\begin{array}{cc} -2a - 2b + 2, \ -2a \\ 1, \quad 1 \end{array} \Bigg| x \right)$	$\dfrac{x^{-2a}}{(1 + x^2)^b}$	$(-\infty, \infty)$, Finite
$S_n^{(\alpha,\beta)}\left(\begin{array}{cc} -2a + 2, \ 2 \\ 1, \quad 0 \end{array} \Bigg| x \right)$	$x^{-2a}\exp(-\frac{1}{x^2})$	$(-\infty, \infty)$, Finite

then (1.203) yields

$$-2a + n + m + 1 + 2\alpha + (\beta - \alpha - 1)(\sigma_n + \sigma_m) \leq 0,$$

which is exactly the same condition as (1.204) for $b = 0$. Therefore, by referring to (1.205) and (1.206), if we define

$$M_{\alpha,\beta}^{(a)} = \min\{a - (\alpha + \beta)/2, \ a - \alpha - 1/2, \ a - \beta + 1/2\},$$

the following corollary is deduced.

Corollary 1.8 *The finite set of symmetric functions* $\left\{S_n^{(\alpha,\beta)}(x; 1, 0, -2a + 2, 2)\right\}_{n=0}^{N}$ *is orthogonal with respect to the weight function* $x^{-2a}\exp(-1/x^2)$ *on* $(-\infty, \infty)$ *if and only if* $N \leq M_{\alpha,\beta}^{(a)}$, $a - \alpha - 1/2 > 0$, $a - \beta - 1/2 > 0$, $(-1)^{2a} = 1$, $(-1)^{\alpha} = 1$, *and* $(-1)^{\beta} = -1$.

Table 1.6 shows the principal properties of the four classes.

1.11 A Generalization of Fourier Trigonometric Series

Although we introduced four main cases of $S_n^{(\theta)}(x; p, q, r, s)$ in Sect. 1.9, further important cases can be still found for the specific values of p, q, r, s, and θ. In this section, using the generalized Sturm–Liouville theorem for symmetric functions, we introduce one of these samples, which generalizes Fourier trigonometric sequences and is orthogonal with respect to the same constant weight function on $[0, \pi]$. One of the advantages of this generalization is its use in finding some new trigonometric series.

1.11.1 A Generalization of Trigonometric Orthogonal Sequences

It is known that the trigonometric sequences $\{\sin nx\}_{n=1}^{\infty}$ and $\{\cos nx\}_{n=0}^{\infty}$ are orthogonal solutions of a usual Sturm–Liouville equation of the form

$$\Phi_n''(x) + n^2\,\Phi_n(x) = 0, \quad x \in [0, \pi], \tag{1.208}$$

where $[0, \pi]$ can be transformed to any other arbitrary interval, say $[-l, l]$, with period $2l$, by a simple linear transformation.

To extend such trigonometric sequences, we start with a special case of Eq. (1.165) as

$$x^2(px^2+q)\Phi_n''(x)+x(rx^2 + s)\Phi_n'(x) + \left(\lambda_n^{(\theta)}x^2 - \theta(s+(\theta - 1)q)\sigma_n\right)\Phi_n(x) = 0, \tag{1.209}$$

for

$$\lambda_n^{(\theta)} = -\left(n + (\theta - 1)\sigma_n\right)\left(r + (n - 1 + (\theta - 1)\sigma_n)p\right),$$

and then assume that $p = -1$ and $q = 1$.

In this case, by the change of variable $x = \cos t$, we obtain the modified differential equation (1.209) as

$$\cos^2 t\ \Phi_n''(t) + \left(\frac{-(r + 1)\cos^3 t - s \cos t}{\sin t}\right)\Phi_n'(t)$$

$$+ \left(\lambda_n^{(\theta)}\cos^2 t - \theta(s + \theta - 1)\sigma_n\right)\Phi_n(t) = 0. \tag{1.210}$$

If $r + 1 = 0$ and $s = 0$ in (1.210), a generalization of differential equation (1.208) is derived for $\theta = 1$. Consequently, by substituting the initial vector

$$(p, q, r, s, \theta) = (-1, 1, -1, 0, \theta)$$

into Eq. (1.209) and applying the change of variable $x = \cos t$, we get

$$\Phi_n''(t) + \left(\left(n + (\theta - 1)\sigma_n\right)^2 - \frac{\theta\,(\theta - 1)}{\cos^2 t}\,\sigma_n\right)\Phi_n(t) = 0. \tag{1.211}$$

By referring to the comments described in Sect. 1.9, Eq. (1.211) has a basic solution as

$$\Phi_n(t) = S_n^{(\theta)}\begin{pmatrix} -1 & 0 \\ -1 & 1 \end{pmatrix}\cos t\bigg) = \prod_{j=0}^{[n/2]-1}\frac{2j + 1 + 2\theta\sigma_n}{2j + n + (2\theta - 1)\sigma_n}\,(\cos t)^{(\theta - 1)\sigma_n}$$

$$\times \sum_{k=0}^{[n/2]} (-1)^k \binom{[n/2]}{k} \left(\prod_{j=0}^{[n/2]-(k+1)} \frac{2j+n+(2\theta-1)\sigma_n}{2j+1+2\theta\sigma_n} \right) \cos^{n-2k} t,$$

which can be simplified to an easier form, because for $n = 2m$ we have

$$S_{2m}^{(\theta)} \left(\left. \begin{matrix} -1 & 0 \\ -1 & 1 \end{matrix} \right| \cos t \right) =$$

$$\prod_{j=0}^{m-1} \frac{j+1/2}{j+m} \sum_{k=0}^{m} (-1)^k \binom{m}{k} \left(\prod_{j=0}^{m-(k+1)} \frac{j+m}{j+1/2} \right) \cos^{2m-2k} t = \frac{1}{2^{2m-1}} \cos 2mt,$$

while for $n = 2m + 1$, it reads as

$$S_{2m+1}^{(\theta)} \left(\left. \begin{matrix} -1 & 0 \\ -1 & 1 \end{matrix} \right| \cos t \right) =$$

$$\prod_{j=0}^{m-1} \frac{j+\theta+1/2}{j+\theta+m} \cos^{(\theta-1)} t \sum_{k=0}^{m} (-1)^k \binom{m}{k} \left(\prod_{j=0}^{m-(k+1)} \frac{j+\theta+m}{j+\theta+1/2} \right) \cos^{2m+1-2k} t,$$

$$(1.212)$$

which still has a complicated form. To simplify (1.212), let us assume that

$$A_j^{(m)}(\theta) = (-1)^j \binom{m}{j} \left(\prod_{i=0}^{m-(j+1)} \frac{i+\theta+m}{i+\theta+1/2} \right).$$

Therefore we have

$$S_{2m+1}^{(\theta)} \left(\left. \begin{matrix} -1 & 0 \\ -1 & 1 \end{matrix} \right| \cos t \right) = \frac{(\theta+1/2)_m}{(\theta+m)_m} \cos^{(\theta-1)} t \sum_{j=0}^{m} A_j^{(m)}(\theta) \cos^{2m+1-2j} t,$$

$$(1.213)$$

where $(a)_m = \prod_{j=0}^{m-1} (a+j)$. On the other hand, since

$$\cos^{2n+1} x = \frac{1}{2^{2n}} \sum_{k=0}^{n} \binom{2n+1}{k} \cos(2n+1-2k) x,$$

the sum in (1.213) is simplified as

$$\sum_{j=0}^{m} A_j^{(m)}(\theta) \cos^{2m+1-2j} t = \sum_{k=0}^{m} A_{m-k}^{(m)}(\theta) \cos^{2k+1} t$$

$$= \sum_{k=0}^{m} A_{m-k}^{(m)}(\theta) \left(\frac{1}{2^{2k}} \sum_{i=0}^{k} \binom{2k+1}{i} \cos(2k+1-2i)t \right) = \sum_{k=0}^{m} B_k^{(m)}(\theta) \cos(2m+1-2k)t,$$

in which

$$B_k^{(m)}(\theta) = \sum_{j=0}^{k} \frac{A_j^{(m)}(\theta)}{2^{2(m-j)}} \binom{2m+1-2j}{k-j}$$

$$= \sum_{j=0}^{k} \frac{(-1)^j}{2^{2(m-j)}} \binom{m}{j} \left(\prod_{i=0}^{m-(j+1)} \frac{i+\theta+m}{i+\theta+1/2} \right) \binom{2m+1-2j}{k-j}.$$

So it remains to compute $B_k^{(m)}(\theta)$. After some calculations, it is simplified as

$$B_k^{(m)}(\theta) = \sum_{j=0}^{k} \frac{4^j \Gamma(\theta+1/2)\Gamma(\theta+2m)\Gamma(2m+2-2j)}{2^{2m}\Gamma(\theta+m)\Gamma(\theta+m+1/2)\Gamma(2m+2-k)k!}$$

$$\times \frac{(-k)_j(k-2m-1)_j(-m)_j(1/2-\theta-m)_j}{(1-\theta-2m)_j\, j!}.$$

Since $B_k^{(m)}(\theta)$ is not still in a hypergeometric form, applying the duplication Legendre formula

$$\Gamma(2z) = \frac{2^{2z-1}}{\sqrt{\pi}} \, \Gamma(z)\, \Gamma\left(z+\frac{1}{2}\right) \tag{1.214}$$

yields

$$B_k^{(m)} = \frac{2}{\sqrt{\pi}} \frac{\Gamma(\theta+1/2)\Gamma(\theta+2m)\,\Gamma(m+3/2)\,m!}{\Gamma(\theta+m)\Gamma(\theta+m+1/2)\Gamma(2m+2-k)k!}$$

$$\times \sum_{j=0}^{k} \frac{(-k)_j(k-2m-1)_j(1/2-\theta-m)_j}{(-m-1/2)_j(1-\theta-2m)_j} \frac{1^j}{j!}$$

$$= \frac{2}{\sqrt{\pi}} \frac{\Gamma(\theta + 1/2)\Gamma(\theta + 2m)\,\Gamma(m + 3/2)\,m!}{\Gamma(\theta + m)\Gamma(\theta + m + 1/2)\Gamma(2m + 2 - k)k!}$$

$$\times \; {}_3F_2 \left(\begin{array}{c} -k,\; k - 2m - 1,\; 1/2 - \theta - m \\ -m - 1/2,\; 1 - \theta - 2m \end{array} \right| \left. 1 \right). \qquad (1.215)$$

The hypergeometric term of (1.215) is a special case of Saalschutz's theorem [1], which says that if c is a negative integer and $a + b + c + 1 = d + e$, then

$$_3F_2 \left(\begin{array}{c} a\,b\,c \\ d\,e \end{array} \right| \left. 1 \right) = \frac{(d - a)_{|c|}(d - b)_{|c|}}{(d)_{|c|}(d - a - b)_{|c|}}.$$

Therefore, after some calculations, (1.213) is simplified as

$$S_{2m+1}^{(\theta)} \left(\begin{array}{cc} -1 & 0 \\ -1 & 1 \end{array} \right| \left. \cos t \right)$$

$$= \frac{1}{2^{2m}} \cos^{(\theta-1)} t \sum_{k=0}^{m} \binom{2m + 1}{k} \frac{(\theta - 1)_k(-m + 1/2)_k}{(1 - \theta - 2m)_k(-m - 1/2)_k} \cos(2m + 1 - 2k)\,t \, .$$

Corollary 1.9 *The trigonometric sequence $\Phi_n^{(\theta)}(t)$ defined by*

$$\Phi_{2n}^{(\theta)}(t) = \frac{1}{2^{2n-1}} \cos 2nt \, ,$$

$$\Phi_{2n+1}^{(\theta)}(t) = \frac{1}{2^{2n}} \cos^{(\theta-1)} t \sum_{k=0}^{n} \binom{2n + 1}{k} \frac{(\theta - 1)_k(-n + 1/2)_k}{(1 - \theta - 2n)_k(-n - 1/2)_k} \cos(2n+1-2k)\,t \, ,$$

satisfies the differential equation (1.211).
Clearly $\Phi_n^{(1)}(t)$ leads to the usual cosine sequence.

To compute the norm square value of this sequence, we should refer to relations (1.109), (1.168), and (1.171). Hence on substituting $(p, q, r, s, \theta) = (-1, 1, -1, 0, \theta)$ and $\alpha = 1$ into (1.168), we get

$$\int_{-1}^{1} \frac{1}{\sqrt{1 - x^2}} \, S_n^{(\theta)} \left(\begin{array}{cc} -1 & 0 \\ -1 & 1 \end{array} \right| \left. x \right) S_m^{(\theta)} \left(\begin{array}{cc} -1 & 0 \\ -1 & 1 \end{array} \right| \left. x \right) dx = \int_0^{\pi} \Phi_n^{(\theta)}(t) \, \Phi_m^{(\theta)}(t) \, dt$$

$$= \left(\prod_{i=1}^{2[n/2]} C_i \left(\begin{array}{cc} -1 - 2\theta\sigma_n, & 2\theta\sigma_n \\ -1 & 1 \end{array} \right) \right.$$

$$\times \int_{-1}^{1} W \left(\begin{array}{cc} -1-2\theta\sigma_n, & 2\theta\sigma_n \\ -1 & 1 \end{array} \middle| x \right) dx \right) \delta_{n,m}. \qquad (1.216)$$

Now let $n \to 2n+1$ in (1.216). In this case, the orthogonality (1.216) changes to

$$\int_0^{\pi} \Phi_{2n+1}^{(\theta)}(t)\, \Phi_{2m+1}^{(\theta)}(t)\, dt = \left(\prod_{i=1}^{2n} C_i \left(\begin{array}{cc} -1-2\theta, & 2\theta \\ -1 & 1 \end{array} \right) \int_{-1}^{1} x^{2\theta} (1-x^2)^{-1/2}\, dx \right) \delta_{n,m}$$

$$= \left(\sqrt{\pi}\, \frac{\Gamma(\theta+1/2)}{\Gamma(\theta+1)} \prod_{i=1}^{2n} \frac{(i+2\theta\sigma_i)(-i+1-2\theta\sigma_i)}{4\,(i+\theta)(i+\theta-1)} \right) \delta_{n,m},$$

if and only if $\theta+1 > 0$ and $(-1)^{\theta} = -1$. This result leads to the following final corollary.

Corollary 1.10 *The modified trigonometric sequence $C_n^{(a)}(x)$ defined as*

$$C_{2n}^{(a)}(x) = \cos 2nx,$$

$$C_{2n+1}^{(a)}(x) = (\cos^a x) \sum_{k=0}^{n} \binom{2n+1}{k} \frac{(a)_k(-n+1/2)_k}{(-a-2n)_k(-n-1/2)_k} \cos(2n+1-2k)\,x,$$

satisfies the orthogonality relation

$$\int_0^{\pi} C_i^{(a)}(x)\, C_j^{(a)}(x)\, dx = \begin{cases} 0 & \text{if } i \neq j, \\ \pi & \text{if } i = j = 0, \\ \pi/2 & \text{if } i = j = 2n, \\ (2^{2n}(2n)!\sqrt{\pi})\, \dfrac{\Gamma(a+3/2+n)\Gamma(a+1+n)}{\Gamma(a+2+2n)\Gamma(a+1+2n)} & \\ & \text{if } i = j = 2n+1, \end{cases}$$

$$(1.217)$$

provided that $a > -2$ and $(-1)^a = 1$.
Note that using (1.214), the last equality of (1.217) is equal to $\pi/2$ if $a = 0$.

1.11.2 Application to Function Expansion Theory

Let $f(x)$ be a periodic function satisfying Dirichlet conditions and suppose for convenience that

$$C_{2n+1}^{(a)}(x) = (\cos^a x)\, D_n^{(a)}(x).$$

According to [2], Courant and Hilbert were probably the first to prove that every sequence of eigenfunctions of a Sturm–Liouville problem is a complete set of orthogonal functions. Hence $\{C_n^{(a)}(x)\}_{n=0}^{\infty}$ is a complete set over $[0, \pi]$, and the periodic function $f(x)$ can be expanded in terms of them as

$$f(x) = \frac{1}{2} b_0 \, C_0^{(a)}(x) + \sum_{k=1}^{\infty} b_k C_k^{(a)}(x) = \frac{1}{2} b_0 + \sum_{k=1}^{\infty} b_{2k} \cos 2kx + (\cos^a x) \sum_{k=0}^{\infty} b_{2k+1} D_k^{(a)}(x),$$

$$(1.218)$$

in which

$$b_{2k} = \frac{2}{\pi} \int_0^{\pi} \cos 2kx \, f(x) \, dx,$$

$$b_{2k+1} = \frac{\Gamma(a+2+2k)\Gamma(a+1+2k)}{\sqrt{\pi} \, 2^{2k}(2k)! \, \Gamma(a+3/2+k)\Gamma(a+1+k)} \int_0^{\pi} f(x) \cos^a x \, D_k^{(a)}(x) \, dx,$$

$$(1.219)$$

and

$$D_k^{(a)}(x) = \sum_{j=0}^{k} d_j^k(a) \cos(2k+1-2j) \, x$$

$$= \sum_{j=0}^{k} \binom{2k+1}{j} \frac{(a)_j(-k+1/2)_j}{(-a-2k)_j(-k-1/2)_j} \cos(2k+1-2j) \, x.$$

As (1.219) shows, the integral in b_{2k+1} can be simplified as

$$R_k^{(a)} = \int_0^{\pi} f(x) \cos^a x \, D_k^{(a)}(x) \, dx = \sum_{j=0}^{k} d_j^k(a) \int_0^{\pi} f(x) \cos^a x \, \cos(2k+1-2j)x \, dx.$$

$$(1.220)$$

One of the interesting choices for a in (1.220), by noting that $a > -2$ and $(-1)^a = 1$, is an even integer, because in this case,

$$\cos^{2m} x = \frac{1}{2^{2m-1}} \sum_{k=0}^{m} \binom{2m}{k} \cos(2m-2k) x \qquad (m \in \mathbb{N}), \qquad (1.221)$$

and therefore we have

$$R_k^{(2m)}$$

$$= \frac{1}{2^{2m-1}} \sum_{j=0}^{k} d_j^k(2m) \left(\sum_{r=0}^{m} \binom{2m}{r} \int_0^{\pi} f(x) \cos(2m-2r)x \, \cos(2k+1-2j)x \, dx \right)$$

$$= \frac{1}{2^{2m}} \sum_{j=0}^{k} d_j^k (2m) \left(\sum_{r=0}^{m} \binom{2m}{r} \left(\int_0^{\pi} f(x) \cos(2m - 2r + 2k + 1 - 2j)x \, dx \right. \right.$$

$$\left. \left. + \int_0^{\pi} f(x) \cos(2m - 2r - 2k - 1 + 2j) x \, dx \right) \right).$$

Let us consider a practical example. It can be straightforwardly verified that the usual cosine series of the function $f(x) = x^2$ over $[0, \pi]$ is as follows:

$$x^2 = \frac{\pi^2}{3} + 4 \sum_{k=1}^{\infty} \frac{(-1)^k}{k^2} \cos kx = \frac{\pi^2}{3} + \sum_{k=1}^{\infty} \frac{1}{k^2} \cos 2kx - 4 \sum_{k=0}^{\infty} \frac{1}{(2k+1)^2} \cos(2k+1)x.$$

$$(1.222)$$

Now, to derive the generalized cosine series of type (1.218), if we suppose $a = 2m$, then we get

$$R_k^{(2m)} = \frac{-\pi}{2^{2m-1}} \times$$

$$\sum_{j=0}^{k} d_j^k (2m) \left(\sum_{r=0}^{m} \binom{2m}{r} \left(\frac{1}{(2m - 2r + 2k + 1 - 2j)^2} + \frac{1}{(2m - 2r - (2k+1 - 2j))^2} \right) \right).$$

Hence the corresponding odd coefficients take the form

$$b_{2k+1}^{(m)} = -2^{2m+2} \frac{(2m + 2k + 1)! \, (2m + 2k)!}{(4m + 2k + 1)! \, (2k)!}$$

$$\times \sum_{j=0}^{k} \binom{2k + 1}{j} \frac{(2m)_j (-k + 1/2)_j}{(-2m - 2k)_j (-k - 1/2)_j} S^{(m)}(j, k),$$

where

$$S^{(m)}(j, k) = \sum_{r=0}^{m} \binom{2m}{r} \left(\frac{1}{(2m - 2r + (2k + 1 - 2j))^2} + \frac{1}{(2m - 2r - (2k + 1 - 2j))^2} \right).$$

This gives the desired series as

$$x^2 = \frac{\pi^2}{3} + \sum_{k=1}^{\infty} \frac{1}{k^2} \cos 2kx + (\cos^{2m} x) \sum_{k=0}^{\infty} b_{2k+1}^{(m)} D_k^{(2m)}(x),$$

$$(1.223)$$

which is a generalization of (1.222) for $m = 0$, because

$$D_k^{(0)}(x) = \cos(2k+1)x \quad \text{and} \quad b_{2k+1}^{(0)}/2 = -2\,S^{(0)}(0,k) = -4/(2k+1)^2.$$

In this sense, note that the identity (1.221) is not valid for $m = 0$, and $b_{2k+1}^{(0)}/2$ should be considered in (1.223) not $b_{2k+1}^{(0)}$, though for the rest of the values of m, (1.223) is valid. For instance, if $m = 1$, then

$$b_{2k+1}^{(1)} = -16\,\frac{(k+1)\,(2k+1)}{(k+2)\,(2k+5)} \sum_{j=0}^{k} \binom{2k+1}{j} \frac{(2)_j(-k+1/2)_j}{(-2-2k)_j(-k-1/2)_j}\,S^{(1)}(j,k),$$

where

$$S^{(1)}(j,k) = \frac{1}{(2k+3-2j)^2} + \frac{4}{(2k+1-2j)^2} + \frac{1}{(2k-1-2j)^2}.$$

There exists a similar case for a generalization of the Fourier sine series such that if we replace the initial data

$$(p,q,r,s,\theta) = (-1,1,-3,0,\theta)$$

in the main orthogonality relation (1.168) for $\alpha = 1$ as

$$\int_{-1}^{1} \sqrt{1-x^2}\,S_n^{(\theta)}\left(\begin{array}{cc} -3 & 0 \\ -1 & 1 \end{array}\middle| x\right) S_m^{(\theta)}\left(\begin{array}{cc} -3 & 0 \\ -1 & 1 \end{array}\middle| x\right) dx$$

$$= \int_0^{\pi} (\sin^2 t)\,S_n^{(\theta)}\left(\begin{array}{cc} -3 & 0 \\ -1 & 1 \end{array}\middle| \cos t\right) S_m^{(\theta)}\left(\begin{array}{cc} -3 & 0 \\ -1 & 1 \end{array}\middle| \cos t\right) dt$$

$$= \left(\prod_{i=1}^{2[n/2]} C_i\left(\begin{array}{cc} -3-2\theta\sigma_n, & 2\theta\sigma_n \\ -1 & 1 \end{array}\right)\right.$$

$$\left. \times \int_{-1}^{1} W\left(\begin{array}{cc} -3-2\theta\sigma_n, & 2\theta\sigma_n \\ -1 & 1 \end{array}\middle| x\right) dx\right) \delta_{n,m},$$

then by defining the trigonometric sequence

$$S_{n+1}^{(a)}(x) = 2^{n+1}\sin x\;S_n^{(a+1)}\left(\begin{array}{cc} -3 & 0 \\ -1 & 1 \end{array}\middle| \cos x\right),$$

we can verify (for example) that

$$S_{2n+1}^{(a)}(x) = 2^{2n+1} \sin x \ S_{2n}^{(a+1)} \left(\begin{vmatrix} -3 & 0 \\ -1 & 1 \end{vmatrix} \cos x \right) = \sin(2n+1)x.$$

1.12 Another Extension for Trigonometric Orthogonal Sequences

As we pointed out in Sect. 1.10, if

$$
\begin{aligned}
A(x) &= x^2(px^2 + q)\,, \\
B(x) &= x(rx^2 + s)\,, \\
C(x) &= x^2 > 0\,, \\
D(x) &= -\alpha\,(s + (\alpha - 1)\,q)\,, \\
E(x) &= (\alpha - \beta)s + (\alpha(\alpha - 1) - \beta(\beta - 1))q\,,
\end{aligned}
$$

and

$$\lambda_n^{(\alpha,\beta)} = (n + \alpha + (\beta - \alpha - 1)\,\sigma_n)\,(r + (n - 1 + \alpha + (\beta - \alpha - 1)\,\sigma_n)p)$$

are substituted into Eq. (1.94), we obtain the second-order equation

$$x^2(px^2 + q)\,\Phi_n''(x) + x(rx^2 + s)\,\Phi_n'(x) - \left(\lambda_n^{(\alpha,\beta)} x^2 + (-1)^n c^* + d^* \right) \Phi_n(x) = 0,$$
$$(1.224)$$

where

$$2c^* = (\alpha - \beta)s + (\alpha(\alpha - 1) - \beta(\beta - 1))q$$

and

$$2d^* = (\alpha + \beta)s + (\alpha(\alpha - 1) + \beta(\beta - 1))q.$$

To introduce a sequence that generalizes $C_n^{(a)}(x)$, we first suppose in Eq. (1.224) that $p = -1$ and $q = 1$. Then by the change of variable $x = \cos t$, we obtain

$$\cos^2 t \ \Phi_n''(t) + \left(\frac{-(r + 1)\cos^3 t - s \cos t}{\sin t} \right) \Phi_n'(t)$$
$$+ \left(\lambda_n^{(\alpha,\beta)} \cos^2 t + (-1)^n c^* + d^* \right) \Phi_n(t) = 0. \quad (1.225)$$

If $r + 1 = 0$ and $s = 0$ in (1.225), an extension of Eq. (1.211) is derived for $\alpha = 0$. Consequently, by substituting the initial vector

$$(p, q, r, s, \alpha, \beta) = (-1, 1, -1, 0, \alpha, \beta)$$

into Eq. (1.224) and applying the change of variable $x = \cos t$, we get

$$\Phi_n''(t)$$
$$+ \left((n + \alpha + (\beta - \alpha - 1)\sigma_n)^2 + \frac{(\alpha(\alpha - 1) - \beta(\beta - 1))\sigma_n - \alpha(\alpha - 1)}{\cos^2 t} \right) \Phi_n(t) = 0.$$
$$(1.226)$$

According to (1.190), Eq. (1.226) has a basis solution as

$$\Phi_n(t) = S_n^{(\alpha,\beta)}\left(\begin{vmatrix} -1 & 0 \\ -1 & 1 \end{vmatrix} \cos t \right)$$

$$= \frac{(\alpha + (\beta - \alpha)\sigma_n + 1/2)_{[n/2]}}{(\alpha + (\beta - \alpha - 1/2)\sigma_n + n/2)_{[n/2]}} (\cos t)^{\alpha + (\beta - \alpha - 1)\sigma_n}$$

$$\times \sum_{k=0}^{[n/2]} (-1)^k \binom{[n/2]}{k} \left(\prod_{j=0}^{[n/2]-(k+1)} \frac{2j + n + 2\alpha + (2\beta - 2\alpha - 1)\sigma_n}{2j + 1 + 2\alpha + 2(\beta - \alpha)\sigma_n} \right) \cos^{n-2k} t,$$

which can be represented as

$$S_n^{(\alpha,\beta)}\left(\begin{vmatrix} -1 & 0 \\ -1 & 1 \end{vmatrix} \cos t \right) =$$

$$(\cos t)^{n+\alpha+(\beta-\alpha-1)\sigma_n} {}_2F_1\left(\begin{array}{c} -[n/2], \ -[n/2] + 1/2 - \alpha + (\alpha - \beta)\sigma_n \\ -n + 1 - \alpha + (\alpha - \beta + 1)\sigma_n \end{array} \Bigg| \ \frac{1}{\cos^2 t} \right).$$
$$(1.227)$$

The above representation can be further simplified. For this purpose, if we take

$$A_k^m(\beta) = \frac{(-m)_k(-m - \beta + 1/2)_k}{(-2m - \beta + 1)_k \, k!},$$

then (1.227) changes to

$$S_{2m}^{(\alpha,\beta)}\left(\begin{vmatrix} -1 & 0 \\ -1 & 1 \end{vmatrix} \cos t \right) = \cos^\alpha t \sum_{k=0}^m A_k^m(\alpha) \cos^{2m-2k} t \qquad (1.228)$$

and

$$S_{2m+1}^{(\alpha,\beta)}\left(\begin{vmatrix} -1 & 0 \\ -1 & 1 \end{vmatrix}\cos t\right) = \cos^{(\beta-1)}t \sum_{k=0}^{m} A_k^{(m)}(\beta)\cos^{2m+1-2k}t .$$ (1.229)

Since in general,

$$\cos^{2n+1} x = \frac{1}{2^{2n}} \sum_{k=0}^{n} \binom{2n+1}{k} \cos(2n+1-2k)x,$$

the sum in (1.229) is simplified as

$$\sum_{k=0}^{m} A_k^{(m)}(\beta)\cos^{2m+1-2k}t = \sum_{k=0}^{m} A_{m-k}^{(m)}(\beta)\cos^{2k+1}t$$

$$= \sum_{k=0}^{m} A_{m-k}^{(m)}(\beta)\left(\frac{1}{2^{2k}} \sum_{j=0}^{k} \binom{2k+1}{j}\cos(2k+1-2j)t\right)$$

$$= \sum_{k=0}^{m} B_k^{(m)}(\beta)\cos(2m+1-2k)t,$$

where

$$B_k^{(m)}(\beta) = \sum_{j=0}^{k} \frac{A_j^{(m)}(\beta)}{2^{2(m-j)}}\binom{2m+1-2j}{k-j}$$

$$= \sum_{j=0}^{k} \frac{(-m)_j(-m-\beta+1/2)_j}{2^{2(m-j)}(-2m-\beta+1)_j\, j!}\binom{2m+1-2j}{k-j},$$

which is equivalent to

$$B_k^{(m)}(\beta) = \frac{2}{\sqrt{\pi}}\frac{m!\,\Gamma(m+3/2)}{\Gamma(2m+2-k)\,k!} \sum_{j=0}^{k} \frac{(-m-\beta+1/2)_j(-k)_j(k-2m-1)_j}{(-m-1/2)_j(-2m-\beta+1)_j\, j!}$$

$$= \frac{2}{\sqrt{\pi}}\frac{m!\,\Gamma(m+3/2)}{\Gamma(2m+2-k)\,k!}\,{}_3F_2\left(\begin{matrix} -k,\ k-2m-1,\ 1/2-\beta-m \\ -m-1/2,\ 1-\beta-2m \end{matrix}\ \middle|\ 1\right).$$ (1.230)

Therefore, (1.229) reads as

$$
S_{2m+1}^{(\alpha,\beta)}\left(\left.\begin{matrix} -1 & 0 \\ -1 & 1 \end{matrix}\right| \cos t\right)
$$

$$
= \frac{\cos^{(\beta-1)}t}{2^{2m}} \sum_{k=0}^{m} \binom{2m+1}{k} \frac{(\beta-1)_k(-m+1/2)_k}{(1-\beta-2m)_k(-m-1/2)_k} \cos(2m+1-2k)t.
$$

Similarly, this method can be applied to the function (1.228), so that by noting the general identity

$$
\cos^{2n}x = \frac{1}{2^{2n}}\left(2\sum_{k=0}^{n}\binom{2n}{k}\cos(2n-2k)x - \binom{2n}{n}\right),
$$

the sum in (1.228) is simplified as

$$
\sum_{k=0}^{m} A_k^{(m)}(\alpha)\cos^{2m-2k}t = \sum_{k=0}^{m} A_{m-k}^{(m)}(\alpha)\cos^{2k}t
$$

$$
= \sum_{k=0}^{m} A_{m-k}^{(m)}(\alpha)\frac{1}{2^{2k}}\left(2\sum_{j=0}^{k}\binom{2k}{j}\cos(2k-2j)t - \binom{2k}{k}\right)
$$

$$
= \sum_{k=0}^{m} C_k^{(m)}(\alpha)\cos(2m-2k)t - \sum_{k=0}^{m} A_{m-k}^{(m)}(\alpha)\frac{1}{2^{2k}}\binom{2k}{k}, \qquad (1.231)
$$

in which

$$
C_k^{(m)}(\alpha) = \sum_{j=0}^{k} \frac{A_j^{(m)}(\alpha)}{2^{2(m-j)-1}} \binom{2m-2j}{k-j}
$$

$$
= \frac{2}{\sqrt{\pi}} \frac{m!\,\Gamma(m+1/2)}{\Gamma(2m+1-k)\,k!} \, _3F_2\left(\left.\begin{matrix} -k & k-2m & 1/2-\alpha-m \\ -m+1/2 & 1-\alpha-2m \end{matrix}\right| 1\right)
$$

$$
= \frac{1}{2^{2m-1}}\binom{2m}{k}\frac{(\alpha)_k}{(1-\alpha-2m)_k},
$$

where we have once again used the Saalschutz identity. Therefore, (1.228) is simplified as

$$
S_{2m}^{(\alpha,\beta)} \left(\begin{matrix} -1 & 0 \\ -1 & 1 \end{matrix} \middle| \cos t \right)
$$

$$
= \frac{\cos^\alpha t}{2^{2m-1}} \left(\sum_{k=0}^{m} \binom{2m}{k} \frac{(\alpha)_k}{(1-\alpha-2m)_k} \cos(2m-2k)t - \frac{1}{2} \binom{2m}{m} \frac{(\alpha)_m}{(1-\alpha-2m)_m} \right).
$$

Corollary 1.11 *The trigonometric sequence*

$$
\Phi_n(t;\alpha,\beta) = (\cos t)^{\alpha+(\beta-1-\alpha)\sigma_n} \times
$$

$$
\left(\frac{1}{2^{n-1}} \sum_{k=0}^{[n/2]} \binom{n}{k} \frac{(\alpha+(\beta-1-\alpha)\sigma_n)_k(-[n/2]+1/2)_k \cos(n-2k)t}{(1-\alpha-n+(\alpha-\beta+1)\sigma_n)_k(-[n/2]+(-1)^n/2)_k} \right.
$$

$$
\left. - \frac{(1/2)_{[n/2]}(\alpha)_{[n/2]}}{[n/2]!\,(1-\alpha-n)_{[n/2]}} \sigma_{n+1} \right)
$$

satisfies the differential equation (1.226).

To compute the norm square value of $\Phi_n(t;\alpha,\beta)$, it is enough to refer to relations (1.194) and (1.109) and substitute $(p,q,r,s,\alpha,\beta) = (-1,1,-1,0,\alpha,\beta)$ and $v=1$ into (1.194) to get

$$
\int_{-1}^{1} \frac{1}{\sqrt{1-x^2}} S_n^{(\alpha,\beta)} \left(\begin{matrix} -1 & 0 \\ -1 & 1 \end{matrix} \middle| x \right) S_m^{(\alpha,\beta)} \left(\begin{matrix} -1 & 0 \\ -1 & 1 \end{matrix} \middle| x \right) dx
$$

$$
= \int_{0}^{\pi} \Phi_n(t;\alpha,\beta)\,\Phi_m(t;\alpha,\beta)\,dt
$$

$$
= \prod_{j=1}^{2[n/2]} C_j \left(\begin{matrix} -1 - 2(\alpha+(\beta-1-\alpha)\sigma_n)\,,\ 2(\alpha+(\beta-1-\alpha)\sigma_n) \\ -1 \qquad\qquad\qquad\qquad 1 \end{matrix} \right)
$$

$$
\times \int_{-1}^{1} W \left(\begin{matrix} -1 - 2(\alpha+(\beta-1-\alpha)\sigma_n)\,,\ 2(\alpha+(\beta-1-\alpha)\sigma_n) \\ -1 \qquad\qquad\qquad\qquad 1 \end{matrix} \middle| x \right) dx\,\delta_{n,m}.
$$

$$
(1.232)
$$

Let $n \to 2n$. Then (1.232) changes to

$$\int_0^\pi \Phi_{2n}(t; \alpha, \beta)\, \Phi_{2m}(t; \alpha, \beta)\, dt$$

$$= \left(\prod_{j=1}^{2n} C_j \begin{pmatrix} -2\alpha - 1, \, 2\alpha \\ -1 \qquad 1 \end{pmatrix} \int_{-1}^1 x^{2\alpha}(1 - x^2)^{-1/2}\, dx \right) \delta_{n,m}$$

$$= \left(\sqrt{\pi}\, \frac{\Gamma(\alpha + 1/2)}{\Gamma(\alpha + 1)} \prod_{j=1}^{2n} \frac{(j + 2\alpha\, \sigma_j)(j + 2\alpha\, \sigma_j - 1)}{4\,(j + \alpha)(j + \alpha - 1)} \right) \delta_{n,m}$$

if and only if $\alpha + 1/2 > 0$ and $(-1)^{2\alpha} = 1$.

Similarly, for $n \to 2n + 1$ in (1.232), we obtain

$$\int_0^\pi \Phi_{2n+1}(t; \alpha, \beta)\, \Phi_{2m+1}(t; \alpha, \beta)\, dt$$

$$= \left(\prod_{j=1}^{2n} C_j \begin{pmatrix} -2\beta + 1, \, 2\beta - 2 \\ -1 \qquad 1 \end{pmatrix} \int_{-1}^1 x^{2\beta-2}(1 - x^2)^{-1/2}\, dx \right) \delta_{n,m}$$

$$= \left(\sqrt{\pi}\, \frac{\Gamma(\beta - 1/2)}{\Gamma(\beta)} \prod_{j=1}^{2n} \frac{(j + 2(\beta - 1)\, \sigma_j)(j + 2(\beta - 1)\sigma_j - 1)}{4\,(j + \beta - 1)(j + \beta - 2)} \right) \delta_{n,m}$$

if and only if $\beta - 1/2 > 0$ and $(-1)^{2\beta} = 1$.

Corollary 1.12   *The modified trigonometric sequence*

$$\bar{\Phi}_{2m}(t; \alpha, \beta) = \cos^\alpha t$$

$$\times \left(-\frac{1}{2} \binom{2m}{m} \frac{(\alpha)_m}{(1 - \alpha - 2m)_m} + \sum_{k=0}^m \binom{2m}{k} \frac{(\alpha)_k}{(1 - \alpha - 2m)_k} \cos(2m - 2k)\, t \right),$$

$$\bar{\Phi}_{2m+1}(t; \alpha, \beta) = \cos^{\beta-1} t$$

$$\times \sum_{k=0}^m \binom{2m + 1}{k} \frac{(\beta - 1)_k(-m + 1/2)_k}{(1 - \beta - 2m)_k(-m - 1/2)_k} \cos(2m + 1 - 2k)\, t,$$

in which $\bar{\Phi}_n(t; 0, 1) = \cos nt$, satisfies the orthogonality relation

$$\int_0^\pi \bar{\Phi}_i(t; \alpha, \beta)\bar{\Phi}_j(t; \alpha, \beta)dt = \begin{cases} 0 & \text{if } i \neq j, \\[2ex] \dfrac{\pi}{2^{2\alpha+1}} \dfrac{\Gamma(2n+2\alpha)\Gamma(2n+1)}{\Gamma(2n+\alpha)\Gamma(2n+\alpha+1)} & \text{if } i = j = 2n, \\[2ex] \dfrac{\pi}{2^{2\beta-1}} \dfrac{\Gamma(2n+2\beta-2)\Gamma(2n+1)}{\Gamma(2n+\beta-1)\Gamma(2n+\beta)} & \text{if } i=j=2n+1, \end{cases}$$

if and only if $\alpha + 1/2 > 0$, $\beta - 1/2 > 0$, $(-1)^{2\alpha} = 1$, and $(-1)^{2\beta} = 1$.

1.13 Incomplete Symmetric Orthogonal Polynomials

In the classical case, systems of orthogonal polynomials are such that the nth polynomial has the exact degree n. Such systems are most often complete and form a basis of the space of arbitrary polynomials. In this section, we introduce some incomplete sets of orthogonal polynomials that do not contain all degrees but are solutions of some symmetric generalized Sturm–Liouville problems. Although such polynomials do not possess all the properties as in the classical cases, they can nevertheless be applied to approximation of functions, since we will compute their explicit norm square values.

1.13.1 Incomplete Symmetric Orthogonal Polynomials of Jacobi Type

Here we introduce incomplete symmetric orthogonal polynomials of Jacobi type as specific solutions of a generalized Sturm–Liouville equation. For this purpose, we first consider the shifted Jacobi polynomials on $[0, 1]$ as

$$P_{n,+}^{(\alpha,\beta)}(x) = \sum_{k=0}^n (-1)^k \binom{n+\alpha+\beta+k}{k}\binom{n+\alpha}{n-k} x^k,$$

which satisfy the equation

$$x(1-x)y'' - ((\alpha+\beta+2)x - (\alpha+1))\, y' + n(n+\alpha+\beta+1)y = 0 \text{ with } y = P_{n,+}^{(\alpha,\beta)}(x),$$
$$(1.233)$$

and the orthogonality relation

$$\int_0^1 x^\alpha (1-x)^\beta\, P_{n,+}^{(\alpha,\beta)}(x)P_{m,+}^{(\alpha,\beta)}(x)\, dx$$

$$= \frac{\Gamma(n+\alpha+1)\Gamma(n+\beta+1)}{(2n+\alpha+\beta+1)\,\Gamma(n+1)\,\Gamma(n+\alpha+\beta+1)}\, \delta_{n,m}. \qquad (1.234)$$

By referring to the main Theorem 1.2 of Sect. 1.6 and noting the Jacobi differential equation (1.233), a differential equation of type (1.94) can be constructed whose solutions are orthogonal with respect to an even weight function, say $|x|^{2a}(1 - x^{2m})^b$ on the symmetric interval $[-1, 1]$. Hence, suppose in the generic equation (1.94) that

$$A(x) = x^2(1 - x^{2m}),$$
$$B(x) = -2x\left((a + mb + 1)x^{2m} - a + m - 1\right),$$
$$C(x) = x^{2m} > 0,$$
$$D(x) = -2s(2s + 2a - 2m + 1),$$
$$E(x) = 2s(2s + 2a - 2m + 1) - 2(2r + 1)(r + a - m + 1),$$

and

$$\lambda_n = (mn + 2s + (2r + 1 - m - 2s)\sigma_n)(mn + 2s + 2a + 1 + 2mb + (2r + 1 - m - 2s)\sigma_n),$$

where $\sigma_n = \frac{1-(-1)^n}{2}$ and a, b, m, r, s are free parameters.

In this case, the following differential equation appears:

$$x^2(1 - x^{2m})\,\Phi_n''(x) - 2x\left((a + mb + 1)x^{2m} - a + m - 1\right)\Phi_n'(x)$$
$$+ (\alpha_n x^{2m} + \beta + \sigma_n \gamma)\,\Phi_n(x) = 0, \qquad (1.235)$$

in which

$$\beta = -2s(2s + 2a - 2m + 1), \quad \gamma = 2s(2s + 2a - 2m + 1) - 2(2r + 1)(r + a - m + 1)$$

and

$$\alpha_n = \left(mn + 2s + (r - s + (m - 1)/2)(1 - (-1)^n)\right)$$
$$\times \left(mn + 2s + 2a + 1 + 2mb + (r - s + (m - 1)/2)(1 - (-1)^n)\right).$$

We prove that Eq. (1.235) has a polynomial solution of the form

$$\Phi_n^{(r,s)}(x; a, b, m) = (x^{2s} + (x^{2r+1} - x^{2s})\sigma_n)\, P_{[n/2],+}^{\left(\frac{a+1-m+s+r}{m}+(-1)^n\frac{2s-2r-1}{2m}, b\right)}(x^{2m}),$$
$$(1.236)$$

which satisfies the orthogonality relation

$$\int_{-1}^{1} x^{2a}(1 - x^{2m})^b\,\Phi_n^{(r,s)}(x; a, b, m)\Phi_k^{(r,s)}(x; a, b, m)\,dx$$
$$= \left(\int_{-1}^{1} x^{2a}(1 - x^{2m})^b\left(\Phi_n^{(r,s)}(x; a, b, m)\right)^2 dx\right)\delta_{n,k}. \qquad (1.237)$$

To prove this claim, we first substitute

$$g(x) = x^\lambda P_{n,+}^{(\alpha,\beta)}(x^\theta) \quad \text{for } \lambda, \theta \in \mathbb{R}$$

into Eq. (1.233) to obtain the differential equation

$$x^2(1 - x^\theta) g''(x) + x \left((2\lambda - (\alpha + \beta + 1)\theta - 1) x^\theta - 2\lambda + \alpha\theta + 1 \right) g'(x)$$
$$+ \left((\theta^2 n(n + \alpha + \beta + 1) + (\alpha + \beta + 1)\theta\lambda - \lambda^2) x^\theta + \lambda^2 - \alpha\theta\lambda \right) g(x) = 0.$$

$$(1.238)$$

For convenience, let

$$\alpha = \frac{q + 2\lambda - 1}{\theta} \quad \text{and} \quad \beta = -\frac{p + q}{\theta} - 1$$

in Eq. (1.238), which changes to

$$x^2(1 - x^\theta) g''(x) + x \left(p x^\theta + q \right) g'(x)$$
$$+ \left((\theta n + \lambda)(\theta n + \lambda - p - 1) x^\theta - \lambda(\lambda + q - 1) \right) g(x) = 0. \quad (1.239)$$

Now, by noting Theorem 1.2, we define the following odd and even polynomial sequences:

$$\Phi_{2n}(x) = x^{2s} P_{n,+}^{(\frac{q+4s-1}{2m}, -\frac{p+q}{2m} - 1)}(x^{2m}) \quad \text{for} \quad \lambda = 2s, \ s \in \mathbb{Z}^+ \text{ and } \theta = 2m, \ m \in \mathbb{N},$$

and

$$\Phi_{2n+1}(x) = x^{2r+1} P_{n,+}^{(\frac{q+4r+1}{2m}, -\frac{p+q}{2m} - 1)}(x^{2m}) \quad \text{for } \lambda = 2r + 1, \ r \in \mathbb{Z}^+ \text{ and } \theta = 2m, \ m \in \mathbb{N}.$$

$$(1.240)$$

Since in general,

$$u + (w - u)\sigma_n = \frac{u + w}{2} + (-1)^n \frac{u - w}{2} = \begin{cases} u & \text{if } n = 2k, \\ w & \text{if } n = 2k + 1, \end{cases}$$

the polynomial sequence $\Phi_n(x)$ defined in (1.240) can be written in a unique form as

$$\Phi_n(x) = \left(x^{2s} + (x^{2r+1} - x^{2s})\sigma_n \right) P_{[n/2],+}^{(\frac{q+2s+2r+(-1)^n(2s-2r-1)}{2m}, -\frac{p+q}{2m} - 1)}(x^{2m}).$$

According to definitions (1.240) and differential equation (1.239), $\Phi_{2n}(x)$ should satisfy the equation

$$x^2(1 - x^{2m})\, \Phi_{2n}''(x) + x\left(px^{2m} + q\right)\Phi_{2n}'(x)$$

$$+ \left((2mn + 2s)(2mn + 2s - p - 1)\, x^{2m} - 2s(2s + q - 1)\right)\Phi_{2n}(x) = 0,$$

and $\Phi_{2n+1}(x)$ should satisfy

$$x^2(1 - x^{2m})\, \Phi_{2n+1}''(x) + x\left(px^{2m} + q\right)\Phi_{2n+1}'(x)$$

$$+ \left((2mn + 2r + 1)(2mn + 2r - p)\, x^{2m} - (2r + 1)(2r + q)\right)\Phi_{2n+1}(x) = 0.$$

By combining these two equations, we finally obtain

$$x^2(1 - x^{2m})\, \Phi_n''(x) + x\left(px^{2m} + q\right)\Phi_n'(x) +$$

$$\left((mn + 2s + (2r + 1 - m - 2s)\,\sigma_n)(mn + 2s - p - 1 + (2r + 1 - m - 2s)\,\sigma_n)\, x^{2m}\right.$$

$$\left. - 2s(2s + q - 1) - ((2r + 1)(2r + q) - 2s(2s + q - 1))\,\sigma_n\right)\Phi_n(x) = 0, \qquad (1.241)$$

which is a special case of the generalized Sturm–Liouville equation (1.94). Also, the weight function corresponding to (1.241) takes the form

$$W(x) = x^{2m} \exp\!\left(\int \frac{x(px^{2m} + q) - (2x - (2m + 2)x^{2m+1})}{x^2(1 - x^{2m})}\, dx\right)$$

$$= K\, x^{2m+q-2}(1 - x^{2m})^{-\frac{p+q}{2m}-1}. \qquad (1.242)$$

Without loss of generality, we can assume that $K = 1$, and since $W(x)$ must be positive, the weight function (1.242) can be considered as

$$W(x) = |x|^{2a}(1 - x^{2m})^b,$$

for $2a = 2m + q - 2$ and $b = -1 - (p + q)/2m$.

To compute the norm square value of (1.237) we directly use the orthogonality relation (1.234), so that for $n = 2j$ we have

$$N_{2j} = \int_{-1}^{1} x^{2a}(1 - x^{2m})^b \left(\Phi_{2j}^{(r,s)}(x; a, b, m)\right)^2 dx$$

$$= \int_{-1}^{1} x^{2a+4s}(1-x^{2m})^b \left(P_{j,+}^{(\frac{2a+4s+1-2m}{2m},b)}(x^{2m}) \right)^2 dx$$

$$= \frac{1}{m}\int_{0}^{1} t^{\frac{2a+4s+1-2m}{2m}}(1-t)^b \left(P_{j,+}^{(\frac{2a+4s+1-2m}{2m},b)}(t) \right)^2 dt$$

$$= \frac{\Gamma(j+\frac{2a+4s+1}{2m})\,\Gamma(j+b+1)}{(2mj+a+2s+\frac{1}{2}+mb)\Gamma(j+1)\,\Gamma(j+\frac{2a+4s+1}{2m}+b)}, \qquad (1.243)$$

and for $n = 2j+1$ we get

$$N_{2j+1} = \int_{-1}^{1} x^{2a}(1-x^{2m})^b \left(\Phi_{2j+1}^{(r,s)}(x;a,b,m) \right)^2 dx$$

$$= \int_{-1}^{1} x^{2a+4r+2}(1-x^{2m})^b \left(P_{j,+}^{(\frac{2a+4r+3-2m}{2m},b)}(x^{2m}) \right)^2 dx$$

$$= \frac{1}{m}\int_{0}^{1} t^{\frac{2a+4r+3-2m}{2m}}(1-t)^b \left(P_{j,+}^{(\frac{2a+4r+3-2m}{2m},b)}(t) \right)^2 dt$$

$$= \frac{\Gamma(j+\frac{2a+4r+3}{2m})\,\Gamma(j+1+b)}{(2mj+a+2r+\frac{3}{2}+mb)\Gamma(j+1)\,\Gamma(j+\frac{2a+4r+3}{2m}+b)}. \qquad (1.244)$$

By combining both relations (1.243) and (1.244), the following value is derived:

$$N_n =$$

$$\frac{\Gamma(\frac{n-\sigma_n}{2} + \frac{2a+4s+1}{2m} + \frac{2r+1-2s}{m}\sigma_n)\,\Gamma(\frac{n-\sigma_n}{2}+b+1)}{(m(n-\sigma_n)+a+2s+\frac{1}{2}+mb+(2r+1-2s)\sigma_n)\,\Gamma(\frac{n-\sigma_n}{2}+1)\,\Gamma(\frac{n-\sigma_n}{2}+\frac{2a+4s+1}{2m}+b+\frac{2r+1-2s}{m}\sigma_n)}.$$

This value shows that the orthogonality (1.237) is valid if and only if

$$b > -1,\ 2a+4s+1 > 0,\ 2a+4s+1+2mb > 0,\ m \in \mathbb{N},\ 2a+4r+3 > 0,\ 2a+4r+3+2mb > 0,$$

and $(-1)^{2a} = 1$.

Example 1.7 Find the standard properties of incomplete symmetric polynomials orthogonal with respect to the weight function $x^4\sqrt{1-x^4}$ on $[-1, 1]$.
To solve the problem, it is sufficient in (1.236) to choose $m = a = 2$ and $b = \frac{1}{2}$ to get the polynomials

$$\Phi_n^{(r,s)}\left(x; 2, \frac{1}{2}, 2\right) = \left(x^{2s} + (x^{2r+1} - x^{2s})\sigma_n\right) P_{[n/2],+}^{(\frac{1+s+r}{2}+(-1)^n\frac{2s-2r-1}{4},\,\frac{1}{2})}(x^4) \qquad r, s \in \mathbb{Z}^+,$$

$$(1.245)$$

that satisfy the differential equation

$$x^2(1 - x^4)\, \Phi_n''(x) + 2x\left(1 - 4x^4\right)\Phi_n'(x)+$$

$$\left((2n + 2s + (2r - 1 - 2s)\,\sigma_n)(2n + 2s + 7 + (2r - 1 - 2s)\,\sigma_n)\, x^4\right.$$

$$\left. - 2s(2s + 1) + (2s(2s + 1) - 2(2r + 1)(r + 1))\,\sigma_n\right)\Phi_n(x) = 0$$

and the orthogonality relation

$$\int_{-1}^{1} x^4\sqrt{1 - x^4}\, \Phi_n^{(r,s)}\left(x; 2, \frac{1}{2}, 2\right) \Phi_k^{(r,s)}\left(x; 2, \frac{1}{2}, 2\right) dx =$$

$$\frac{\Gamma(\frac{n-\sigma_n}{2} + s + \frac{5}{4} + \frac{2r-2s+1}{2}\,\sigma_n)\,\Gamma(\frac{n-\sigma_n+3}{2})}{(2(n - \sigma_n) + \frac{7}{2} + 2s + (2r - 2s + 1)\,\sigma_n)\,\Gamma(\frac{n-\sigma_n}{2} + 1)\,\Gamma(\frac{n-\sigma_n}{2} + s + \frac{7}{4} + \frac{2r-2s+1}{2}\,\sigma_n)}\, \delta_{n,k}.$$

As (1.245) shows, $\Phi_n^{(r,s)}(x; 2, 1/2, 2)$ are incomplete symmetric polynomials with degrees respectively $\{2s, 2r + 1, 2s + 4, 2r + 5, 2s + 8, 2r + 9, \ldots\}$. For instance, we have

$$\text{degrees of } \Phi_n^{(1,0)}\left(x; 2, \frac{1}{2}, 2\right) = \{0, 3, 4, 7, 8, 11, \ldots\},$$

$$\text{degrees of } \Phi_n^{(2,0)}\left(x; 2, \frac{1}{2}, 2\right) = \{0, 5, 4, 9, 8, 13, \ldots\},$$

and

$$\text{degrees of } \Phi_n^{(1,1)}\left(x; 2, \frac{1}{2}, 2\right) = \{2, 3, 6, 7, 10, 11, \ldots\}.$$

Example 1.8 If $m = 1$ is considered in (1.236), a generalization of generalized ultraspherical polynomials as

$$\Phi_n^{(r,s)}(x; a, b, 1) = (x^{2s} + (x^{2r+1} - x^{2s})\,\sigma_n)\, P_{[n/2],+}^{(a+s+r+(-1)^n(s-r-\frac{1}{2}),\, b)}(x^2) \qquad (1.246)$$

is derived for $r = s = 0$ that satisfies the differential equation

$$x^2(1 - x^2)\, \Phi_n''(x) - 2x\left((a + b + 1)\, x^2 - a\right)\Phi_n'(x)+$$

$$\left((n + 2s + (2r - 2s)\sigma_n)(n + 2s + 2a + 1 + 2b + (2r - 2s)\,\sigma_n)\, x^2\right.$$

$$\left. - 2s(2s + 2a - 1) + (2s(2s + 2a - 1) - 2(2r + 1)(r + a))\,\sigma_n\right)\Phi_n(x) = 0$$

and has an orthogonality relation of the form

$$\int_{-1}^{1} x^{2a}(1-x^2)^b \, \Phi_n^{(r,s)}(x;a,b,1) \, \Phi_k^{(r,s)}(x;a,b,1) \, dx = \delta_{n,k} \times$$

$$\frac{\Gamma(\frac{n-\sigma_n+2a+4s+1}{2}+(2r+1-2s)\,\sigma_n)\,\Gamma(\frac{n-\sigma_n}{2}+b+1)}{(n-\sigma_n+a+2s+\frac{1}{2}+b+(2r+1-2s)\,\sigma_n)\,\Gamma(\frac{n-\sigma_n}{2}+1)\,\Gamma(\frac{n-\sigma_n+2a+4s+1}{2}+b+(2r+1-2s)\,\sigma_n)}.$$

As (1.246) shows, $\Phi_n^{(r,s)}(x;a,b,1)$ are incomplete symmetric polynomials with degrees respectively $D^{(r,s)} = \{2s, 2r+1, 2s+2, 2r+3, 2s+4, 2r+5, \ldots\}$ though it is complete for $r = s = 0$, because in this case we have $D^{(0,0)} = \{0, 1, 2, 3, \ldots\}$.

1.13.2 Incomplete Symmetric Orthogonal Polynomials of Laguerre Type

In this part, we introduce incomplete symmetric orthogonal polynomials of Laguerre type as specific solutions of a generalized Sturm–Liouville equation. First, we suppose in the generic equation (1.94) that

$$A(x) = x^2,$$
$$B(x) = -2x \left(m \, x^{2m} - a + m - 1\right),$$
$$C(x) = x^{2m} > 0,$$
$$D(x) = -2s(2s + 2a - 2m + 1),$$
$$E(x) = 2s(2s + 2a - 2m + 1) - 2(2r+1)(r + a - m + 1),$$

and

$$\lambda_n = 2m \left(mn + 2s + (2r + 1 - m - 2s)\,\sigma_n\right),$$

where a, m, r, s are free parameters. Replacing these options in (1.94) leads to the differential equation

$$x^2 \Phi_n''(x) - 2x \left(m \, x^{2m} - a + m - 1\right) \Phi_n'(x) + \left(\alpha_n x^{2m} + \beta + \frac{1 - (-1)^n}{2} \gamma\right) \Phi_n(x) = 0,$$

$$(1.247)$$

in which

$$\beta = -2s(2s + 2a - 2m + 1), \quad \gamma = 2s(2s + 2a - 2m + 1) - 2(2r + 1)(r + a - m + 1)$$

and

$$\alpha_n = 2m \left(mn + 2s + (r - s + (m - 1)/2)(1 - (-1)^n)\right).$$

Once again, we prove that Eq. (1.247) has a polynomial solution as

$$\Phi_n^{(r,s)}(x; a, m) = (x^{2s} + (x^{2r+1} - x^{2s}) \sigma_n) L_{[n/2]}^{(\frac{a+1-m+s+r}{m} + (-1)^n \frac{2s-2r-1}{2m})}(x^{2m}), \qquad (1.248)$$

which satisfies the orthogonality relation

$$\int_{-\infty}^{\infty} x^{2a} \exp(-x^{2m}) \Phi_n^{(r,s)}(x; a, m) \Phi_k^{(r,s)}(x; a, m) \, dx$$

$$= \left(\int_{-\infty}^{\infty} x^{2a} \exp(-x^{2m}) \left(\Phi_n^{(r,s)}(x; a, m) \right)^2 dx \right) \delta_{n,k}. \qquad (1.249)$$

For this purpose, we first substitute

$$g(x) = x^\lambda L_n^{(\alpha)}(x^\theta) \quad \text{for } \lambda, \theta \in \mathbb{R}$$

into Eq. (1.40) to obtain the differential equation

$$x^2 g''(x) + x \left(-2\lambda + \alpha\theta + 1 - \theta x^\theta\right) g'(x) + \left(\theta(n\theta + \lambda) x^\theta + \lambda(\lambda - \alpha\theta)\right) g(x) = 0. \qquad (1.250)$$

If

$$-2\lambda + \alpha\theta + 1 = q, \qquad \text{or equivalently} \qquad \alpha = \frac{q + 2\lambda - 1}{\theta},$$

Equation (1.250) is transformed to

$$x^2 g''(x) + x \left(q - \theta x^\theta\right) g'(x) + \left(\theta(\theta n + \lambda) x^\theta - \lambda(\lambda + q - 1)\right) g(x) = 0. \qquad (1.251)$$

By referring to Theorem 1.2, we now define the following odd and even polynomial sequences:

$$\Phi_{2n}(x) = x^{2s} L_n^{(\frac{q+4s-1}{2m})}(x^{2m}) \quad \text{for } \lambda = 2s, \ s \in \mathbb{Z}^+, \text{ and } \theta = 2m, \ m \in \mathbb{N},$$

and

$$\Phi_{2n+1}(x) = x^{2r+1} L_n^{(\frac{q+4r+1}{2m})}(x^{2m}) \quad \text{for } \lambda = 2r + 1, \ r \in \mathbb{Z}^+, \text{ and } \theta = 2m, \ m \in \mathbb{N}, \qquad (1.252)$$

which can be written in the unique form

$$\Phi_n(x) = (x^{2s} + (x^{2r+1} - x^{2s}) \sigma_n) L_{[n/2]}^{(\frac{q+2s+2r+(-1)^n(2s-2r-1)}{2m})}(x^{2m}).$$

According to the two definitions of (1.252) and Eq. (1.251), $\Phi_{2n}(x)$ should satisfy the equation

$$x^2 \Phi_{2n}''(x) - x \left(2m\, x^{2m} - q\right) \Phi_{2n}'(x) + \left(4m\,(m\,n + s)\,x^{2m} - 2s(2s + q - 1)\right) \Phi_{2n}(x) = 0,$$

while $\Phi_{2n+1}(x)$ should satisfy

$$x^2\,\Phi_{2n+1}''(x) - x\left(2m\,x^{2m} - q\right) \Phi_{2n+1}'(x)$$
$$+ \left(2m\,(2mn + 2r + 1)\,x^{2m} - (2r + 1)(2r + q)\right)\Phi_{2n+1}(x) = 0.$$

Combining these two equations finally gives

$$x^2 \Phi_n''(x) - x\left(2m\,x^{2m} - q\right) \Phi_n'(x) + \left(2m\,(mn + 2s + (2r + 1 - m - 2s)\,\sigma_n)\,x^{2m}\right.$$
$$\left. - 2s(2s + q - 1) - ((2r + 1)(2r + q) - 2s(2s + q - 1))\,\sigma_n\right)\Phi_n(x) = 0, \qquad (1.253)$$

which is a special case of the generalized Sturm–Liouville equation (1.94).

The weight function corresponding to Eq. (1.253) takes the form

$$W(x) = x^{2m}\,\exp\!\left(\int \frac{-x(2m\,x^{2m} - q) - 2x}{x^2}\,dx\right) = K\,x^{2m+q-2}\exp(-x^{2m}),$$

with $K = 1$ without loss of generality. Moreover, since $W(x)$ is positive, it can be considered as $|x|^{2a}\exp(-x^{2m})$ for $2a = 2m + q - 2$.

To compute the norm square value of (1.249) we use the Laguerre orthogonality relation in (1.41), so that for $n = 2j$ we have

$$N_{2j} = \int_{-\infty}^{\infty} x^{2a}\,\exp(-x^{2m})\left(\Phi_{2j}^{(r,s)}(x; a, m)\right)^2 dx$$

$$= \int_{-\infty}^{\infty} x^{2a+4s}\,\exp(-x^{2m})\left(L_j^{\left(\frac{2a+4s+1-2m}{2m}\right)}(x^{2m})\right)^2 dx$$

$$= \frac{1}{m}\int_0^{\infty} t^{\frac{2a+4s+1-2m}{2m}}\,e^{-t}\left(L_j^{\left(\frac{2a+4s+1-2m}{2m}\right)}(t)\right)^2 dt = \frac{1}{m\,j!}\,\Gamma\!\left(j + \frac{2a + 4s + 1}{2m}\right),$$

$$(1.254)$$

and for $n = 2j + 1$ we get

$$N_{2j+1} = \int_{-\infty}^{\infty} x^{2a}\,\exp(-x^{2m})\left(\Phi_{2j+1}^{(r,s)}(x; a, m)\right)^2 dx$$

$$= \int_{-\infty}^{\infty} x^{2a+4r+2} \exp(-x^{2m}) \left(L_j^{(\frac{2a+4r+3-2m}{2m})}(x^{2m}) \right)^2 dx$$

$$= \frac{1}{m} \int_0^{\infty} t^{\frac{2a+4r+3-2m}{2m}} e^{-t} \left(L_j^{(\frac{2a+4r+3-2m}{2m})}(t) \right)^2 dt = \frac{1}{m\, j!} \Gamma\left(j + \frac{2a+4r+3}{2m} \right).$$

(1.255)

Combining relations (1.254) and (1.255) finally gives

$$N_n = \frac{1}{m\,((n-\sigma_n)/2)!} \Gamma\left(\frac{n-\sigma_n}{2} + \frac{2a+4s+1}{2m} + \frac{2r+1-2s}{m}\, \sigma_n \right),$$

which shows that the orthogonality (1.249) is valid if and only if

$$2a + 4s + 1 > 0, \quad 2a + 4r + 3 > 0 \quad \text{and} \quad (-1)^{2a} = 1.$$

Example 1.9 Find the standard properties of incomplete symmetric polynomials orthogonal with respect to the weight function $x^4 \exp(-x^4)$ on $(-\infty, \infty)$.
To find the solution, it is sufficient in (1.248) to choose $m = a = 2$ to get the polynomials

$$\Phi_n^{(r,s)}(x; 2, 2) = (x^{2s} + (x^{2r+1} - x^{2s})\sigma_n) L_{[n/2]}^{(\frac{1+s+r}{2} + (-1)^n \frac{2s-2r-1}{4})}(x^4) \quad r, s \in \mathbb{Z}^+,$$

(1.256)

which satisfy the equation

$$x^2 \Phi_n''(x) - 2x \left(2x^4 - 1 \right) \Phi_n'(x) + \left(4\,(2n + 2s + (2r - 1 - 2s)\,\sigma_n) x^4 \right.$$

$$\left. - 2s\,(2s + 1) + (2s\,(2s + 1) - 2(2r + 1)(r + 1))\,\sigma_n \right) \Phi_n(x) = 0$$

and the orthogonality relation

$$\int_{-\infty}^{\infty} x^4 \exp(-x^4)\, \Phi_n^{(r,s)}(x; 2, 2)\, \Phi_k^{(r,s)}(x; 2, 2)\, dx$$

$$= \frac{1}{2\,((n-\sigma_n)/2)!} \Gamma\left(\frac{n-\sigma_n}{2} + s + \frac{5}{4} + \frac{2r - 2s + 1}{2}\, \sigma_n \right) \delta_{n,k}.$$

As (1.256) shows, $\Phi_n^{(r,s)}(x; 2, 2)$ are incomplete symmetric polynomials with degrees respectively $\{2s, 2r + 1, 2s + 4, 2r + 5, 2s + 8, 2r + 9, \ldots\}$. For instance, we have

$$\text{degrees of } \Phi_n^{(1,0)}(x; 2, 2) = \{0, 3, 4, 7, 8, 11, \ldots\},$$

$$\text{degrees of } \Phi_n^{(2,0)}(x; 2, 2) = \{0, 5, 4, 9, 8, 13, \ldots\},$$

and

$$\text{degrees of } \Phi_n^{(1,1)}(x; 2, 2) = \{2, 3, 6, 7, 10, 11, \ldots\}.$$

Example 1.10 If we set $m = 1$ in (1.248), a generalization of generalized Hermite polynomials as

$$\Phi_n^{(r,s)}(x; a, 1) = (x^{2s} + (x^{2r+1} - x^{2s})\sigma_n) L_{[n/2]}^{(a+s+r+(-1)^n(s-r-\frac{1}{2}))}(x^2) \qquad (1.257)$$

is derived for $r = s = 0$ that satisfies the equation

$$x^2 \Phi_n''(x) - 2x \left(x^2 - a \right) \Phi_n'(x) + \left(2\,(n + 2s + (2r - 2s)\sigma_n)\,x^2 - 2s(2s + 2a - 1) \right.$$

$$\left. + (2s(2s + 2a - 1) - 2(2r + 1)(r + a))\,\sigma_n \right) \Phi_n(x) = 0$$

and the orthogonality relation

$$\int_{-\infty}^{\infty} x^{2a} \exp(-x^2)\, \Phi_n^{(r,s)}(x; a, 1)\, \Phi_k^{(r,s)}(x; a, 1)\, dx$$

$$= \frac{\Gamma(\frac{n - \sigma_n + 2a + 4s + 1}{2}) + (2r + 1 - 2s)\,\sigma_n)}{((n - \sigma_n)/2)!}\, \delta_{n,k}\,.$$

As (1.257) shows, $\Phi_n^{(r,s)}(x; a, 1)$ are incomplete symmetric polynomials with degrees respectively $D^{(r,s)} = \{2s, 2r + 1, 2s + 2, 2r + 3, 2s + 4, 2r + 5, \ldots\}$ though they are complete for $r = s = 0$, because in this case we have $D^{(0,0)} = \{0, 1, 2, 3, \ldots\}$.

Remark 1.8 For the finite sequences of hypergeometric orthogonal polynomials, we can also construct their incomplete type just as in the previous two sections. The weight functions corresponding to the finite type of incomplete symmetric orthogonal polynomials are clearly symmetric on the real line and can be defined as

$$W_1(x; m, p) = x^{-2p} \exp\left(-\frac{1}{x^{2m}} \right), \qquad x \in (-\infty, \infty),$$

and

$$W_2(x; m, p, q) = \frac{x^{2q}}{(1 + x^{2m})^p}, \qquad x \in (-\infty, \infty).$$

1.14   A Class of Hypergeometric-Type Orthogonal Functions

Here we introduce a sequence of hypergeometric series that satisfies a specific type of Sturm–Liouville equation. For this purpose, by recalling that the Gauss hypergeometric function

$$y(z) = {}_2F_1 \left(\begin{array}{cc} a, & b \\ & c \end{array} \middle| z \right) = \sum_{k=0}^{\infty} \frac{(a)_k (b)_k}{(c)_k} \frac{z^k}{k!} \tag{1.258}$$

satisfies the differential equation

$$z(z-1)y'' + \big((a+b+1)z - c\big)y' + aby = 0, \tag{1.259}$$

we first define the following hypergeometric function:

$$y(x) = y\big(x; a(x), b(x), p, q, r\big) = a(x)\,{}_2F_1 \left(\begin{array}{c} p+iq, \ p-iq \\ r \end{array} \middle| b(x) \right), \tag{1.260}$$

in which $a(x)$ and $b(x)$ are two predetermined functions, $p, q, r \in \mathbb{R}$, and $i = \sqrt{-1}$.
Since

$$(p+iq)_k (p-iq)_k = \prod_{j=0}^{k-1} \big((p+j)^2 + q^2\big) \geq 0$$

is always a positive real value, the function (1.260) is real, so that we have

$$y\big(x; a(x), b(x), p, q, r\big) = a(x) \sum_{k=0}^{\infty} \left(\prod_{j=0}^{k-1} (p+j)^2 + q^2 \right) \frac{(b(x))^k}{(r)_k\, k!}. \tag{1.261}$$

To obtain the differential equation of the function (1.260), we can use relations (1.258) and (1.259) such that if $z = b(x), a = p+iq, b = p-iq$, and $c = r$ are replaced in (1.259), then

$$\frac{b(x)\big(b(x)-1\big)}{\big(b'(x)\big)^2} \frac{d^2}{dx^2}\, {}_2F_1\big(p+iq, p-iq, r\,;\, b(x)\big)$$

$$+ \left(\frac{(2p+1)b(x) - r}{b'(x)} - \frac{b(x)\big(b(x)-1\big)}{\big(b'(x)\big)^3} \right) \frac{d}{dx}\, {}_2F_1\big(p+iq, p-iq, r\,;\, b(x)\big)$$

$$+ (p^2+q^2)\, {}_2F_1\big(p+iq, p-iq, r\,;\, b(x)\big) = 0. \tag{1.262}$$

On the other hand, relation (1.260) shows that

$$
{}_2F_1\left(\begin{array}{c}p+iq,\ p-iq\\r\end{array}\bigg|\ b(x)\right)=\frac{y\big(x;\,a(x),b(x),p,q,r\big)}{a(x)}.
\tag{1.263}
$$

Therefore, by substituting (1.263) into (1.262), we finally obtain

$$
\frac{b(b-1)}{b'^2}y''-\left(\frac{b(b-1)}{b'^3}\left(b''+2\frac{a'b'}{a}\right)-\frac{(2p+1)b-r}{b'}\right)y'
$$
$$
+\left(\frac{b(b-1)}{b'^3}\left(\frac{a'b''-a''b'}{a}+2\frac{a'^2b'}{a^2}\right)-\frac{(2p+1)b-r}{b'}\frac{a'}{a}+p^2+q^2\right)y=0,
$$

which is a particular case of the Sturm–Liouville equation

$$
A(x)y_q''(x)+B(x)y_q'(x)+\big(\lambda_q+C(x)\big)y_q(x)=0,
\tag{1.264}
$$

for

$$
A(x)=\frac{b(x)\big(b(x)-1\big)}{\big(b'(x)\big)^2},
$$

$$
B(x)=\frac{(2p+1)b(x)-r}{b'(x)}-\frac{b(x)\big(b(x)-1\big)}{\big(b'(x)\big)^3}\left(b''(x)+2\frac{a'(x)b'(x)}{a(x)}\right),
$$

$$
C(x)=\frac{b(x)\big(b(x)-1\big)}{\big(b'(x)\big)^3}\left(\frac{a'(x)b''(x)-a''(x)b'(x)}{a(x)}+2b'(x)\left(\frac{a'(x)}{a(x)}\right)^2\right)
$$
$$
-\frac{(2p+1)b(x)-r}{b'(x)}\frac{a'(x)}{a(x)},
$$

and

$$
\lambda_q=q^2+p^2.
$$

This means that if $A(x)$ is a real function that vanishes at the boundary points of an interval, say $[\alpha,\beta]$, then the solution of Eq. (1.264), which is indeed the series (1.261), is orthogonal with respect to the weight function

$$
w(x)=\exp\int\frac{B(x)-A'(x)}{A(x)}\,dx
$$

$$
=\exp\int\left((2p+1)\frac{b'(x)}{b(x)-1}-r\frac{b'(x)}{b(x)\big(b(x)-1\big)}-\frac{b''(x)}{b'(x)}-2\frac{a'(x)}{a(x)}-\frac{A'(x)}{A(x)}\right)dx
$$

$$= \exp\left((2p+1)\ln(b(x)-1) - r\ln\frac{b(x)-1}{b(x)} - \ln\frac{b(x)(b(x)-1)(a(x))^2}{b'}\right)$$

$$= \frac{b'(x)}{(a(x))^2}(b(x)-1)^{2p-r}(b(x))^{r-1},$$

which must be positive on the interval $[\alpha, \beta]$.

Example As the simplest case, assume in (1.262) that $a(x) = 1$ and $b(x) = x$. Hence the series

$$y_q(x) = {}_2F_1\left(\begin{array}{cc} p+iq, & p-iq \\ & r \end{array}\bigg| x\right) \tag{1.265}$$

satisfies the equation

$$x(x-1)y_q'' + \left((2p+1)x - r\right)y_q' + (p^2+q^2)y_q = 0. \tag{1.266}$$

Equation (1.266) shows that the orthogonality interval is $[0, 1]$, and the corresponding weight function is denoted by

$$w(x) = x^{r-1}(1-x)^{2p-r},$$

which is a particular case of the shifted beta distribution on $[0, 1]$.

Therefore we have

$$\int_0^1 x^{r-1}(1-x)^{2p-r} y_{q_1}(x)y_{q_2}(x)\,dx = \left(\int_0^1 x^{r-1}(1-x)^{2p-r}y_{q_1}^2(x)\,dx\right)\delta_{q_1,q_2}. \tag{1.267}$$

In order to obtain the norm square value of the series (1.265), we need to evaluate the general integral

$$I^* = \int_0^1 x^{u-1}(1-x)^{v-1}{}_2F_1^2\left(\begin{array}{cc} a, & b \\ & c \end{array}\bigg| x\right)dx.$$

For this purpose, we first have

$$I^* = \sum_{k=0}^\infty \frac{(a)_k(b)_k}{(c)_k\,k!}\int_0^1 x^{u-1+k}(1-x)^{v-1}{}_2F_1\left(\begin{array}{cc} a, & b \\ & c \end{array}\bigg| x\right)dx.$$

Since

$$
\int_0^1 x^{u-1+k}(1-x)^{v-1}\,_2F_1\left(\begin{array}{c} a,b \\ c \end{array}\middle|\, x\right) dx = \sum_{j=0}^\infty \frac{(a)_j(b)_j}{(c)_j\, j!}\int_0^1 x^{u-1+k+j}(1-x)^{v-1}\,dx
$$

$$
= B(u,v)\frac{(u)_k}{(u+v)_k}\,_3F_2\left(\begin{array}{c} a,b,u+k \\ c,u+v+k \end{array}\middle|\, 1\right),
$$

we eventually obtain

$$
\int_0^1 x^{u-1}(1-x)^{v-1}\,_2F_1{}^2\left(\begin{array}{c} a,b \\ c \end{array}\middle|\, x\right) dx
$$

$$
= B(u,v)\sum_{k=0}^\infty \frac{(a)_k(b)_k(u)_k}{(c)_k(u+v)_k k!}\,_3F_2\left(\begin{array}{c} a,b,u+k \\ c,u+v+k \end{array}\middle|\, 1\right). \qquad (1.268)
$$

Now, it is just enough to replace

$$
u = r,\ v = 2p - r + 1,\ a = p + iq,\ b = p - iq,\ \text{and}\ c = r
$$

in (1.268) to get

$$
\int_0^1 x^{r-1}(1-x)^{2p-r}y_q^2(x)\,dx
$$

$$
= B(r,2p+1-r)\sum_{k=0}^\infty \frac{(p+iq)_k\,(p-iq)_k}{(2p+1)_k\,k!}\,_3F_2\left(\begin{array}{c} p+iq,p-iq,r+k \\ r,2p+1+k \end{array}\middle|\, 1\right),
$$

which is positive if $2p + 1 > r > 0$ and $q \in \mathbb{R}$.

Hence the orthogonality relation (1.267) reads as

$$
\int_0^1 x^{r-1}(1-x)^{2p-r}y_{q_1}(x)y_{q_2}(x)\,dx
$$

$$
= \frac{\Gamma(r)\Gamma(2p+1-r)}{\Gamma(2p+1)}\sum_{k=0}^\infty \frac{(p+iq)_k(p-iq)_k}{(2p+1)_k\,k!}\,_3F_2\left(\begin{array}{c} p+iq,p-iq,r+k \\ r,2p+1+k \end{array}\middle|\, 1\right)\delta_{q_1,q_2}.
$$

For $q = k \in \mathbb{Z}^+$, one of the advantages of the above relation is the following expansion:

$$
f(x) = \sum_{k=0}^\infty c_k\,_2F_1\left(\begin{array}{c} p+ik,p-ik \\ r \end{array}\middle|\, x\right),
$$

where

$$c_k = \frac{\int_0^1 x^{r-1}(1-x)^{2p-r} f(x) \, {}_2F_1\left(\begin{array}{c} p+ik, \, p-ik \\ r \end{array} \bigg| \, x\right) dx}{B(r, 2p+1-r) \sum_{j=0}^{\infty} \frac{(p+ik)_j (p-ik)_j}{(2p+1)_j \, j!} \, {}_3F_2\left(\begin{array}{c} p+ik, \, p-ik, \, r+j \\ r, \, 2p+1+j \end{array} \bigg| \, 1\right)}.$$

This expansion is especially useful when $f(x)$ is considered as a generalized hypergeometric function.

The symmetric case of the orthogonal series (1.265) is defined as

$$g_n(x) = x^{\sigma_n} \, {}_2F_1\left(\begin{array}{c} p+\frac{1}{2}\sigma_n + i[\frac{n}{2}], \; p+\frac{1}{2}\sigma_n - i[\frac{n}{2}] \\ r + \sigma_n \end{array} \bigg| \, x^2 \right),$$

which satisfies the equation

$$x^2(x^2-1)g_n'' + x\big((4p+1)x^2 + 1 - 2r\big)g_n' + \Big(4\big(p^2 + [\tfrac{n}{2}]^2\big)x^2 + (2r-1)\sigma_n\Big)g_n = 0. \tag{1.269}$$

Equation (1.269) shows that the orthogonality interval is $[-1, 1]$, and the corresponding weight function is given by

$$w(x) = x^2 \exp \int \frac{x\big((4p+1)x^2 + 1 - 2r\big) - 2x(2x^2 - 1)}{x^2(x^2 - 1)} \, dx$$

$$= x^2 \exp \int \left(\frac{(4p+1)x}{x^2 - 1} + \frac{3 - 2r}{x(x^2 - 1)} - \frac{4x}{x^2 - 1} \right) dx$$

$$= x^2 \exp \left(\frac{4p+1}{2} \ln |x^2 - 1| + (3 - 2r)\big(-\ln|x| + \frac{1}{2}\ln|x^2 - 1|\big) - 2\ln|x^2 - 1| \right)$$

$$= x^2 \exp \left(\ln \big((1 - x^2)^{2p-r}|x|^{2r-3}\big) \right),$$

which can be denoted by

$$w(x) = x^{2r-1}(1 - x^2)^{2p-r} = w(-x) \quad \Leftrightarrow \quad (-1)^{2r} = -1.$$

Consequently, according to the generalized Sturm–Liouville theorem for symmetric functions, we have

$$\int_{-1}^{1} x^{2r-1}(1 - x^2)^{2p-r} g_n(x) g_m(x) \, dx = \left(\int_{-1}^{1} x^{2r-1}(1 - x^2)^{2p-r} g_n^2(x) \, dx \right) \delta_{n,m}.$$

1.15 Application of Zero Eigenvalue in Solving Some Sturm–Liouville Problems

Usually, the zero eigenvalue is ignored in the solution of some Sturm–Liouville problems. This case also occurs in calculating the eigenvalues of a matrix, so that we would often like to find nonzero solutions of the linear system

$$AX = \lambda X \quad \text{when} \quad \lambda \neq 0.$$

But if in the above system we take $\lambda = 0$, then $\det A = 0$ for $X \neq 0$, and the dimension of the matrix A is reduced by at least one. Now, if in a boundary value problem we similarly take one of the eigenvalues of the equations related to the main problem to be equal to zero, then one of the solutions will be specified in advance. In view of this fact, in this section we define a class of special functions and apply them to the classical potential, heat, and wave equations in spherical coordinates.

We start by defining the two sequences

$$C_n(z; a(z)) = \frac{(a(z))^n + (a(z))^{-n}}{2} = \cosh(n\ln(a(z))),$$

$$S_n(z; a(z)) = \frac{(a(z))^n - (a(z))^{-n}}{2} = \sinh(n\ln(a(z))), \tag{1.270}$$

in which $a(z)$ can be a complex (or real) function and n is a positive integer.

It is not difficult to verify that both above sequences satisfy a unique second-order differential equation of the form

$$a^2(z)a'(z)y'' + \left(a(z)\big(a'(z)\big)^2 - a^2(z)a''(z)\right)y' - n^2\big(a'(z)\big)^3 y = 0, \tag{1.271}$$

provided that $a^2(z)a'(z) \neq 0$.

Clearly, Eq. (1.271) is a particular case of the Sturm–Liouville equation (1.264) for

$$A(z) = \frac{a^2(z)}{(a'(z))^2},$$

$$B(z) = \frac{a(z)(a'(z))^2 - a^2(z)a''(z)}{(a'(z))^3},$$

$$C(z) = 0,$$

and

$$\lambda_n = -n^2.$$

There are various cases for sequences (1.270). The first case may lead to the Chebyshev polynomials if one takes $a(z) = \exp(i \arccos z)$ and uses the well-known Euler identity. In other words, we have

$$C_n(z; \exp(i\arccos z)) = \cos(n\arccos z) = T_n(z),$$

$$S_n(z; \exp(i\arccos z)) = i \, \sin(n\arccos z) = i\sqrt{1 - z^2} \, U_{n-1}(z).$$

Also replacing $a(z) = \exp(i \arccos z)$ in (1.271) gives the well-known differential equation of Chebyshev polynomials of the first kind

$$(1 - z^2)y'' - zy' + n^2 y = 0.$$

Rational Chebyshev functions are another case that can be derived by choosing $a(z) = \exp(i \arccot z)$ in (1.270), leading to

$$C_n(z; \exp(i\arccot z)) = \cos(n\arccot z),$$

$$S_n(z; \exp(i\arccot z)) = i \, \sin(n\arccot z). \tag{1.272}$$

Also, since in this case

$$a^2(z)a'(z) = \frac{-i \exp(3i\arccot z)}{1 + z^2},$$

$$a(z)(a'(z))^2 = \frac{- \exp(3i\arccot z)}{(1 + z^2)^2},$$

$$a^2(z)a''(z) = \frac{(2iz - 1) \exp(3i\arccot z)}{(1 + z^2)^2},$$

and

$$(a'(z))^3 = \frac{i \exp(3i\arccot z)}{(1 + z^2)^3},$$

the functions (1.272) satisfy the equation

$$(1 + z^2)^2 y'' + 2z(1 + z^2)y' + n^2 y = 0.$$

Notice that the explicit forms of the real functions (1.272) can be directly obtained by de Moivre's formula, because it is enough to substitute $\theta = \arccot x$ into the above formula to get

$$\frac{(x + i)^n}{(\sqrt{1 + x^2})^n} = \cos(n \arccot x) + i \, \sin(n \arccot x).$$

Consequently, we have

$$C_n(x; \exp(i\,\text{arccot}\,x)) = \left(\sum_{k=0}^{[n/2]} (-1)^k \binom{n}{2k} x^{n-2k} \right) / (\sqrt{1+x^2})^n = T_n^*(x),$$

$$-i\,S_{n+1}(x; \exp(i\,\text{arccot}\,x)) = \left(\sum_{k=0}^{[n/2]} (-1)^k \binom{n+1}{2k+1} x^{n-2k} \right) / (\sqrt{1+x^2})^{n+1} = U_n^*(x).$$

We can prove that the Chebyshev rational functions $T_n^*(x)$ and $U_n^*(x)$ are orthogonal with respect to the weight function

$$W(x) = \frac{1}{1+x^2} \quad \text{on} \quad (-\infty, \infty),$$

so that we have

$$\int_{-\infty}^{\infty} \frac{T_n^*(x)T_n^*(x)}{1+x^2}\,dx = \begin{cases} 0 & (m \neq n), \\ \pi/2 & (m = n), \\ \pi & (m = n = 0), \end{cases}$$

and

$$\int_{-\infty}^{\infty} \frac{U_n^*(x)U_n^*(x)}{1+x^2}\,dx = \begin{cases} 0 & (m \neq n), \\ \pi/2 & (m = n). \end{cases}$$

As the last particular example for sequences (1.270), we know that the associated Legendre differential equation

$$(1-x^2)y''(x) - 2xy'(x) + \left(p - \frac{q}{1-x^2} \right) y(x) = 0 \tag{1.273}$$

has been solved for the following three cases up to now:

(a) $p \neq 0$, $q \neq 0$ (generating the associated Legendre functions),
(b) $p \neq 0$, $q = 0$ (generating the Legendre polynomials),
(c) $p = 0$, $q = 0$, which is reduced to the simple equation

$$(1-x^2)y''(x) - 2xy'(x) = 0,$$

having the general solution

$$y(x) = c_1 \ln \frac{1+x}{1-x} + c_2.$$

Hence, the fourth case $p = 0$, $q \neq 0$ remains to be studied. To derive the solution of the corresponding equation, we first substitute

$$a(z) = \left(\frac{1-z}{1+z}\right)^{\frac{1}{2}}$$

into Eq. (1.271) to get

$$(1 - z^2)y'' - 2zy' - \frac{n^2}{1 - z^2}y = 0, \tag{1.274}$$

which is a particular case of Eq. (1.273) for $p = 0$, $q = n^2$. By noting (1.270), the solutions of Eq. (1.274) are of the form

$$C_n\left(z; \left(\frac{1-z}{1+z}\right)^{\frac{1}{2}}\right) = \frac{1}{2}\left(\left(\frac{1-z}{1+z}\right)^{\frac{n}{2}} + \left(\frac{1-z}{1+z}\right)^{\frac{-n}{2}}\right),$$

$$S_n\left(z; \left(\frac{1-z}{1+z}\right)^{\frac{1}{2}}\right) = \frac{1}{2}\left(\left(\frac{1-z}{1+z}\right)^{\frac{n}{2}} - \left(\frac{1-z}{1+z}\right)^{\frac{-n}{2}}\right),$$

and will appear in Helmholtz's equation in spherical coordinates. In other words, if the equation

$$\nabla^2 U(r, \theta, \Phi) = k^2 U(r, \theta, \Phi)$$

is separated to three ordinary differential equations, one of the equations takes the form

$$\frac{1}{\sin\theta}\frac{d}{d\theta}\left(\sin\theta\frac{dy}{d\theta}\right) + \left(m(m+1) - \frac{n^2}{\sin^2\theta}\right)y(\theta) = 0, \tag{1.275}$$

which is the same as Eq. (1.273) for $x = \cos\theta$, $p = m(m+1)$, and $q = n^2$.

Therefore, if $a(z) = tg\frac{z}{2}$ in (1.271), a special case of Eq. (1.275) for $m = 0$, i.e.,

$$y'' + (\cot z)\, y' - \frac{n^2}{\sin^2 z}y = 0, \tag{1.276}$$

would have the following solutions:

$$C_n\left(z; \tan\frac{z}{2}\right) = \frac{1}{2}\left(\left(\tan\frac{z}{2}\right)^n + \left(\tan\frac{z}{2}\right)^{-n}\right),$$

$$S_n\left(z; \tan\frac{z}{2}\right) = \frac{1}{2}\left(\left(\tan\frac{z}{2}\right)^n - \left(\tan\frac{z}{2}\right)^{-n}\right). \tag{1.277}$$

- **Application of the functions** (1.277) **for the potential, heat, and wave equations in spherical coordinates**

Most of boundary value problems related to the wave, heat, and potential equations are somehow connected to the Helmholtz partial differential equation

$$\nabla^2 U(r, \theta, \Phi) = k^2 U(r, \theta, \Phi)$$

in spherical coordinates. For instance, $k = 0$ in this equation leads to the potential equation

$$\frac{\partial^2 U}{\partial r^2} + \frac{2}{r}\frac{\partial U}{\partial r} + \frac{1}{r^2}\frac{\partial^2 U}{\partial \theta^2} + \frac{\cot \theta}{r^2}\frac{\partial U}{\partial \theta} + \frac{1}{r^2 \sin^2 \theta}\frac{\partial^2 U}{\partial \Phi^2} = 0, \qquad (1.278)$$

and by separating the variables as

$$U(r, \theta, \Phi) = R(r) A(\theta) B(\Phi),$$

Equation (1.278) is transformed into three ordinary differential equations

$$(i)\ r^2 R'' + 2r R' - \lambda_1 R = 0,$$

$$(ii)\ B'' - \lambda_2 B = 0,$$

$$(iii)\ A'' + (\cot \theta) A' + \left(\lambda_1 + \frac{\lambda_2}{\sin^2\theta}\right) A = 0. \qquad (1.279)$$

Since the solution of a potential equation is generally determined when the boundary conditions are known, if $\lambda_1 = 0$ and $\lambda_2 = -k^2 \neq 0$ are assumed in (1.279), then for the variable r we get

$$R(r) = -c_1^2 \frac{1}{r} + c_2 \qquad (c_1 \text{ and } c_2 \text{ are constant}),$$

and for the third equation, we get the same form as Eq. (1.276). Hence, the general solution corresponding to third equation is as

$$A(\theta) = A_1 C_k\left(\theta; \tan\frac{\theta}{2}\right) + A_2 S_k\left(\theta; \tan\frac{\theta}{2}\right) = a_1 \tan^k\left(\frac{\theta}{2}\right) + a_2 \tan^{-k}\left(\frac{\theta}{2}\right),$$

where A_1 and A_2 and a_1 and a_2 are all constant values.

This means that the general solution of the potential equation $\nabla^2 U(r, \theta, \Phi) = 0$ with the predetermined condition $R(r) = \frac{-c_1^2}{r} + c_2$ is

$$U(r, \theta, \Phi) = \left(\frac{-c_1^2}{r} + c_2\right)(b_1 \cos k\Phi + b_2 \sin k\Phi)\left(a_1 \tan^k\left(\frac{\theta}{2}\right) + a_2\tan^{-k}\left(\frac{\theta}{2}\right)\right).$$

$$(1.280)$$

The solution (1.280) reveals the sensitivity of the potential equation with respect to the variable r, because $\lim\limits_{r \to 0} U(r, \theta, \Phi) = \infty$.

For example, let us consider the equation

$$\nabla^2 U(r, \theta, \Phi) = 0 \quad \text{for} \quad 0 < r < a$$

in spherical coordinates when the variable r takes the preassigned form

$$R(r) = \frac{-c_1^2}{r} + c_2$$

and the following initial and boundary conditions hold:

$$\lim_{r \to 0} U(r, \theta, \Phi) = \infty,$$

$$U(\tfrac{a}{2}, \theta, \Phi) = 0,$$

$$U(r, \tfrac{\pi}{2}, \Phi) = 0,$$

$$U(a, \tfrac{\pi}{3}, \Phi) = \Phi.$$

The general solution of this problem, according to the given conditions and assuming $A_k = a_1 b_1 c_2$ and $B_k = a_2 b_1 c_2$ for simplicity, is denoted by

$$U(r, \theta, \Phi) = \sum_k U_k(r, \theta, \Phi),$$

$$U_k(r, \theta, \Phi) = \left(1 - \frac{a}{2r}\right)\left(\tan^k\left(\frac{\theta}{2}\right) - \tan^{-k}\left(\frac{\theta}{2}\right)\right)(A_k \cos k\Phi + B_k \sin k\Phi).$$

Now by employing the last condition in the above solution, we get

$$2\Phi = \sum_k \left(\sqrt{3}^{-k} - \sqrt{3}^{+k}\right)(A_k \cos k\Phi + B_k \sin k\Phi),$$

where A_k and B_k are calculated as

$$\left(\sqrt{3}^{-k} - \sqrt{3}^{+k}\right) A_k = \frac{4}{\pi}\int_0^\pi \Phi \cos k\Phi \, d\Phi = \frac{4((-1)^k - 1)}{\pi k^2},$$

$$\left(\sqrt{3}^{-k} - \sqrt{3}^{+k}\right) B_k = \frac{4}{\pi}\int_0^\pi \Phi \sin k\Phi \, d\Phi = \frac{4(-1)^k}{k}.$$

Therefore, the solution under the given conditions is

$$U(r, \theta, \Phi) = \left(1 - \frac{a}{2r}\right)$$

$$\times \sum_{k=1}^{\infty} \left(\frac{4((-1)^k - 1)k^{-2}}{\pi \, (3^{-k/2} - 3^{k/2})} \cos k \, \Phi + \frac{4(-1)^k k^{-1}}{(3^{-k/2} - 3^{k/2})} \sin k \, \Phi\right) \left(\tan^k \frac{\theta}{2} - \tan^{-k} \frac{\theta}{2}\right).$$

This approach can be similarly stated for the classical heat and wave equations. For instance, if the heat equation

$$\nabla^2 U = \frac{\partial \, U}{\partial \, t}$$

is considered, separating the variables as

$$U(r, \theta, \Phi, t) = S(r, \theta, \Phi)T(t),$$

where

$$S(r, \theta, \Phi) = R(r)A(\theta)B(\Phi),$$

yields

$$\begin{cases} \nabla^2 U = \nabla^2(S\,T) = T \, \nabla^2 S \\ \dfrac{\partial \, U}{\partial \, t} = \dfrac{\partial \, (S\,T)}{\partial \, t} = S \dfrac{\partial \, T}{\partial \, t} \end{cases} \quad \Rightarrow \quad \frac{\nabla^2 S}{S} = \frac{T'(t)}{T(t)} = \alpha.$$

Hence the corresponding differential equations appear as

$$(i) \ T' - \alpha \, T = 0,$$

$$(ii) \ r^2 R'' + 2r R' - (\alpha \, r^2 + \lambda_1)R = 0,$$

$$(iii) \ B'' - \lambda_2 B = 0,$$

$$(iv) \ A'' + (\cot \theta) \, A' + \left(\lambda_1 + \frac{\lambda_2}{\sin^2\theta}\right)A = 0.$$

(1.281)

Once again, if

$$\lambda_1 = 0, \quad \lambda_2 = -n^2 \neq 0, \quad \text{and} \quad \alpha = -k^2 \neq 0$$

are considered in (1.281), then the general solution, when

$$R(r) = \frac{c_1 J_{1/2}(kr) + c_2 J_{-1/2}(kr)}{\sqrt{kr}},$$

takes the form

$$U(r, \theta, \Phi, t) = \frac{e^{-k^2 t}}{\sqrt{kr}} (c_1 J_{1/2}(kr) + c_2 J_{-1/2}(kr)) \left(b_1 \tan^n \left(\frac{\theta}{2} \right) + b_2 \tan^{-n} \left(\frac{\theta}{2} \right) \right)$$

$$\times (a_1 \cos n\Phi + a_2 \sin n\Phi), \qquad (1.282)$$

where $J_{1/2}(x)$ and $J_{-1/2}(x)$ are two particular cases of the well-known Bessel functions $J_p(x)$.

For example, let us consider the heat equation

$$\nabla^2 U(r, \theta, \Phi, t) = \frac{\partial U}{\partial t} \quad \text{for} \quad 0 < r < a$$

in spherical coordinates when the variable r takes the preassigned form

$$R(r) = \frac{c_1 J_{1/2}(kr) + c_2 J_{-1/2}(kr)}{\sqrt{kr}},$$

and the following initial and boundary conditions hold:

$$\lim_{r \to 0} U(r, \theta, \Phi, t) < M,$$

$$U\left(r, \frac{\pi}{2}, \Phi, t\right) = 0,$$

$$U(r, \theta, 0, t) = 0,$$

$$U\left(r, \frac{\pi}{3}, \Phi, 0\right) = f(r, \Phi).$$

By referring to the general solution of the problem and using the given conditions, we first conclude in (1.282) that $c_2 = a_1 = 0$ and $b_1 + b_2 = 0$. Hence if $A_{k,n} = a_2 b_1 c_1$, then

$$U(r, \theta, \Phi, t) = \sum_k \sum_n U_{k,n}(r, \theta, \Phi, t),$$

$$U_{k,n}(r, \theta, \Phi, t) = A_{k,n} \frac{e^{-k^2 t}}{\sqrt{kr}} J_{1/2}(kr) \left(\tan^n \frac{\theta}{2} - \tan^{-n} \frac{\theta}{2} \right) \sin n\Phi,$$

is the general solution.

On the other hand, by noting the orthogonality relation of the Bessel functions $J_p(x)$ as

$$\int_0^a J_p\left(Z_{(p,m)}\frac{x}{a}\right) J_p\left(Z_{(p,n)}\frac{x}{a}\right) x\, dx = \frac{a^2}{2} J_{p+1}^2\left(Z_{(p,m)}\right) \delta_{n,m},$$

where $Z_{(p,m)}$ is the mth zero of $J_p(x)$, i.e., $J_p(Z_{(p,m)}) = 0$, it is better for the eigenvalues k to be considered as $k = \frac{Z_{(1/2,m)}}{a}$. Thus, the general solution becomes

$$U(r,\theta,\Phi,t)$$

$$= \sum_n \sum_m A_{n,m}^* \frac{\exp(-(\frac{Z_{(1/2,m)}}{a})^2 t)}{\sqrt{Z_{(1/2,m)}r/a}} J_{\frac{1}{2}}(Z_{(1/2,m)}r/a)\left(\tan^n\frac{\theta}{2} - \tan^{-n}\frac{\theta}{2}\right)\sin n\,\Phi,$$

in which $A_{n,m}^* = A_{n,\frac{Z_{(1/2,m)}}{a}}$.

In the sequel, it is sufficient to compute the coefficients $A_{n,m}^*$. To this end, substituting the last given condition into the above solution yields

$$\sqrt{\frac{r}{a}}\, f(r,\Phi) = \sum_n \sum_m \frac{A_{n,m}^*(3^{\frac{-n}{2}} - 3^{\frac{n}{2}})}{\sqrt{Z_{(1/2,m)}}} J_{1/2}\left(Z_{(1/2,m)}\frac{r}{a}\right)\sin(n\,\Phi).$$

Therefore, using the orthogonality relation of the Bessel functions and also the orthogonality property of the sequence $\{\sin(n\Phi)\}_{n=1}^\infty$ on $[0,\pi]$, $A_{n,m}^*$ is computed as

$$A_{n,m}^* = \frac{4\sqrt{Z_{(1/2,m)}}\int_0^a\int_0^\pi f(r,\Phi)\, J_{1/2}(Z_{(1/2,m)}\frac{r}{a})\,\sin(n\,\Phi) r^{\frac{3}{2}}\, dr\, d\Phi}{\pi\,(3^{\frac{-n}{2}} - 3^{\frac{n}{2}})a^{\frac{3}{2}} J_{3/2}^2(Z_{(1/2,m)})}.$$

A similar problem can be stated for the classical wave equation

$$\nabla^2 U = \frac{\partial^2 U}{\partial t^2}.$$

First we have

$$\begin{cases} \nabla^2 U = \nabla^2(S\,T) = T\,\nabla^2 S \\[2mm] \dfrac{\partial^2 U}{\partial t^2} = \dfrac{\partial^2(S\,T)}{\partial t^2} = S\dfrac{\partial^2 T}{\partial t^2} \end{cases} \Rightarrow \frac{\nabla^2 S}{S} = \frac{T''(t)}{T(t)} = \alpha,$$

which results in four ordinary differential equations

$$(i)\ T'' - \alpha\, T = 0,$$

$$(ii)\ r^2 R'' + 2r R' - (\alpha\, r^2 + \lambda_1) R = 0,$$

$$(iii)\ B'' - \lambda_2 B = 0,$$

$$(iv)\ A'' + (\cot \theta) A' + \left(\lambda_1 + \frac{\lambda_2}{\sin^2\theta}\right) A = 0.$$

(1.283)

Now, if

$$\lambda_1 = 0,\ \ \lambda_2 = -k^2 \neq 0 \quad \text{and} \quad \alpha = -n^2 \neq 0$$

in (1.283), then the general solution of the wave equation when

$$R\,(r) = \frac{c_1\, J_{1/2}(n\,r) + c_2\, J_{-1/2}(n\,r)}{\sqrt{nr}},$$

is

$$U\,(r, \theta, \Phi, t) = (d_1 \cos n\,t + d_2 \sin n\,t)\,(b_1 \cos k\,\Phi + b_2 \sin\, k\,\Phi)$$

$$\times \left(a_1 \tan^k \frac{\theta}{2} + a_2 \tan^{-k} \frac{\theta}{2}\right)\left(\frac{c_1\, J_{1/2}(n\,r) + c_2\, J_{-1/2}(n\,r)}{\sqrt{nr}}\right). \qquad (1.284)$$

For example, let us consider the classical wave equation in spherical coordinates when the variable r has the preassigned form

$$R\,(r) = \frac{c_1\, J_{1/2}(n\,r) + c_2\, J_{-1/2}(n\,r)}{\sqrt{nr}},$$

together with the following initial and boundary conditions:

$$\lim_{r \to 0} U(r, \theta,\ \Phi, t) < M,$$

$$U\left(r, \frac{\pi}{2},\ \Phi, t\right) = 0,$$

$$U(r, \theta,\, 0, t) = 0,$$

$$U(r, \theta,\ \Phi, 0) = 0,$$

$$U\left(r, \frac{\pi}{3},\ \Phi, q\right) = g(r, \Phi).$$

To solve this problem, substituting the given conditions into the general solution (1.284) gives $c_2 = b_1 = d_1 = 0$ and $a_1 + a_2 = 0$. Now if $B_{k,n} = a_1 b_2 c_1 d_2$, then (1.284) is simplified as

$$U(r, \theta, \Phi, t) = \sum_k \sum_n U_{k,n}(r, \theta, \Phi, t),$$

$$U_{k,n}(r, \theta, \Phi, t) = B_{k,n} \sin(nt) \frac{J_{1/2}(nr)}{\sqrt{nr}} \left(\tan^k \frac{\theta}{2} - \tan^{-k} \frac{\theta}{2} \right) \sin k\, \Phi.$$

Similar to the previous example, if $n = \frac{Z_{(1/2,m)}}{a}$, then

$$U(r, \theta, \Phi, t) = \sum_k \sum_m B_{k,m}^* \sin\left(Z_{(1/2,m)} \frac{t}{a} \right) \frac{J_{1/2}(Z_{(1/2,m)} r/a)}{\sqrt{Z_{(1/2,m)} r/a}} \left(\tan^k \frac{\theta}{2} - \tan^{-k} \frac{\theta}{2} \right) \sin k\, \Phi$$

is the general solution of the given problem, where $B_{k,m}^* = B_{k, \frac{Z_{(1/2,m)}}{a}}$.

Finally, by replacing the last condition in the above solution, i.e.,

$$\sqrt{\frac{r}{a}} g(r, \Phi) = \sum_k \sum_m \frac{B_{k,m}^* (3^{\frac{-k}{2}} - 3^{\frac{k}{2}}) \sin(Z_{(1/2,m)} \frac{q}{a})}{\sqrt{Z_{(1/2,m)}}} J_{1/2}\left(Z_{(1/2,m)} \frac{r}{a} \right) \sin(k\, \Phi),$$

and using the orthogonality relation of the Bessel functions $J_{1/2}(Z_{(1/2,m)} \frac{r}{a})$ on $[0, a]$, the unknown coefficients $B_{k,m}^*$ will be derived as

$$B_{k,m}^* = \frac{4\sqrt{Z_{(1/2,m)}} \int_0^a \int_0^\pi g(r, \Phi) J_{1/2}(Z_{(1/2,m)} \frac{r}{a}) \sin(k\, \Phi) r^{\frac{3}{2}} \, dr \, d\Phi}{\pi \, \sin(Z_{(1/2,m)} \frac{q}{a}) (3^{\frac{-k}{2}} - 3^{\frac{k}{2}}) a^{\frac{3}{2}} J_{3/2}^2(Z_{(1/2,m)})}.$$

1.15.1 A Relationship Between the Chebyshev Polynomials of the Third and Fourth Kinds and the Associated Legendre Differential Equation

In this section, we provide a short survey on the Chebyshev polynomials of the third and fourth kinds and show that their differential equations are particular cases of Eq. (1.273). Let us start with the Jacobi polynomials $y = P_n^{(\alpha, \beta)}(x)$ that are orthogonal with respect to the weight function $(1 - x)^\alpha (1 + x)^\beta$ for $\alpha, \beta > -1$ on $[-1, 1]$ and satisfy the differential equation

$$(1 - x^2) y'' - ((\alpha + \beta + 2)x + (\alpha - \beta)) y' + n(n + \alpha + \beta + 1) y = 0. \tag{1.285}$$

If

$$A_n(x) = (1 - x)^{\frac{\alpha}{2}} (1 + x)^{\frac{\beta}{2}} P_n^{(\alpha,\beta)}(x),$$

Then Eq. (1.285) changes to the self-adjoint form

$$(1 - x^2)^2 A'' - 2x(1 - x^2) A' - \left(\frac{(\alpha + \beta)(\alpha + \beta + 2) + 4n(n + \alpha + \beta + 1)}{4} x^2 \right.$$

$$\left. + \frac{(\alpha + \beta)(\alpha - \beta)}{2} x - \frac{2(\alpha + \beta) - (\alpha - \beta)^2 + 4n(n + \alpha + \beta + 1)}{4} \right) A = 0.$$

$$(1.286)$$

There are two cases in Eq. (1.286) that are transformed to the associated Legendre differential equation

$$(1 - x^2) y''(x) - 2x\, y'(x) + (p - \frac{q}{1 - x^2}) y(x) = 0. \tag{1.287}$$

The first case is $\alpha = \beta$, which leads to the particular equation

$$(1 - x^2) A'' - 2x A' + \left((n + \alpha)(n + \alpha + 1) - \frac{\alpha^2}{1 - x^2} \right) A = 0,$$

where

$$A_n(x) = (1 - x^2)^{\frac{\alpha}{2}} P_n^{(\alpha,\alpha)}(x) \quad \text{for} \quad \alpha > -1.$$

The second case, which generates the third and fourth kinds of Chebyshev polynomials, is $\alpha = -\beta$. In this case, the equation

$$(1 - x^2) A'' - 2x A' + \left(n(n + 1) - \frac{\alpha^2}{1 - x^2} \right) A = 0 \tag{1.288}$$

is a special case of Eq. (1.287) for $p = n(n + 1)$ and $q = \alpha^2$, where

$$A_n(x) = \left(\frac{1 - x}{1 + x} \right)^{\frac{\alpha}{2}} P_n^{(\alpha,-\alpha)}(x) \quad \text{for} \quad -1 < \alpha < 1.$$

Note that $-1 < \alpha < 1$ is a necessary condition for orthogonality of the polynomials $P_n^{(\alpha,-\alpha)}(x)$, because we must simultaneously have $\alpha + \beta = 0$ and $\alpha, \beta > -1$. This means that the parameter $q = \alpha^2$ should lie in $[0, 1)$, and we can, for instance, consider the eigenvalues $q = 1/m$, $m \in \mathbb{N}$, corresponding to this case. Hence Eq. (1.288) can be directly applied in the potential theory when $\alpha + \beta = 0$. It is clear that the polynomials generated by this condition are orthogonal on $[-1, 1]$, since

$$\int_{-1}^{1} \left(\frac{1-x}{1+x}\right)^{\alpha} P_n^{(\alpha,-\alpha)}(x) P_m^{(\alpha,-\alpha)}(x)\, dx = \frac{2}{2n+1} \frac{(n+\alpha)!(n-\alpha)!}{(n!)^2} \delta_{n,m},$$

and they satisfy the differential equation

$$(1-x^2)y'' - 2(x+\alpha)y' + n(n+1)y = 0 \quad \text{for} \quad y = P_n^{(\alpha,-\alpha)}(x).$$

There are two subsequences of polynomials $P_n^{(\alpha,-\alpha)}(x)$ that have trigonometric forms just like the ultraspherical polynomials $P_n^{(\alpha,\alpha)}(x)$ that generate the first and second kinds of Chebyshev polynomials for $\alpha = -\frac{1}{2}$ and $\alpha = \frac{1}{2}$ respectively. To introduce these two subsequences, it is enough to substitute $2\theta = \arccos x$ and $m = 2n + 1$ into de Moivre's formula to get

$$\left(\cos\left(\frac{\arccos x}{2}\right) + i\,\sin\left(\frac{\arccos x}{2}\right)\right)^{2n+1} = \cos\left(\left(n+\frac{1}{2}\right)\arccos x\right) + i\,\sin\left(\left(n+\frac{1}{2}\right)\arccos x\right).$$

$$(1.289)$$

Since

$$\cos\left(\frac{\arccos x}{2}\right) = \sqrt{\frac{1+x}{2}} \quad \text{and} \quad \sin\left(\frac{\arccos x}{2}\right) = \sqrt{\frac{1-x}{2}},$$

relation (1.289) changes to

$$\left(\sqrt{\frac{1+x}{2}} + i\sqrt{\frac{1-x}{2}}\right)^{2n+1} = \cos\left(\left(n+\frac{1}{2}\right)\arccos x\right) + i\,\sin\left(\left(n+\frac{1}{2}\right)\arccos x\right).$$

Consequently, we have

$$\sqrt{\frac{1+x}{2}} \sum_{k=0}^{n} (-1)^k \binom{2n+1}{2k} \left(\frac{1-x}{2}\right)^k \left(\frac{1+x}{2}\right)^{n-k}$$

$$+ i\sqrt{\frac{1-x}{2}} \sum_{k=0}^{n} (-1)^k \binom{2n+1}{2k+1} \left(\frac{1-x}{2}\right)^k \left(\frac{1+x}{2}\right)^{n-k}$$

$$= \cos\left(\left(n+\frac{1}{2}\right)\arccos x\right) + i\,\sin\left(\left(n+\frac{1}{2}\right)\arccos x\right),$$

which yields

$$V_n(x) = \sqrt{\frac{2}{1+x}} \cos\left(\left(n+\frac{1}{2}\right)\arccos x\right) = \sum_{k=0}^{n}(-1)^k\binom{2n+1}{2k}\left(\frac{1-x}{2}\right)^k\left(\frac{1+x}{2}\right)^{n-k},$$

and

$$W_n(x) = \sqrt{\frac{2}{1-x}} \sin\left(\left(n+\frac{1}{2}\right)\arccos x\right) = \sum_{k=0}^{n}(-1)^k\binom{2n+1}{2k+1}\left(\frac{1-x}{2}\right)^k\left(\frac{1+x}{2}\right)^{n-k},$$

as the explicit representations of the third and fourth kinds of Chebyshev polynomials. In this direction, note that since

$$P_n^{(\alpha,-\alpha)}(-x) = (-1)^n P_n^{(-\alpha,\alpha)}(x),$$

it follows that

$$V_n(-x) = (-1)^n W_n(x).$$

Corollary 1.13 *The two sequences*

$$A_n(x) = \left(\frac{1-x}{1+x}\right)^{-\frac{1}{4}} V_n(x) \quad and \quad A_n(-x) = (-1)^n \left(\frac{1-x}{1+x}\right)^{\frac{1}{4}} W_n(x)$$

satisfy the following self-adjoint equation according to the general Eq. (1.288):

$$(1-x^2)A'' - 2xA' + \left(n(n+1) - \frac{1}{4(1-x^2)}\right)A = 0.$$

Moreover, the norm square value of the two sequences is equal to π, *because*

$$\int_{-1}^{1}\sqrt{\frac{1+x}{1-x}}\, V_n(x)V_m(x)dx = 2\int_{-1}^{1}\frac{\cos((n+1/2)\arccos x)\cos((m+1/2)\arccos x)}{\sqrt{1-x^2}}dx$$

$$= 2\int_{0}^{\pi}\cos(n+1/2)\theta\,\cos(m+1/2)\theta\,d\theta$$

$$= \int_{0}^{\pi}\left(\cos(n+m+1)\theta + \cos(n-m)\theta\right)d\theta = \pi\,\delta_{m,n},$$

and for $W_n(x)$ we similarly have

$$\int_{-1}^{1} \sqrt{\frac{1-x}{1+x}} \, W_n(x) W_m(x) dx = 2 \int_{0}^{\pi} \sin(n+1/2)\theta \, \sin(m+1/2)\theta \, d\theta$$

$$= \int_{0}^{\pi} \left(\cos(n-m)\theta - \cos(n+m+1)\theta \right) d\theta = \pi \, \delta_{m,n}.$$

Therefore

$$\int_{-1}^{1} V_n(x) V_m(x) \sqrt{\frac{1+x}{1-x}} \, dx = \int_{-1}^{1} W_n(x) W_m(x) \sqrt{\frac{1-x}{1+x}} \, dx = \pi \, \delta_{m,n}. \qquad (1.290)$$

From the result (1.290), one can conclude that the sequences

$$f_n(x) = \cos\left(n+\frac{1}{2}\right)\frac{\pi}{l}x \quad \text{and} \quad g_n(x) = \sin\left(n+\frac{1}{2}\right)\frac{\pi}{l}x$$

are orthogonal on $[0, l]$, so that we have

$$\int_{0}^{l} f_n(x) f_m(x) dx = \frac{1}{2} \int_{0}^{l} \left(\cos(n+m+1)\frac{\pi}{l}x + \cos(n-m)\frac{\pi}{l}x \right) dx = \frac{l}{2}\delta_{m,n}$$

and

$$\int_{0}^{l} g_n(x) g_m(x) dx = \frac{1}{2} \int_{0}^{l} \left(\cos(n-m)\frac{\pi}{l}x - \cos(n+m+1)\frac{\pi}{l}x \right) dx = \frac{l}{2}\delta_{m,n}.$$

The two latter relations show that

$$A = \{f_n(x)\}_{n=0}^{\infty} \quad \text{and} \quad B = \{g_n(x)\}_{n=0}^{\infty}$$

are complete orthogonal sets in $[0, l]$. On the other hand, by recalling that the sequences

$$A_n(x) = \cos\frac{n\pi}{l}x \quad \text{and} \quad B_n(x) = \sin\frac{n\pi}{l}x$$

are also orthogonal on $[0, l]$, the following expansions can be derived, which are various combinations of the above four trigonometric sequences:

$$f(x) = \sum_{k=0}^{\infty} a_k^{(\frac{1}{2})} \cos\left(k+\frac{1}{2}\right)\frac{\pi x}{l},$$

where

$$a_k^{\left(\frac{1}{2}\right)} = \frac{2}{l} \int_0^l f(x) \cos\left(k + \frac{1}{2}\right)\frac{\pi x}{l}\, dx,$$

$$f(x) = \sum_{k=0}^{\infty} b_k^{\left(\frac{1}{2}\right)} \sin\left(k + \frac{1}{2}\right)\frac{\pi x}{l},$$

where

$$b_k^{\left(\frac{1}{2}\right)} = \frac{2}{l} \int_0^l f(x) \sin\left(k + \frac{1}{2}\right)\frac{\pi x}{l}\, dx,$$

$$f(x) = \sum_{k=0}^{\infty} a_k^{\left(\frac{1}{2}\right)} \cos\left(k + \frac{1}{2}\right)\frac{\pi x}{l} + b_k^{\left(\frac{1}{2}\right)} \sin\left(k + \frac{1}{2}\right)\frac{\pi x}{l}, \qquad (1.291)$$

where

$$\begin{cases} a_k^{\left(\frac{1}{2}\right)} = \frac{1}{l} \int_{-l}^{l} f(x) \cos(k + \frac{1}{2})\frac{\pi x}{l}\, dx, \\[2mm] b_k^{\left(\frac{1}{2}\right)} = \frac{1}{l} \int_{-l}^{l} f(x) \sin(k + \frac{1}{2})\frac{\pi x}{l}\, dx, \end{cases}$$

$$f(x) = \sum_{k=0}^{\infty} a_k^{\left(\frac{1}{2}\right)} \cos\left(k + \frac{1}{2}\right)\frac{\pi x}{l} + b_k \sin\frac{k\pi x}{l}, \qquad (1.292)$$

where

$$\begin{cases} a_k^{\left(\frac{1}{2}\right)} = \frac{1}{l} \int_0^{2l} f(x) \cos(k + \frac{1}{2})\frac{\pi x}{l}\, dx, \\[2mm] b_k = \frac{1}{l} \int_{-l}^{l} f(x) \sin\frac{k\pi x}{l}\, dx, \end{cases}$$

$$f(x) = \sum_{k=0}^{\infty} a_k \cos\frac{k\pi x}{l} + b_k^{\left(\frac{1}{2}\right)} \sin(k + \frac{1}{2})\frac{\pi x}{l}, \qquad (1.293)$$

where

$$\begin{cases} a_0 = \frac{1}{2l} \int_{-l}^{l} f(x)\, dx, \\[2mm] a_k = \frac{1}{l} \int_{-l}^{l} f(x) \cos\frac{k\pi x}{l}\, dx, \\[2mm] b_k^{\left(\frac{1}{2}\right)} = \frac{1}{l} \int_{-l}^{l} f(x) \sin(k + \frac{1}{2})\frac{\pi x}{l}\, dx. \end{cases}$$

Note that relations (1.291)–(1.293) also have complex forms. For instance, the complex form of relation (1.291) is

$$f(x) = C_{-0}^{(\frac{1}{2})} e^{-i\frac{\pi x}{2l}} + \sum_{k=-\infty}^{\infty} C_k^{(\frac{1}{2})} e^{i(k+\frac{1}{2})\frac{\pi x}{l}},$$

where

$$\begin{cases} C_k^{(\frac{1}{2})} = \frac{1}{2l} \int_{-l}^{l} f(x) e^{-i(k+\frac{1}{2})\frac{\pi x}{l}} dx, \\ \\ C_{-0}^{(\frac{1}{2})} = \frac{1}{2l} \int_{-l}^{l} f(x) e^{i\frac{\pi x}{2l}} dx. \end{cases}$$

Furthermore, Bessel's inequality, Parseval's identity, and the Dirikhlet kernel are some other items that can be applied to relations (1.291)–(1.293).

For example, the Dirikhlet kernel corresponding to relation (1.291) for $l = \pi$ is derived as

$$S_n(x) = \frac{1}{\pi} \sum_{k=0}^{n} \int_{-\pi}^{\pi} f(t) \left(\cos\left(k + \frac{1}{2}\right) t \, \cos\left(k + \frac{1}{2}\right) x + \sin\left(k+\frac{1}{2}\right) t \, \sin\left(k+\frac{1}{2}\right) x \right) dt$$

$$= \frac{1}{\pi} \int_{-\pi}^{\pi} f(t) \left(\sum_{k=0}^{n} \cos(k + \frac{1}{2})(t - x) \right) dt = \frac{1}{2\pi} \int_{-\pi}^{\pi} f(t) \frac{\sin(n + 1)(t - x)}{\sin \frac{t-x}{2}} dt.$$

Hence

$$D_{n+1/2}(x) = \frac{\sin(n + 1)x}{\sin \frac{x}{2}}$$

is the Dirichlet kernel corresponding to relation (1.291). Moreover, since

$$D_n(x) = \sum_{k=-n}^{n} e^{ikx} = \frac{\sin(n + \frac{1}{2})x}{\sin \frac{x}{2}},$$

it follows that

$$D_n(\arccos x) = \frac{\sin\left((n + \frac{1}{2}) \arccos x\right)}{\sin(\frac{\arccos x}{2})} = W_n(x) \quad \Leftrightarrow \quad W_n(\cos x) = D_n(x).$$

It is not difficult to verify that

$$V_n(x) =_2 F_1\left(-n, n + 1, \frac{1}{2}; \frac{1 - x}{2}\right)$$

and

$$W_n(x) = (2n + 1)\, {}_2F_1\left(-n, n+1, \frac{3}{2}; \frac{1-x}{2}\right),$$

in which ${}_2F_1(a, b, c; x)$ denotes the hypergeometric function of order $(2, 1)$.

To compute the generating functions of the polynomials $V_n(x)$ and $W_n(x)$, we can start with the relation

$$\sum_{n=0}^{\infty} t^n \cos\left(n + \frac{1}{2}\right)\theta + i \sum_{n=0}^{\infty} t^n \sin\left(n + \frac{1}{2}\right)\theta = e^{i\frac{\theta}{2}} \sum_{n=0}^{\infty} \left(te^{i\theta}\right)^n = \frac{e^{i\frac{\theta}{2}}}{1 - te^{i\theta}}$$

$$= \frac{(1 - t\cos\theta)\cos(\theta/2) - t\sin\theta\sin(\theta/2)}{1 + t^2 - 2t\cos\theta} + i\frac{(1 - t\cos\theta)\sin(\theta/2) + t\sin\theta\cos(\theta/2)}{1 + t^2 - 2t\cos\theta},$$

which yields

$$\sum_{n=0}^{\infty} \frac{\cos(n + 1/2)\theta}{\cos(\theta/2)} t^n = \frac{1 - t}{1 + t^2 - 2t\cos\theta} \qquad (1.294)$$

and

$$\sum_{n=0}^{\infty} \frac{\sin(n + 1/2)\theta}{\sin(\theta/2)} t^n = \frac{1 + t}{1 + t^2 - 2t\cos\theta}. \qquad (1.295)$$

Thus, by setting $x = \cos\theta$ in (1.294) and (1.295), the generating functions of the third and fourth kinds of Chebyshev polynomials will be respectively derived as follows:

$$\sum_{n=0}^{\infty} V_n(x)\, t^n = \frac{1 - t}{1 + t^2 - 2tx}$$

and

$$\sum_{n=0}^{\infty} W_n(x)\, t^n = \frac{1 + t}{1 + t^2 - 2tx},$$

which are special cases of the Jacobi polynomial generating function for $\alpha = -\beta = -\frac{1}{2}$ and $\alpha = -\beta = \frac{1}{2}$.

Finally, since the Chebyshev polynomials of the first and second kinds have trigonometric forms, by applying different trigonometric identities, we can derive various relationships. For example, the identity

$$\frac{1}{2}\arccos x = \arccos\sqrt{\frac{1+x}{2}}$$

implies that

$$V_n(x) = \sqrt{\frac{2}{1+x}}\, T_{2n+1}\left(\sqrt{\frac{1+x}{2}}\right)$$

and

$$W_n(x) = U_{2n}\left(\sqrt{\frac{1+x}{2}}\right),$$

where $T_n(x)$ and $U_n(x)$ denote the first and second kinds of Chebyshev polynomials respectively.

References

1. G.E. Andrews, R. Askey, R. Roy, Special functions, in *Encyclopedia of Mathematics and Its Applications*, vol. 71 (Cambridge University Press, Cambridge, 1999)
2. G. Arfken, *Mathematical Methods for Physicists* (Academic, Cambridge, 1985)
3. A. Erdelyi, W. Magnus, F. Oberhettinger, F.G. Tricomi, *Tables of Integral Transforms*, vol. 1 (McGraw-Hill, New York, 1954)
4. W. Lesky, Endliche und unendliche systeme von kontinuierlichen klassischen orthogonalpolynomen. Z. Angew. Math. Mech. **76**, 181–184 (1996)
5. P. Lesky, *Eine Charakterisierung der Klassischen Kontinuierlichen, Diskreten und q-Orthogonalpolynome* (Shaker, Aachen, 2005)
6. M. Masjed-Jamei, Three finite classes of hypergeometric orthogonal polynomials and their application in functions approximation. Integral Transform. Spec. Funct. **13**(2), 169–190 (2002)
7. M. Masjed-Jamei, Classical orthogonal polynomials with weight function $((ax+b)^2 + (cx+d)^2)^{-p}\exp(q\arctan\frac{ax+b}{cx+d})$; $x \in (-\infty, \infty)$ and a generalization of T and F distributions. Integral Transform. Spec. Funct. **15**(2), 137–153 (2004)
8. V. Romanovski, Sur quelques classes nouvelles de polynômes orthogonaux. Comptes Rendues Acad. Sci. Paris. **188**, 1023–1025 (1929)

Further Reading

- M. Abramowitz, I.A. Stegun, *Handbook of Mathematical Functions with Formulas, Graphs, and Mathematical Tables, 9th Printing* (Dover, New York, 1972)
- W. Al-Salam, W.R. Allaway, R. Askey, Sieved ultraspherical polynomials. Trans. Am. Math. Soc. **284**, 39–55 (1984).
- R. Askey, Orthogonal polynomials old and new, and some combinatorial connections, in *Enumeration and Design*, ed. by D.M. Jacson, S.A. Vanstone (Academic, New York, 1984)
- R. Askey, Continuous Hahn polynomials. J. Phys. A **18**(16), L1017–L1019 (1985)

- R. Askey, An integral of Ramanujan and orthogonal polynomials. J. Indian Math. Soc. **51**, 27–36 (1987)
- S. Bochner, Über Sturm–Liouvillesche polynomsysteme. Math. Zeit. **29**, 730–736 (1929)
- A. Branquinho, F. Marcellán, J.C. Petronilho, Classical orthogonal polynomials: a functional approach. Acta Appl. Math. **34** , 283–303 (1994)
- A.L. Cauchy, Sur les integrales définies prises entre des limites imaginaires, Bulletin de Ferussoc, T. III (1825), 214–221, in Oeuvres de A. L. Cauchy, 2 serie, T. II, Gauthier-Villars. Paris, 59–65 (1958)
- J.A. Charris, M.E.H. Ismail, On sieved orthogonal polynomials II: random walk polynomials. Canad. J. Math. **38**, 397–415 (1986)
- J.A. Charris, M.E.H. Ismail, Sieved orthogonal polynomials VII: generalized polynomial mappings. Trans. Am. Math. Soc. **340**, 71–93 (1993)
- S.A. Chihara, *An Introduction to Orthogonal Polynomials* (Gordon and Breach, New York, 1978)
- H. Dette, Characterizations of generalized Hermite and sieved ultraspherical polynomials. Proc. Am. Math. Soc. **384**, 691–711 (1996)
- Digital Library of Mathematical Functions, DLMF. Available at: http://dlmf.nist.gov/
- L. Fox, I.B. Parker, *Chebyshev Polynomials in Numerical Analysis* (Oxford University Press, London, 1968)
- E. Grosswald, *Bessel Polynomials*. Lecture Notes Mathematics, vol. 698 (Springer, New York, 1973)
- M.E.H. Ismail, *Classical and Quantum Orthogonal Polynomials in One Variable*. Encyclopedia of Mathematics, vol. 98 (Cambridge University Press, Cambridge, 2005)
- R. Koekoek, P.A. Lesky, R.F. Swarttouw, *Hypergeometric Orthogonal Polynomials and Their q-Analogues*. Springer Monographs in Mathematics (Springer, Berlin, 2010)
- W. Koepf, *Hypergeometric Summation* (Vieweg, Braunschweig/Wiesbaden, 1988)
- W. Koepf, M. Masjed-Jamei, A generic formula for the values at the boundary points of monic classical orthogonal polynomials. J. Comput. Appl. Math. **191**, 98–105 (2006)
- W. Koepf, M. Masjed-Jamei, A generic polynomial solution for the differential equation of hypergeometric type and six sequences of classical orthogonal polynomials related to it. Integral Transform. Spec. Funct. **17**, 559–576 (2006)
- W. Koepf, M. Masjed-Jamei, Two classes of special functions using Fourier transforms of some finite classes of classical orthogonal polynomials. Proc. Am. Math. Soc. **135**, 3599–3606 (2007)
- W. Koepf, D. Schmersau, Representation of orthogonal polynomials. J. Comput. Appl. Math. **90**, 57–94 (1998)
- W. Koepf, D. Schmersau, Recurrence equations and their classical orthogonal polynomial solutions. Appl. Math. Comput. **128**(2–3), 303–327 (2002)
- H.T. Koelink, On Jacobi and continuous Hahn polynomials. Proc. Am. Math. Soc. **124**, 997–898 (1996)

- T. H. Koornwinder, Special orthogonal polynomial systems mapped onto each other by the Fourier–Jacobi transform, in *Polynomes Orthogonaux et Applications*, ed. by C. Brezinski, A. Draux, A.P. Magnus, P. Maroni, A. Ronveaux. Lecture Notes Mathematics, vol. 1171 (Springer, Berlin, 1985), pp. 174–183
- H.L. Krall, O. Frink, A new class of orthogonal polynomials: the Bessel polynomials. Trans. Am. Math. Soc. **65**, 100–115 (1949)
- W. Lesky, Uber Polynomlösungen von Differentialgleichungen und Differenzengleichungen zweiter Ordnung. Anz. Österr. Akad. Wiss., math. nat. Kl. **121**, 29–33 (1985)
- F. Marcellán, R. Sfaxi, Orthogonal polynomials and second-order pseudo-spectral linear differential equations. Integral Transform. Spec. Funct. **21**, 487–501 (2010)
- M. Masjed-Jamei, A generalization of classical symmetric orthogonal functions using a symmetric generalization of Sturm–Liouville problems. Integral Transform. Spec. Funct. **18**, 871–883 (2007)
- M. Masjed-Jamei, A basic class of symmetric orthogonal polynomials using the extended Sturm–Liouville theorem for symmetric functions. J. Math. Anal. Appl. **325**, 753–775 (2007)
- M. Masjed-Jamei, A basic class of symmetric orthogonal functions using the extended Sturm–Liouville theorem for symmetric functions. J. Comput. Appl. Math. **216**, 128–143 (2008)
- M. Masjed-Jamei, Biorthogonal exponential sequences with weight function $\exp(ax^2 + ibx)$ on the real line and an orthogonal sequence of trigonometric functions. Proc. Am. Math. Soc. **136**, 409–417 (2008)
- M. Masjed-Jamei, W. Koepf, Two classes of special functions using Fourier transforms of generalized ultraspherical and generalized Hermite polynomials. Proc. Am. Math. Soc. **140**, 2053–2063 (2012)
- M. Masjed-Jamei, F. Marcellan, E.J. Huertas, A finite class of orthogonal functions generated by Routh–Romanovski polynomials. Complex Var. Elliptic Equ. **59**(2), 162–171 (2014)
- A.F. Nikiforov, V.B. Uvarov, *Special Functions of Mathematical Physics* (Birkhäuser, Berlin, 1988)
- W.H. Press, B.P. Flannery, S.A. Teukolsky, W.T. Vetterling, Beta Function, T Student distribution and F-Distribution, in *Numerical Recipes in Fortran: The Art of Scientific Computing*, Section 6.2, 2nd edn. (Cambridge University Press, Cambridge, 1992), pp. 219–223
- S. Ramanujan, A class of definite integrals. Quarterly J. Math, **48**, 294–310 (1920)
- T. Rivlin, *Chebyshev Polynomials: From Approximation Theory to Algebra and Number Theory*, 2nd edn. (Wiley, New York, 1990)
- E. Routh, On some properties of certain solutions of a differential equation of the second order. Proc. Lond. Math. Soc. **16**, 245–261 (1884)
- J. Shohat, A differential equation for orthogonal polynomials. Duke Math. J. **5**, 401–417 (1939)

- Y. Sosov, C.E. Theodosiou, On the complete solution of the Sturm–Liouville problem $d^2X/dx^2 + \lambda^2 X = 0$ over a closed interval. J. Math. Phys. **43**, 2831–2843 (2002)
- C. Truesdell, *An Assay Toward a Unified Theory of Special Functions* (Princeton University Press, Princeton, 1948)
- E.W. Weisstein, Saalschütz's Theorem From MathWorld: A Wolfram Web Resource. http://mathworld.wolfram.com/SaalschuetzsTheorem.html
- E.T. Whittaker, G.N. Watson, *A Course of Modern Analysis*, 4th edn. (Cambridge University Press, Cambridge, 1927)

Special Functions Generated by Generalized Sturm–Liouville Problems in Discrete Spaces

2.1 Introduction

Orthogonal functions of a discrete variable may be solutions of a discrete-type Sturm–Liouville problem of the form

$$\Delta\Big(k^*(x)\nabla y_n(x)\Big) + \Big(\lambda_n w(x) - q^*(x)\Big)y_n(x) = 0 \quad \text{for} \quad k^*(x) > 0 \quad \text{and} \quad w(x) > 0,$$

$$(2.1)$$

where

$$\Delta f(x) = \nabla f(x+1) = f(x+1) - f(x),$$

and Eq. (2.1) satisfies a set of discrete boundary conditions as

$$\alpha_1 y(a) + \beta_1 \nabla y(a) = 0,$$

$$\alpha_2 y(b) + \beta_2 \nabla y(b) = 0.$$

This means that if $y_n(x)$ and $y_m(x)$ are two eigenfunctions of Eq. (2.1), then they are orthogonal with respect to the weight function $w(x)$ on a discrete set of points.

Similar to the continuous case, many special functions of a discrete variable are solutions of a regular or singular Sturm–Liouville problem of type (2.1).

In this chapter, we extend discrete-type Sturm–Liouville problems to symmetric functions whose corresponding solutions preserve the orthogonality property.

Let us start with the second-order difference equation of hypergeometric type

$$\sigma(x)\Delta\nabla y(x) + \tau(x)\Delta y(x) + \lambda y(x) = 0, \qquad (2.2)$$

© Springer Nature Switzerland AG 2020
M. Masjed-Jamei, *Special Functions and Generalized Sturm–Liouville Problems*,
Frontiers in Mathematics, https://doi.org/10.1007/978-3-030-32820-7_2

where

$$\sigma(x) = ax^2 + bx + c \quad \text{and} \quad \tau(x) = px + q \quad \text{with} \quad p \neq 0$$

are polynomials of degree at most 2 and 1, respectively, and λ is a constant.
   Then Eq. (2.2) can be written as

$$\Delta\Big(\sigma(x)w(x)\nabla y(x)\Big) + \lambda w(x)y(x) = 0,$$

in which $w(x)$ satisfies the Pearson-type difference equation

$$\Delta\Big(\sigma(x)w(x)\Big) = \tau(x)w(x). \tag{2.3}$$

The solutions of Eq. (2.2) with

$$\lambda \equiv \lambda_n = -n\big((n-1)a + d\big)$$

are polynomials of degree n, say $y(x) = y_n(x)$, and usually called hypergeometric-type discrete polynomials. They are orthogonal with respect to the weight function $w(x)$ on the counter set $x = A, \ A+1, \ \ldots, \ B$ as

$$\sum_{x=A}^{B} y_n(x)y_m(x)w(x) = \left(\sum_{x=A}^{B} y_n^2(x)w(x)\right)\delta_{n,m},$$

provided that $w(x) > 0$ for $A \leq x \leq B$ and

$$\sigma(x)w(x)x^k\Big|_{x=A,B+1} = 0, \ \forall k \geq 0.$$

For $y(x) = y_n(x)$, Eqs. (2.2) and (2.3) can be represented as

$$\sigma_1(x)(\Delta^2 y_n)(x) + \tau_1(x)(\Delta y_n)(x) + \lambda_n \, y_n(x+1) = 0$$

and

$$\Delta\Big(\sigma_1(x-1)w(x)\Big) = \tau_1(x)w(x+1),$$

where

$$\sigma_1(x) = \sigma(x+1) + \tau(x+1) \quad \text{and} \quad \tau_1(x) = \tau(x+1),$$

or as

$$\sigma_2(x)(\nabla^2 y_n)(x) + \tau_2(x)(\nabla y_n)(x) + \lambda_n \, y_n(x-1) = 0$$

and

$$\Delta\Big(\sigma_2(x+1)w(x)\Big) = \tau_2(x+1)w(x),$$

where

$$\sigma_2(x) = \sigma(x-1) \quad \text{and} \quad \tau_2(x) = \tau(x-1).$$

According to Hahn's classification, Eq. (2.2) has four orthogonal polynomial solutions, which are known in the literature, respectively, as follows:

- The CHARLIER polynomials:

$$C_n(x; a) = {}_2F_0\left(\begin{array}{c} -n, -x \\ - \end{array}\middle|\, -\frac{1}{a}\right),$$

  orthogonal with respect to the Poisson weight function as

$$\sum_{k=0}^{\infty} \frac{a^k}{k!} C_n(k; a)C_m(k; a) = a^{-n}e^a n! \, \delta_{n,m} \quad (a > 0).$$

- The MEIXNER polynomials:

$$M_n(x; \beta, c) = {}_2F_1\left(\begin{array}{c} -n, -x \\ \beta \end{array}\middle|\, 1 - \frac{1}{c}\right),$$

  orthogonal with respect to the Pascal weight function as

$$\sum_{k=0}^{\infty} \frac{(\beta)_k c^k}{k!} M_n(k; \beta, c)M_m(k; \beta, c) = \frac{c^{-n}n!}{(\beta)_n(1-c)^{\beta}} \delta_{n,m} \quad (\beta > 0, \, 0 < c < 1).$$

- The KRAVCHUK polynomials:

$$K_n(x; p, N) = {}_2F_1\left(\begin{array}{c} -n, -x \\ -N \end{array}\middle|\, \frac{1}{p}\right), \quad n = 0, 1, 2, \ldots, N,$$

orthogonal with respect to the binomial weight function as

$$\sum_{k=0}^{N} \binom{N}{k} p^k (1-p)^{N-k} K_n(k; p, N) K_m(k; p, N) = \frac{(-1)^n n!}{(-N)_n} \left(\frac{1-p}{p}\right)^n \delta_{n,m} \quad (0 < p < 1),$$

- the HAHN polynomials:

$$Q_n(x; \alpha, \beta, N) = {}_3F_2 \left(\begin{array}{c} -n, -x, n+\alpha+\beta+1 \\ -N, \alpha+1 \end{array} \middle| 1 \right), \quad n = 0, 1, 2, \ldots, N,$$

orthogonal with respect to the hypergeometric weight function as

$$\sum_{k=0}^{N} \binom{\alpha+k}{k} \binom{\beta+N-k}{N-k} Q_n(k; \alpha, \beta, N) Q_m(k; \alpha, \beta, N)$$

$$= \frac{(-1)^n (n+\alpha+\beta+1)_{N+1} (\beta+1)_n n!}{(2n+\alpha+\beta+1)(\alpha+1)_n (-N)_n N!} \delta_{n,m} \quad (\alpha > -1, \beta > -1).$$

Table 2.1 shows the data $\sigma(x)$ and $\tau(x)$ for each of the above four polynomial families. By referring to this table, we observe that all polynomial coefficients σ and τ have real zeros. However, there is still the main case of the real difference equation (2.2), whose coefficients have complex zeros with four free parameters. To obtain this case, we first expand the Pearson equation (2.3) as follows:

$$\frac{w(x+1)}{w(x)} = \frac{\sigma(x)+\tau(x)}{\sigma(x+1)} = \frac{\sigma_1(x-1)}{\sigma_1(x)-\tau_1(x)}. \tag{2.4}$$

It is clear that if in (2.4) $\sigma(x)$ has degree < 2, it leads to the Charlier, Meixner, and Kravchuk polynomials, while if $\sigma(x)$ has the exact degree 2, the Hahn polynomials are derived if both the numerator $\sigma(x)+\tau(x)$ and the denominator $\sigma(x+1)$ have real factorizations.

Another case in (2.4) is that in which the polynomials $\sigma(x)+\tau(x)$ and $\sigma(x+1)$ are real and have complex zeros. In other words, consider Eq. (2.4) in the expanded form

$$\frac{w(x+1)}{w(x)} = \frac{ax^2+(b+d)x+c+e}{ax^2+(2a+b)x+a+b+c}. \tag{2.5}$$

Table 2.1 Four cases of classical discrete orthogonal polynomials

Symbol	$Q_n(x; \alpha, \beta, N)$	$M_n(x; \beta, c)$	$K_n(x; p, N)$	$C_n(x; a)$
$\sigma(x)$	$x(N+\alpha-x)$	x	x	x
$\tau(x)$	$(\beta+1)(N-1)-(\alpha+\beta+2)x$	$(c-1)x+\beta c$	$\dfrac{Np-x}{1-p}$	$a-x$

Then only two cases can occur for the parameter a in (2.5), i.e., $a = 0$ or $a \neq 0$. If $a = 0$, the result is well known. So by assuming $a = 1 \neq 0$, we will obtain the simplified equation

$$\frac{w(x+1)}{w(x)} = \frac{x^2 + (b+d)x + c + e}{x^2 + (b+2)x + b + c + 1}. \tag{2.6}$$

Now, four cases can occur for Eq. (2.6):

1. Both numerator and denominator have real zeros, namely

$$\frac{w(x+1)}{w(x)} = \frac{(x+p)(x+q)}{(x+r)(x+s)} \quad (p, q, r, s \in \mathbb{R}). \tag{2.7}$$

2. The numerator has real zeros but the denominator has complex roots, namely

$$\frac{w(x+1)}{w(x)} = \frac{(x+p)(x+q)}{(x+r+is)(x+r-is)} \quad (p, q, r, s \in \mathbb{R}). \tag{2.8}$$

3. The numerator has complex zeros but the denominator has real roots, namely

$$\frac{w(x+1)}{w(x)} = \frac{(x+p+iq)(x+p-iq)}{(x+r)(x+s)} \quad (p, q, r, s \in \mathbb{R}). \tag{2.9}$$

4. Finally, both numerator and denominator have complex zeros, namely

$$\frac{w(x+1)}{w(x)} = \frac{(x+p+iq)(x+p-iq)}{(x+r+is)(x+r-is)} \quad (p, q, r, s \in \mathbb{R}). \tag{2.10}$$

Here we should remark that the solutions of Eqs. (2.7)–(2.10) are expressible in terms of the gamma function

$$\Gamma(z) = \lim_{n \to \infty} \frac{n! \, n^z}{\prod\limits_{k=0}^{n} (z+k)}, \tag{2.11}$$

which implies that

$$\Gamma(p+iq)\,\Gamma(p-iq) = \Gamma^2(p) \prod_{k=0}^{\infty} \frac{(p+k)^2}{(p+k)^2 + q^2} \tag{2.12}$$

is always a real positive value for all $p > 0$ and $q \in \mathbb{R}$. One of the consequences of (2.12) is that the multiplicative term

$$(p + iq)_n \, (p - iq)_n = \prod_{k=0}^{n-1} \left(q^2 + (p + k)^2 \right) \qquad (p, q \in \mathbb{R})$$

is also a real positive value. Moreover, when $p = m \in \mathbb{N}$, we have

$$\Gamma(m + iq) \, \Gamma(m - iq) = \frac{\pi q}{\sinh(\pi q)} \prod_{k=1}^{m-1} \left(q^2 + (m - k)^2 \right).$$

By noting the above remark, the solutions of Eqs. (2.7)–(2.10) can be, respectively, represented as

$$w(x) = w_1(x; p, q, r, s) = \frac{\Gamma(x + p)\Gamma(x + q)}{\Gamma(x + r)\Gamma(x + s)}, \tag{2.13}$$

$$w(x) = w_2(x; p, q, r, s) = \frac{\Gamma(x + p)\Gamma(x + q)}{\Gamma(x + r + is)\Gamma(x + r - is)}, \tag{2.14}$$

$$w(x) = w_3(x; p, q, r, s) = \frac{\Gamma(x + p + iq)\Gamma(x + p - iq)}{\Gamma(x + r)\Gamma(x + s)}, \tag{2.15}$$

and

$$w(x) = w_4(x; p, q, r, s) = \frac{\Gamma(x + p + iq)\Gamma(x + p - iq)}{\Gamma(x + r + is)\Gamma(x + r - is)}. \tag{2.16}$$

On the other hand, since a weight function must be positive, relation (2.13) implies that $p = q$ and $r = s$, leading to

$$w_1(x; p, p, r, r) = \frac{\Gamma^2(x + p)}{\Gamma^2(x + r)} > 0. \tag{2.17}$$

Also, by referring to the important relation (2.12), the function (2.14) is positive when $p = q$, i.e.,

$$w_2(x; p, p, r, s) = \frac{\Gamma^2(x + p)}{\Gamma(x + r + is)\Gamma(x + r - is)} > 0. \tag{2.18}$$

Similarly, for the case (2.15), we must have $r = s$, i.e.,

$$w_3(x; p, q, r, r) = \frac{\Gamma(x + p + iq)\Gamma(x + p - iq)}{\Gamma^2(x + r)} > 0. \tag{2.19}$$

Finally, for the case (2.16), we observe that no restriction is required and $w_4(x; p, q, r, s)$ is always positive. A short look at relations (2.17)–(2.19) shows that they are just particular cases of the positive weight function (2.16). This means that we are dealing with a main sequence of hypergeometric orthogonal polynomials of a discrete variable with four free parameters that is finitely orthogonal on the real line. In the next section, we introduce such a sequence and consider three particular cases of it corresponding to the weight functions (2.17)–(2.19).

2.2   A Finite Sequence of Hahn-Type Orthogonal Polynomials

For $p, q, r, s \in \mathbb{R}$, suppose that

$$\sigma(x) + \tau(x) = (x + p)^2 + q^2$$

and

$$\sigma(x + 1) = (x + r)^2 + s^2,$$

which leads to the real equation

$$\left((x + p + 1)^2 + q^2\right)(\Delta^2 y_n)(x) + \left(2(p + 1 - r)x + (p + 1)^2 + q^2 - r^2 - s^2\right)(\Delta y_n)(x)$$
$$- n(n + 1 + 2p - 2r)y_n(x + 1) = 0, \qquad (2.20)$$

or

$$\left((x + r - 2)^2 + s^2\right)(\nabla^2 y_n)(x)$$
$$+ \left(2(p + 1 - r)x + p^2 + q^2 - r^2 - s^2 - 2p + 4r - 3\right)(\nabla y_n)(x)$$
$$- n(n + 1 + 2p - 2r)y_n(x - 1) = 0. \qquad (2.21)$$

We look for a polynomial solution of Eq. (2.20) of the form

$$y_n(x) = \sum_{k=0}^{n} a_{n,k}\binom{x - 1 + r + is}{k}, \quad a_{n,n} \neq 0,$$

which is a particular case of the general solution

$$y_n(x) = \sum_{k=0}^{\infty} a_{n,k}\binom{x + c}{k}.$$

Since

$$\binom{x+c}{k} = \frac{(-1)^k(-x-c)_k}{k!} = \frac{1}{k!}(x+c)^{[k]}$$

for

$$x^{[k]} = x(x-1)(x-2)\cdots(x-k+1)$$

and we have the relations

$$\Delta(x+c)^{[n]} = n(x+c)^{[n-1]} \quad \text{and} \quad x(x+c)^{[n]} = (x+c)^{[n+1]} + (n-c)(x+c)^{[n]}, \quad (2.22)$$

we can directly prove the following theorem.

Theorem 2.1 *The monic polynomial solution of the difference equation* (2.20) *is*

$$y_n(x) = \bar{R}_n(x; p, q, r, s) = \frac{(-r+p+1-i(q+s))_n(-r+p+1+i(q-s))_n}{(n+2p-2r+1)_n}$$

$$(2.23)$$

$$\times {}_3F_2\left(\begin{array}{c} -n, n+1+2p-2r, -x+1-r-is \\ -r+p+1-i(q+s), -r+p+1+i(q-s) \end{array}\middle|\, 1\right)$$

$$= (-1)^n \frac{(-r+p+1+i(q+s))_n(-r+p+1+i(q-s))_n}{(n+2p-2r+1)_n}$$

$$\times {}_3F_2\left(\begin{array}{c} -n, n+1+2p-2r, x+p+iq \\ -r+p+1+i(q+s), -r+p+1+i(q-s) \end{array}\middle|\, 1\right).$$

Proof For $c = r + is - 1$, if in (2.20) we take

$$y_n(x) = \bar{R}_n(x; p, q, r, s) = \sum_{k=0}^{\infty} a_{n,k}(x+c)^{[k]}$$

and use properties (2.22), we eventually obtain the recurrence equation

$$(k-n)(k+n+1+2p-2r)a_{n,k} + (k+1)\left(-s^2 + 2i(p+k-r+1)s + p^2\right.$$

$$+ (4k-2n-2r+4)p + 2k^2 - 4kr - n^2 + 2nr + q^2 + r^2 + 5k - n - 4r + 3\Big)a_{n,k+1}$$

$$- (k+1)(k+2)(iq+is+k+p-r+2)(iq-is-k-p+r-2)a_{n,k+2} = 0,$$

which has a solution of the hypergeometric form (2.23).

The second representation of $\bar{R}_n$ comes from (2.21) following the same approach and using the basis $(-x+c)^{[k]}$ for $c = -p - iq$ with the property

$$\nabla(-x+c)^{[k]} = -k(-x+c)^{[k-1]}.$$

We can show that both representations of $\bar{R}_n$ are solutions of the following three-term recurrence relation:

$$\bar{R}_{n+1}(x; p, q, r, s) = (x - c_n)\bar{R}_n(x; p, q, r, s) - d_n \bar{R}_{n-1}(x; p, q, r, s),$$

where

$$c_n = -\frac{(p+r-1)n^2 + (p+r-1)(1+2p-2r)n + (-r+p)(p^2+q^2-r^2-s^2+2r-1)}{2(n+p-r)(n+1+p-r)}$$

and

$$d_n = \frac{n(2r-2p-n)\left((n+p-r)^2+(q-s)^2\right)\left((n+p-r)^2+(q+s)^2\right)}{4(n+p-r)^2(2n+1+2p-2r)(2n-1+2p-2r)},$$

(2.24)

with the unique initial conditions

$$\bar{R}_0(x; p, q, r, s) = 1 \text{ and } \bar{R}_1(x; p, q, r, s) = x + \frac{p^2+q^2-r^2-s^2+2r-1}{2(p+1-r)}. \quad \square$$

There is a direct relationship between $\bar{R}_n(x; p, q, r, s)$ and continuous Hahn polynomials defined by

$$P_n(x; a, b, c, d) = \frac{i^n(a+c)_n(a+d)_n}{n!} {}_3F_2\left(\begin{array}{c} -n, n+a+b+c+d-1, a+ix \\ a+c, a+d \end{array}\middle| 1\right),$$

so that we have

$$\bar{R}_n(x; p, q, r, s) = \bar{P}_n(ix; 1-r-is, 1-r+is, p-iq, p+iq).$$

We now prove that the monic polynomials (2.23) are finitely orthogonal on the real line. For this purpose, we first reconsider Eq. (2.20) and write it in a self-adjoint form to obtain

$$\Big[w_4(x; p, q, r, s)(x+p+iq)(x+p-iq)$$

$$\times \left(\bar{R}_m(x; p, q, r, s)\bar{R}_n(x+1; p, q, r, s) - \bar{R}_n(x; p, q, r, s)\bar{R}_m(x+1; p, q, r, s)\right)\Big]_{-\infty}^{\infty}$$

$$= (n(n+1+2p-2r) - m(m+1+2p-2r))$$

$$\times \sum_{x=-\infty}^{\infty} w_4(x; p, q, r, s)\bar{R}_m(x; p, q, r, s)\bar{R}_n(x; p, q, r, s). \quad (2.25)$$

In order to show that the left-hand side of (2.25) is equal to zero when $m \neq n$, we use the following limit relations:

$$\lim_{x \to \infty} \frac{\Gamma(x+a)}{\Gamma(x)x^a} = 1 \quad \text{and} \quad \lim_{x \to -\infty} \frac{\Gamma(x+a)}{\Gamma(x)x^a} = (-1)^a \quad (\forall a \in \mathbb{C}), \tag{2.26}$$

which can be proved directly via the limit definition (2.11). Since

$$\deg\left(\bar{R}_m(x; p,q,r,s)\bar{R}_n(x+1; p,q,r,s) - \bar{R}_n(x; p,q,r,s)\bar{R}_m(x+1; p,q,r,s)\right) = n+m-1,$$

the left-hand side of (2.25) is equal to zero for $n \neq m$ if and only if

$$\lim_{x \to \pm\infty} \frac{\Gamma(x+p+iq)\Gamma(x+p-iq)}{\Gamma(x+r+is)\Gamma(x+r-is)} x^{k+2} = 0 \text{ for all } k = 0, 1, \ldots, n+m-1. \tag{2.27}$$

By noting (2.26), it is now straightforward to verify that (2.27) is equivalent to

$$\lim_{x \to \infty} x^{k+2-2r+2p} = 0 \text{ and } \lim_{x \to -\infty} (-x)^{k+2-2r+2p} = 0 \text{ for any } k = 0, 1, \ldots, n+m-1. \tag{2.28}$$

Finally, if in (2.28) we take $\max\{m, n\} = N$, the left-hand side of (2.25) will be equal to zero if and only if

$$2N + 1 - 2r + 2p < 0 \quad \Leftrightarrow \quad N < r - p - \frac{1}{2}. \tag{2.29}$$

An interesting point is that if the condition (2.29) is satisfied, the coefficient d_n in (2.24) is automatically positive, and Favard's theorem can therefore be applied to conclude that the monic polynomial family $\{\bar{R}_n(x; p,q,r,s)\}$ is orthogonal with respect to the weight function $w_4(x; p,q,r,s)$.

By noting these comments, it now remains to compute the norm square value

$$\sum_{x=-\infty}^{\infty} w_4(x; p,q,r,s)\bar{R}_n^2(x; p,q,r,s) = \left(\sum_{x=-\infty}^{\infty} w_4(x; p,q,r,s)\right)\prod_{k=1}^{n} d_k.$$

From (2.24), first we have

$$\prod_{k=1}^{n} d_k = \left\{ n!\,(2r-2p-n)_n\,(p-r+1+i(q-s))_n\,(p-r+1-i(q-s))_n \right.$$
$$(p-r+1+i(q+s))_n\,(p-r+1-i(q+s))_n \Big\}$$
$$\left. \Big/ \Big\{ 2^{4n}\,(p-r+1)_n^2\,(p-r+3/2)_n\,(p-r+1/2)_n \Big\}. \right.$$

On the other hand, using Dougall's bilateral sum

$$\sum_{n=-\infty}^{\infty} \frac{\Gamma(a+n)\Gamma(b+n)}{\Gamma(c+n)\Gamma(d+n)} = \frac{\Gamma(a)\Gamma(1-a)\Gamma(b)\Gamma(1-b)\Gamma(c+d-a-b-1)}{\Gamma(c-a)\Gamma(c-b)\Gamma(d-a)\Gamma(d-b)}, \quad a, b \notin \mathbb{Z},$$

the moment of order zero can be computed as

$$\sum_{x=-\infty}^{\infty} w_4(x; p, q, r, s) = \sum_{x=-\infty}^{\infty} \frac{\Gamma(x+p+iq)\Gamma(x+p-iq)}{\Gamma(x+r+is)\Gamma(x+r-is)}$$

$$= \frac{\Gamma(p+iq)\Gamma(p-iq)\Gamma(1-p+iq)\Gamma(1-p-iq)\Gamma(2r-2p-1)}{\Gamma(r-p+i(s-q))\Gamma(r-p-i(s-q))\Gamma(r-p-i(s+q))\Gamma(r-p+i(s+q))} > 0.$$

Theorem 2.2   *The polynomial set $\{\bar{R}_n(x; p, q, r, s)\}_{n=0}^{N<r-p-1/2}$ is finitely orthogonal with respect to the weight function $w_4(x; p, q, r, s)$ on the real line, so that we have*

$$\sum_{x=-\infty}^{\infty} \frac{\Gamma(x+p+iq)\Gamma(x+p-iq)}{\Gamma(x+r+is)\Gamma(x+r-is)} \bar{R}_n(x; p, q, r, s) \bar{R}_m(x; p, q, r, s)$$

$$= \left\{ n!\, (2r-2p-n)_n\, (p-r+1+i(q-s))_n\, (p-r+1-i(q-s))_n\, (p-r+1+i(q+s))_n \right.$$

$$\left. (p-r+1-i(q+s))_n\, \Gamma(p+iq)\Gamma(p-iq)\Gamma(1-p+iq)\Gamma(1-p-iq)\Gamma(2r-2p-1) \right\} \delta_{m,n}$$

$$\Big/ \left\{ 2^{4n}\, (p-r+1)_n^2\, (p-r+3/2)_n\, (p-r+1/2)_n\, \Gamma(r-p+i(s-q))\Gamma(r-p-i(s-q)) \right.$$

$$\left. \Gamma(r-p-i(s+q))\Gamma(r-p+i(s+q)) \right\}.$$

As we said, there are three particular cases of the weight function $w_4(x; p, q, r, s)$ in relations (2.17)–(2.19). Hence, three corollaries of the main Theorem 2.2 can be deduced as follows.

Corollary 2.1   *The polynomial set $\{\bar{R}_n(x; p, 0, r, 0)\}_{n=0}^{N<r-p-1/2}$ is finitely orthogonal with respect to the weight function $w_1(x; p, p, r, r)$ on the real line, so that we have*

$$\sum_{x=-\infty}^{\infty} \frac{\Gamma^2(x+p)}{\Gamma^2(x+r)} \bar{R}_n(x; p, 0, r, 0) \bar{R}_m(x; p, 0, r, 0)$$

$$= \left\{ n!\, (2r-2p-n)_n\, (p-r+1)_n^2\, \Gamma^2(p)\Gamma^2(1-p)\Gamma(2r-2p-1) \right\} \delta_{m,n}$$

$$\Big/ \left\{ 2^{4n}\, (p-r+3/2)_n\, (p-r+1/2)_n\, \Gamma^4(r-p) \right\}.$$

Corollary 2.2 *The polynomial set* $\{\bar{R}_n(x; p, 0, r, s)\}_{n=0}^{N<r-p-1/2}$ *is finitely orthogonal with respect to the weight function* $w_2(x; p, p, r, s)$ *on the real line, so that we have*

$$\sum_{x=-\infty}^{\infty} \frac{\Gamma^2(x+p)}{\Gamma(x+r+is)\Gamma(x+r-is)} \bar{R}_n(x; p, 0, r, s)\bar{R}_m(x; p, 0, r, s)$$

$$= \Big\{ n!\,(2r-2p-n)_n\,(p-r+1-is)_n\,(p-r+1+is)_n\,(p-r+1+is)_n$$

$$(p-r+1-is)_n\,\Gamma^2(p)\Gamma^2(1-p)\Gamma(2r-2p-1) \Big\}\,\delta_{m,n}$$

$$\Big/ \Big\{ 2^{4n}\,(p-r+1)_n^2\,(p-r+3/2)_n\,(p-r+1/2)_n\,\Gamma(r-p+is)\Gamma(r-p-is)$$

$$\Gamma(r-p-is)\Gamma(r-p+is) \Big\}.$$

Corollary 2.3 *The polynomial set* $\{\bar{R}_n(x; p, q, r, 0)\}_{n=0}^{N<r-p-1/2}$ *is finitely orthogonal with respect to the weight function* $w_3(x; p, q, r, r)$ *on the real line, so that we have*

$$\sum_{x=-\infty}^{\infty} \frac{\Gamma(x+p+iq)\Gamma(x+p-iq)}{\Gamma^2(x+r)} \bar{R}_n(x; p, q, r, 0)\bar{R}_m(x; p, q, r, 0)$$

$$= \Big\{ n!\,(2r-2p-n)_n\,(p-r+1+iq)_n\,(p-r+1-iq)_n\,(p-r+1+iq)_n$$

$$(p-r+1-iq)_n\,\Gamma(p+iq)\Gamma(p-iq)\Gamma(1-p+iq)\Gamma(1-p-iq)\Gamma(2r-2p-1) \Big\}\,\delta_{m,n}$$

$$\Big/ \Big\{ 2^{4n}\,(p-r+1)_n^2\,(p-r+3/2)_n\,(p-r+1/2)_n\,\Gamma(r-p-iq)\Gamma(r-p+iq)$$

$$\Gamma(r-p-iq)\Gamma(r-p+iq) \Big\}.$$

2.3 Classical Symmetric Orthogonal Polynomials of a Discrete Variable

Let us reconsider the second-order difference equation of hypergeometric type

$$\sigma(x)\Delta\nabla y(x) + \tau(x)\Delta y(x) + \lambda_n y(x) = 0, \tag{2.30}$$

in which

$$\sigma(x) = ax^2 + bx + c,$$

$$\tau(x) = px + q,$$

and

$$\lambda_n = -n\left(p + (n-1)a\right),$$

for $a, b, c, p, q \in \mathbb{R}$.

In this section, we classify symmetric polynomial solutions of the above equation that are orthogonal with respect to a symmetric weight function $\varrho(x)$ supported on a finite set $\{x_k\} \subset \mathbb{R}$ satisfying $x_{k+1} = x_k + 1$ and $\varrho(x_k) > 0$.

Assuming that the polynomial solutions of Eq. (2.30) satisfy a three-term recurrence relation

$$P_{n+1}(x) = (x - B_n)P_n(x) - C_n P_{n-1}(x) \qquad (n \geq 0), \tag{2.31}$$

with

$$P_{-1}(x) = 0, \qquad P_0(x) = 1,$$

then we get [1, 2]

$$-B_n = \frac{q(p - 2a) + n(p + 2b)(p + (n-1)a)}{(p + 2(n-1)a)(p + 2na)}$$

and

$$-C_n = \frac{n(p + (n-2)a)}{(p + (2n-3)a)(p + (2n-1)a)}\left(\sigma\left(\eta_{n-1}\right) + \tau\left(\eta_{n-1}\right)\right),$$

where

$$\eta_n = -\frac{q + nb - n^2 a}{p + 2na}.$$

Note in (2.30) that $\lambda_n \neq \lambda_m$ for fixed $n \in \mathbb{N}$ and $m = 0, 1, 2, \ldots, n - 1$ in order to have a unique polynomial solution of degree exactly n in x, up to a multiplicative constant. Taking into account this property, we conclude that $p \neq 0$.

For symmetric polynomial $(P_n(x) = (-1)^n P_n(-x))$ satisfying (2.31) we clearly have $B_n = 0$ for all $n = 0, 1, 2, \ldots$. Since $B_0 = -q/p$, with $p \neq 0$, it follows that $q = 0$. In this situation, if $p = -2a$, then the trivial polynomial family solution of Eq. (2.30) is $\{1, x, x^2 - c/b\}$. Therefore, we assume that $p \neq -2a$.

Moreover, for $q = 0$, we have $B_1 = -(2b + p)/(2a + p)$. Hence $p = -2b$ gives $B_n = 0$ for all $n = 0, 1, 2, \ldots$.

Consequently, we have $q = 0$, $p = -2b$, $p \neq 0$, and $p \neq -2a$. In this case,

$$
\begin{aligned}
\sigma(x) &= ax^2 + bx + c, \\
\tau(x) &= -2bx, \\
\lambda_n &= n\,(2\,b + a\,(1-n)), \\
B_n &= 0,
\end{aligned}
$$
(2.32)

and

$$
C_n = \frac{n\,(2b + a(2-n))\,(a + 2b + 4c - 2\,(a+b)\,n + a\,n^2)}{4\,(2b + a\,(3-2n))\,(2b + a(1-2n))}.
$$
(2.33)

If Eq. (2.30) is written in the expanded form

$$
\mathcal{T}[y] := \Big(\sigma(x) + \tau(x)\Big)y(x+1) - \Big(2\sigma(x) + \tau(x)\Big)y(x) + \sigma(x)y(x-1) = -\lambda y(x),
$$
(2.34)

the symmetry requirement can be restated as that whenever $y(x)$ is an eigenfunction of $\mathcal{T}$, so is $y(-x)$, which implies that

$$
\sigma(x) + \tau(x) = \sigma(-x), \quad \text{or} \quad \tau(x) = \sigma(-x) - \sigma(x).
$$

For the symmetric polynomial solutions of Eq. (2.30) to be orthogonal on $[A, B-1]$ with weight $\varrho(x)$, i.e.,

$$
\sum_{x_i=A}^{B-1} P_m(x_i)\, P_n(x_i)\, \varrho(x_i) = d_n^2\, \delta_{m,n} \qquad (x_{i+1} = x_i + 1),
$$

where $\varrho(x_i) > 0$ for $A \le x_i \le B-1$, the condition $C_n > 0$ for all $n \ge 1$ must hold. Since $C_1 = c/(2b-a)$, if $a \neq 2b$, then c cannot be zero. Finally, if $a = 2b$ (with $q = 0$, $p = -2b$), then $\lambda_n = 2b(2-n)n$, giving rise to the orthogonal polynomial family $\{1, x\}$.

Summarizing, for nontrivial symmetric orthogonal polynomial families as solutions of Eq. (2.30), the following conditions have to be considered: the coefficients of the difference equation (2.30) and three–term recurrence relation (2.31) are given in (2.32) and (2.33), respectively, with $b \neq 0$, $a \neq b$, $a \neq 2b$, and $c \neq 0$. Under these conditions, we will be able to obtain the factorial moments

$$
\sum_{k=A}^{B-1} (k-A)^{[n]}\varrho(k) \qquad n \in (\mathbb{N}),
$$

where $x^{[n]} = x(x-1)\cdots(x-n+1)$ and $x^{[0]} = 1$.

2.3.1   Classification

Having in mind the aforementioned conditions, the recurrence relation (2.31) is simplified as

$$P_{n+1}(x) = x P_n(x) - C_n P_{n-1}(x) \quad (n \geq 1),\tag{2.35}$$

with

$$P_0(x) = 1, \quad P_1(x) = x,$$

where

$$C_n = \frac{n\,(2\,b + a(2-n))\,\big(a + 2\,b + 4\,c - 2\,(a+b)\,n + a\,n^2\big)}{4\,(2\,b + a\,(3 - 2\,n))\,(2\,b + a(1 - 2n))}.$$

And the Pearson equation

$$\Delta\Big(\sigma(x)\varrho(x)\Big) = \tau(x)\varrho(x)\tag{2.36}$$

is equivalent to

$$\frac{\varrho(x+1)}{\varrho(x)} = \frac{\sigma(x) + \tau(x)}{\sigma(x+1)} = \frac{\sigma(-x)}{\sigma(x+1)} = \frac{a x^2 - b x + c}{c + b(x+1) + a(x+1)^2}.\tag{2.37}$$

We are looking for explicit expressions for $\varrho(x)$ corresponding to different degrees of the polynomial $\sigma(x)$.

For infinite cases of hypergeometric discrete orthogonal polynomials, it is proved in [3] that there exists no solution for Eq. (2.36) with x ranging from $-\infty$ to $+\infty$, i.e., the discrete support of the measure is finite.

Let A be the lower bound of the support and let $B - 1$ be the upper bound. It is known that

$$\sigma(A) = 0 \quad \text{and} \quad \sigma(1 - B) = 0,\tag{2.38}$$

which can be proved in the following way: if we write the operator $\mathscr{T}$ defined in (2.34) in a self-adjoint form, then we get

$$\sigma(-x)\varrho(x) - \sigma(x+1)\varrho(x+1) = 0,$$

for all

$$x = A, A+1, \ldots, B-2,$$

as well as (2.38). This means that $B - 1 - A \in \mathbb{N}$.

Moreover, in order to have $\varrho(x_k) > 0$ for

$$x_k \in \{A, A+1, \ldots, B-2, B-1\},$$

it is necessary and sufficient that

$$\varrho(A) > 0 \quad \text{and} \quad \varrho(x+1)/\varrho(x) > 0 \quad \text{for} \quad x = A, A+1, \ldots, B-2.$$

First Case: $a = 0$

When $a = 0$ in (2.32), we can assume, without loss of generality, that $b = 1$ and therefore

$$\sigma(x) = x + c, \qquad \tau(x) = -2x, \quad \text{and} \quad \lambda_n = 2n \qquad (c \neq 0).$$

Hence relation (2.37) can be rewritten as

$$\frac{\varrho(x+1)}{\varrho(x)} = \frac{c-x}{x+c+1},$$

with the explicit solution

$$\varrho(x) = \frac{1}{\Gamma(c-x+1)\,\Gamma(c+x+1)},$$

for $x \in \{-c, -c+1, \ldots, c-1, c\}$ as support of orthogonality.

Moreover, the condition $2c \in \mathbb{N}$ implies $c \in \mathbb{N}$ or $c + 1/2 \in \mathbb{N}$.

The corresponding monic symmetric polynomial solutions of Eq. (2.30) is obtained from (2.35), with

$$C_n = \frac{(1+2c-n)\,n}{4} \qquad (1 \leq n \leq 2c),$$

or from its representation in terms of hypergeometric series as

$$P_n(x) = 2^{-n}\,(-2c)_n \; {}_2F_1\left(\begin{array}{c} -n, -c-x \\ -2c \end{array}\bigg| 2\right) = k_n^{(1/2)}(x+c; 2c) \qquad (0 \leq n \leq 2c),$$

where $k_n^{(p)}(x; N)$ are monic Kravchuk polynomials for $0 < p < 1$ and $N \in \mathbb{N}$.

In this sense, the factorial moments are computed as

$$\sum_{k=-c}^{c} \frac{(k+c)^{[n]}}{\Gamma(c-k+1)\,\Gamma(c+k+1)} = \frac{2^{2c-n}}{\Gamma(1+2c)}\,(2c)^{[n]} \qquad (0 \leq n \leq 2c).$$

Second Case: $a \neq 0$

In this case, we can assume, without loss of generality, that $a = 1$ in (2.32). As shown in (2.38), the roots of $\sigma(x)$ fix the lower bound of the support A. In order to have real roots for $\sigma(x)$, the condition $b^2 - 4c \geq 0$ is needed. Representation of the polynomials in terms of hypergeometric series must therefore be given in each particular case.

- If $b^2 - 4c = 0$ ($b \neq 0$), then we have

$$\sigma(x) = \left(x + \frac{b}{2}\right)^2, \quad \tau(x) = -2bx, \quad \text{and} \quad \lambda_n = n\,(1 + 2b - n)\,.$$

Hence (2.37) takes the form

$$\frac{\varrho(x+1)}{\varrho(x)} = \frac{(x - b/2)^2}{(x + b/2 + 1)^2},$$

with the symmetric solution

$$\varrho(x) = \frac{1}{\left(\Gamma\left(x + \frac{b}{2} + 1\right) \Gamma\left(-x + \frac{b}{2} + 1\right)\right)^2},$$

and the support of orthogonality is $\{-b/2, -b/2 + 1, \ldots, b/2 - 1, b/2\}$ where $b \in \mathbb{N}$.

Now, the symmetric polynomial solutions corresponding to Eq. (2.30) are obtained from (2.35), with

$$C_n = \frac{(1 + b - n)^2\,(2 + 2b - n)\,n}{4\,(1 + 2b - 2n)\,(3 + 2b - 2n)} \qquad (1 \leq n \leq b),$$

and they can be represented as

$$P_n(x) = \frac{((-b)_n)^2}{(n - 2b - 1)_n}\; {}_3F_2\left(\begin{array}{c} -n,\, n - 2b - 1,\, -\dfrac{b}{2} - x \\ -b,\, -b \end{array}\middle|\,1\right)$$

$$= \tilde{h}_n^{(0,0)}(x + b/2;\, b + 1) \qquad (0 \leq n \leq b)\,,$$

where $\tilde{h}_n^{(\mu,\nu)}(x;\, N)$ are monic Hahn–Eberlein polynomials [1] for $\mu > -1$, $\nu > -1$, and $N \in \mathbb{N}$, generally defined by

$$\tilde{h}_n^{(\mu,\nu)}(x;\, N) = \frac{(-1)^n\,(N - n)_n\,(1 - N - \nu)_n}{(n + 1 - 2N - \mu - \nu)_n}$$

$$\times\; {}_3F_2\left(\begin{array}{c} -n,\, -x,\, n - 2N - \mu - \nu + 1 \\ 1 - N - \nu,\, 1 - N \end{array}\middle|\,1\right) \qquad (0 \leq n \leq N - 1)\,. \qquad (2.39)$$

In this sense, the factorial moments are computed as

$$
\sum_{k=-b/2}^{b/2} \frac{(k+b/2)^{[n]}}{\left(\Gamma\,(k+b/2+1)\;\Gamma\,(-k+b/2+1) \right)^2}
$$

$$
= \frac{2^{2b-n}\;\Gamma(b+1/2)\;b^{[n]}}{\sqrt{\pi}\,(\Gamma(b+1))^3} \frac{(b-E((n+1)/2))^{[E(n/2)]}}{(b-1/2)^{[E(n/2)]}} \qquad (0 \le n \le b),
$$

where $E(x)$ denotes the largest integer less than or equal to x.

- If $b^2 - 4c > 0$ $(b \ne 0)$, it is possible to write

$$
\sigma(x) = x^2 + bx + c = (x - \delta_1)(x - \delta_2),
$$

where

$$
\delta_1 = -\frac{b+d}{2}, \qquad \delta_2 = \frac{d-b}{2},
$$

and $d = \sqrt{b^2 - 4c} > 0$. Thus $\delta_1 < \delta_2$.

Moreover, we have

$$
\tau(x) = 2\,(\delta_1 + \delta_2)\,x,
$$

$$
\lambda_n = -n\,(n + 2(\delta_1 + \delta_2) - 1),
$$

and

$$
\frac{\varrho(x+1)}{\varrho(x)} = \frac{(x + \delta_1)(x + \delta_2)}{(x - \delta_1 + 1)(x - \delta_2 + 1)},
$$

as well as

$$
C_n = \frac{-(n\,(-1 + 2\,\delta_1 + n)\,(-1 + 2\,\delta_2 + n)\,(2\,(-1 + \delta_1 + \delta_2) + n))}{4\,(-3 + 2\,\delta_1 + 2\,\delta_2 + 2\,n)\,(-1 + 2\,\delta_1 + 2\,\delta_2 + 2\,n)}.
$$

Conditions (2.38) and $\varrho(x_k) > 0$ with $x_k \in \{A, A+1, \dots, B-2, B-1\}$ will be satisfied in some cases. The combination of the four roots of $\sigma(x)$ and $\sigma(-x)$ (δ_1, δ_2 and $-\delta_1$, $-\delta_2$, respectively) as endpoints of the support generates five situations, in which the case $[\delta_1, -\delta_1]$ gives two different possibilities.

For the case $\{[\delta_1, -\delta_1], \varrho_1\}$, the support of orthogonality is

$$\{\delta_1, \delta_1 + 1, \ldots, -\delta_1 - 1, -\delta_1\} \quad \text{with} \quad \delta_1 < \delta_2 < 0 \quad \text{and} \quad \delta_2 < \delta_1 + 1.$$

In this situation, $-2\delta_1 \in \mathbb{N}$, and the orthogonality weight appears as

$$\varrho_1(x) = \frac{1}{\Gamma(-x - \delta_1 + 1) \, \Gamma(-x - \delta_2 + 1) \, \Gamma(x - \delta_1 + 1) \, \Gamma(x - \delta_2 + 1)}.$$

For the case $\{[\delta_1, -\delta_1], \varrho_2\}$, the support of orthogonality is

$$\{\delta_1, \delta_1 + 1, \ldots, -\delta_1 - 1, -\delta_1\} \quad \text{with} \quad \delta_2 > -\delta_1.$$

In this case, $-2\delta_1 \in \mathbb{N}$, and the orthogonality weight is denoted by

$$\varrho_2(x) = \frac{\Gamma(-x + \delta_2) \, \Gamma(x + \delta_2)}{\Gamma(-x - \delta_1 + 1) \, \Gamma(x - \delta_1 + 1)}.$$

For the case $\{[\delta_2, -\delta_2], \varrho_1\}$, the support is

$$\{\delta_2, \delta_2 + 1, \ldots, -\delta_2 - 1, -\delta_2\}$$

with the weight function $\varrho_1(x)$. Also $-2\delta_2 \in \mathbb{N}$.
For the case $\{[\delta_2, -\delta_1], \varrho_1\}$, the support of orthogonality is

$$\{\delta_2, \delta_2 + 1, \ldots, -\delta_1 - 1, -\delta_1\} \quad \text{with} \quad \delta_1 - \delta_2 + 1 > 0,$$

and the weight $\varrho_1(x)$, where $-\delta_1 - \delta_2 \in \mathbb{N}$.
For the case $\{[\delta_1, -\delta_2], \varrho_1\}$, the support of orthogonality is

$$\{\delta_1, \delta_1 + 1, \ldots, -\delta_2 - 1, -\delta_2\} \quad \text{with} \quad \delta_1 - \delta_2 + 1 > 0.$$

In this situation, $-\delta_1 - \delta_2 \in \mathbb{N}$, and the orthogonality weight is $\varrho_1(x)$.
Note that the inequality conditions in each case come from a careful analysis of the positivity of

$$\varrho(x + 1)/\varrho(x) \quad \text{for} \quad x \in \{A, A + 1, \ldots, B - 2\}$$

and an appropriate choice of the argument in the gamma function inside the eight possible representations of $\varrho(x)$.

In order to finish the classification, we present in each of these five cases the factorial moments for the weights as well as the representation in terms of hypergeometric series of the polynomial solutions of Eq. (2.30) assuming that $a = 1$ and $b^2 - 4c > 0$.

For the case $\{[\delta_1, -\delta_1], \varrho_1\}$, the symmetric polynomial solutions of Eq. (2.30) can be written in terms of a hypergeometric series as

$$P_n(x) = \frac{(\delta_1 + \delta_2)_n \, (2\delta_1)_n}{(n + 2(\delta_1 + \delta_2) - 1)_n} \; {}_3F_2 \left(\begin{array}{c} -n, \, n + 2(\delta_1 + \delta_2) - 1, \, \delta_1 - x \\ \delta_1 + \delta_2, \, 2\delta_1 \end{array} \middle| \, 1 \right)$$

$$= \tilde{h}_n^{(\delta_1 - \delta_2, \delta_1 - \delta_2)}(x - \delta_1; 1 - 2\delta_1) \qquad (0 \le n \le -2\delta_1),$$

where $\tilde{h}_n^{(\mu,\nu)}(x; N)$ are monic Hahn–Eberlein polynomials defined in (2.39).

Also, the factorial moments are computed as

$$\sum_{k=\delta_1}^{-\delta_1} \varrho_1(k)(k - \delta_1)^{[n]} = \frac{\Gamma(1/2 - \delta_1 - \delta_2)}{2^{2(\delta_1 + \delta_2)} \sqrt{\pi} \, \Gamma(1 - 2\delta_2) \, \Gamma(1 - \delta_1 - \delta_2) \, \Gamma(1 - 2\delta_1)}$$

$$\times \frac{(-1)^n \, \delta_1 \, (1 + 2\delta_1)_{n-1} \, (\delta_1 + \delta_2 + E((n+1)/2))_{E(n/2)}}{(\delta_1 + \delta_2 + 1/2)_{E(n/2)} \, 2^{n-1}},$$

valid for $n = 0, 1, 2, \ldots, -2\delta_1$.

For the case $\{[\delta_1, -\delta_1], \varrho_2\}$, the symmetric polynomial solutions of Eq. (2.30) can be written as

$$P_n(x) = \frac{(\delta_1 + \delta_2)_n \, (2\delta_1)_n}{(n + 2(\delta_1 + \delta_2) - 1)_n} \; {}_3F_2 \left(\begin{array}{c} -n, \, n + 2(\delta_1 + \delta_2) - 1, \, \delta_1 - x \\ \delta_1 + \delta_2, \, 2\delta_1 \end{array} \middle| \, 1 \right)$$

$$= h_n^{(\delta_1 + \delta_2 - 1, \delta_1 + \delta_2 - 1)}(x - \delta_1; 1 - 2\delta_1) \qquad (0 \le n \le -2\delta_1), \qquad (2.40)$$

where $h_n^{(\alpha,\beta)}(x; N)$ are monic Hahn polynomials for $\alpha > -1, \beta > -1$, and $N \in \mathbb{N}$ defined by

$$h_n^{(\alpha,\beta)}(x; N) = \frac{(-1)^n \, (N - n)_n \, (\beta + 1)_n}{(\alpha + \beta + n + 1)_n}$$

$$\times {}_3F_2 \left(\begin{array}{c} -n, \, -x, \, n + \alpha + \beta + 1 \\ 1 - N, \, \beta + 1 \end{array} \middle| \, 1 \right) \qquad (0 \le n \le N - 1).$$

In this sense, note that if $\delta_1 + \delta_2 = 1$, the polynomials (2.40) are reduced to the monic Gram polynomials.

The moments corresponding to this case are computed as

$$\sum_{k=\delta_1}^{-\delta_1} \varrho_2(k)(k-\delta_1)^{[n]} = \frac{\sqrt{\pi}\,\Gamma(2\delta_2)\,\Gamma(\delta_1+\delta_2)}{2^{2(\delta_1+\delta_2)-1}\,\Gamma(1-2\delta_1)\,\Gamma(\delta_1+\delta_2+1/2)}$$

$$\times \frac{(-1)^n\,\delta_1\,(1+2\delta_1)_{n-1}\,(\delta_1+\delta_2+E((n+1)/2))_{E(n/2)}}{(\delta_1+\delta_2+1/2)_{E(n/2)}\,2^{n-1}},$$

valid for $n = 0, 1, 2, \ldots, -2\delta_1$.

For the case $\{[\delta_2, -\delta_2], \varrho_1\}$, the symmetric polynomial solutions of Eq. (2.30) can be represented as

$$P_n(x) = \frac{(\delta_1+\delta_2)_n\,(2\delta_2)_n}{(n+2(\delta_1+\delta_2)-1)_n}\, {}_3F_2\left(\begin{array}{c} -n,\ n+2(\delta_1+\delta_2)-1,\ \delta_2-x \\ \delta_1+\delta_2,\ 2\delta_2 \end{array}\middle| 1\right)$$

$$= \tilde{h}_n^{(\delta_2-\delta_1,\delta_2-\delta_1)}(x-\delta_2;1-2\delta_2) \qquad (0 \le n \le -2\delta_2),$$

where $\tilde{h}_n^{(\mu,\nu)}(x; N)$ are monic Hahn–Eberlein polynomials.

The factorial moments are computed as

$$\sum_{k=\delta_2}^{-\delta_2} \varrho_1(k)(k-\delta_2)^{[n]} = \frac{\Gamma(1/2-\delta_1-\delta_2)}{2^{\delta_1+\delta_2}\,\sqrt{\pi}\,\Gamma(1-2\delta_2)\,\Gamma(1-\delta_1-\delta_2)\,\Gamma(1-2\delta_1)}$$

$$\times \frac{(-1)^n\,\delta_2\,(1+2\delta_2)_{n-1}\,(\delta_1+\delta_2+E((n+1)/2))_{E(n/2)}}{(\delta_1+\delta_2+1/2)_{E(n/2)}\,2^{n-1}},$$

valid for $n = 0, 1, 2, \ldots, -2\delta_2$.

For the case $\{[\delta_2, -\delta_1], \varrho_1\}$, the symmetric polynomial solutions of Eq. (2.30) can be represented as

$$P_n(x) = \frac{(\delta_1+\delta_2)_n\,(2\delta_1)_n}{(n+2(\delta_1+\delta_2)-1)_n}\, {}_3F_2\left(\begin{array}{c} -n,\ n+2(\delta_1+\delta_2)-1,\ \delta_2-x \\ \delta_1+\delta_2,\ 2\delta_1 \end{array}\middle| 1\right)$$

$$= \tilde{h}_n^{(\delta_1-\delta_2,\delta_2-\delta_1)}(x-\delta_2;1-\delta_1-\delta_2) \qquad (0 \le n \le -\delta_1-\delta_2).$$

Also, the factorial moments are computed as

$$\sum_{k=\delta_2}^{-\delta_1} \varrho_1(k)(k-\delta_2)^{[n]} = \frac{\Gamma(1/2-\delta_1-\delta_2)}{2^{2(\delta_1+\delta_2)}\,\sqrt{\pi}\,\Gamma(1-2\delta_2)\,\Gamma(1-\delta_1-\delta_2)\,\Gamma(1-2\delta_1)}$$

$$\times \frac{(-1)^n\,\delta_2\,(1+2\delta_2)_{n-1}\,(\delta_1+\delta_2+E((n+1)/2))_{E(n/2)}}{(\delta_1+\delta_2+1/2)_{E(n/2)}\,2^{n-1}},$$

valid for $n = 0, 1, 2, \ldots, -\delta_1-\delta_2$.

Finally, for the case $\{[\delta_1, -\delta_2], \varrho_1\}$, the symmetric polynomial solutions of Eq. (2.30) can be written in terms of a hypergeometric series as

$$P_n(x) = \frac{(\delta_1 + \delta_2)_n \, (2\delta_1)_n}{(n + 2(\delta_1 + \delta_2) - 1)_n} \; {}_3F_2 \left(\begin{array}{c} -n \, , \, n + 2(\delta_1 + \delta_2) - 1 \, , \, \delta_1 - x \\ \delta_1 + \delta_2 \, , \, 2\delta_1 \end{array} \middle| 1 \right)$$

$$= \tilde{h}_n^{(\delta_1 - \delta_2, \delta_2 - \delta_1)}(x - \delta_1; 1 - \delta_1 - \delta_2) \qquad (0 \le n \le -\delta_1 - \delta_2),$$

and we have

$$\sum_{k=\delta_1}^{-\delta_2} \varrho_1(k)(k - \delta_1)^{[n]} = \frac{\Gamma(1/2 - \delta_1 - \delta_2)}{2^{2(\delta_1 + \delta_2)} \sqrt{\pi} \, \Gamma(1 - 2\delta_2) \, \Gamma(1 - \delta_1 - \delta_2) \, \Gamma(1 - 2\delta_1)}$$

$$\times \frac{(-1)^n \, \delta_1 \, (1 + 2\delta_1)_{n-1} \, (\delta_1 + \delta_2 + E((n+1)/2))_{E(n/2)}}{(\delta_1 + \delta_2 + 1/2)_{E(n/2)} \, 2^{n-1}},$$

which is valid for $n = 0, 1, 2, \ldots, -\delta_1 - \delta_2$.

2.4   A Symmetric Generalization of Sturm–Liouville Problems in Discrete Spaces

In this section, we generalize the difference equation (2.1) symmetrically and show that the corresponding orthogonality property is preserved. For this purpose, we first consider the following limit relations:

$$y''(x) = \lim_{h \to 0} \frac{y(x + h) - 2y(x) + y(x - h)}{h^2}$$

and

$$y'(x) = \lim_{h \to 0} \frac{y(x + h) - y(x)}{h}.$$

Suppose in Eq. (1.94) that $\Phi_n(x) = y(x)$. We can approximate it up to the second order in terms of h in order to obtain a second-order difference equation as

$$A(x) \frac{y(x + h) - 2y(x) + y(x - h)}{h^2} + B(x) \frac{y(x + h) - y(x)}{h}$$

$$+ \left(\lambda_n C(x) + D(x) + \sigma_n E(x) \right) y(x) = 0. \qquad (2.41)$$

In general, two cases can be considered for the step size $h = x_{i+1} - x_i$ in (2.41). They are respectively uniform step size for h, which can be fixed equal to one without loss of generality, and nonuniform lattices.

Although the extension of Sturm–Liouville problems is possible for both cases, we here consider the first one.

If we assume in what follows that $h = 1$, then for $y(x) = \phi_n(x)$, (2.41) takes the form

$$A(x)\Delta\nabla\phi_n(x) + B(x)\Delta\phi_n(x) + (\lambda_n C(x) + D(x) + \sigma_n E(x))\,\phi_n(x) = 0, \qquad (2.42)$$

where

$$\Delta\nabla = \Delta - \nabla \quad \text{and} \quad \sigma_n = \left(1 - (-1)^n\right)/2.$$

In order to have a symmetric solution for the difference equation (2.42), some initial conditions should be considered for its coefficients. If Eq. (2.42) is expanded as

$$\left(A(x) + B(x)\right)\phi_n(x + 1) - \left(2A(x) + B(x)\right)\phi_n(x) + A(x)\phi_n(x - 1)$$
$$= \left(\lambda_n C(x) + D(x) + \sigma_n E(x)\right)\phi_n(x), \qquad (2.43)$$

then by taking into account that the solution is symmetric, i.e., $\phi_n(-x) = (-1)^n \phi_n(x)$, we conclude from (2.43) that

$$A(x) + B(x) = A(-x),$$
$$C(-x) = C(x),$$
$$D(-x) = D(x), \qquad (2.44)$$

and

$$E(-x) = E(x).$$

The first condition of (2.44) implies that $B(x) = A(-x) - A(x)$ is an odd function. Therefore, Eqs. (2.42) and (2.43) are respectively simplified as

$$A(x)\Delta\nabla\phi_n(x) + \left(A(-x) - A(x)\right)\Delta\phi_n(x) + \left(\lambda_n C(x) + D(x) + \sigma_n E(x)\right)\phi_n(x) = 0 \qquad (2.45)$$

and

$$A(-x)\phi_n(x + 1) + A(x)\phi_n(x - 1)$$
$$= -\left(\lambda_n C(x) + D(x) - A(x) - A(-x) + \sigma_n E(x)\right)\phi_n(x), \qquad (2.46)$$

leading to the following theorem.

Theorem 2.3 *Let* $\phi_n(x) = (-1)^n \phi_n(-x)$ *be a sequence of symmetric functions that satisfies the difference equation (2.45) (or equivalently (2.46)). If $A(x)$ is a free real function and $C(x)$, $D(x)$, and $E(x)$ are real even functions, then*

$$\sum_{x=\alpha}^{\beta-1} W^*(x)\phi_n(x)\phi_m(x) = \left(\sum_{x=\alpha}^{\beta-1} W^*(x)\phi_n^2(x)\right)\delta_{n,m},$$

where

$$W^*(x) = C(x)W(x) \tag{2.47}$$

and $W(x)$ is a symmetric solution of the Pearson difference equation

$$\Delta\Big(A(x)W(x)\Big) = \Big(A(-x) - A(x)\Big)W(x), \tag{2.48}$$

which is equivalent to

$$\frac{W(x+1)}{W(x)} = \frac{A(-x)}{A(x+1)}. \tag{2.49}$$

Moreover, the weight function defined in (2.47) must be even over one of the following symmetric counter sets:

(i) $S_1 = \{-a - n, -a - n + 1, \dots, -a - 1, -a, a, a + 1, a + n - 1, a + n\},\ a \in \mathbb{R}$,
(ii) $S_2 = S_1 \cup \{0\}$ *(since every odd function is equal to zero at $x = 0$),*
(iii) $S_3 = \{\dots, -a - n, -a - n + 1, \dots, -a - 1, -a, a, a + 1, a + n - 1, a + n, \dots\}$
(iv) $S_4 = S_3 \cup \{0\}$,

and the function $A(x)W(x)$ must also vanish at $x = \alpha$ and $x = 1 - \beta$, where $[\alpha, \beta - 1] \in \{S_1, S_2, S_3, S_4\}$.

Proof If Eq. (2.45) is written in a self-adjoint form, then

$$\Delta\Big(A(x)W(x)\nabla\phi_n(x)\Big) + \Big(\lambda_n C(x) + D(x) + \sigma_n E(x)\Big)W(x)\phi_n(x) = 0 \tag{2.50}$$

and

$$\Delta\Big(A(x)W(x)\nabla\phi_m(x)\Big) + \Big(\lambda_m C(x) + D(x) + \sigma_m E(x)\Big)W(x)\phi_m(x) = 0. \tag{2.51}$$

By multiplying (2.50) by $\phi_m(x)$ and (2.51) by $\phi_n(x)$ and subtracting one from the other and also noting that

$$\phi_m(x)\Delta\Big(A(x)W(x)\nabla\phi_n(x)\Big) - \phi_n(x)\Delta\Big(A(x)W(x)\nabla\phi_m(x)\Big)$$
$$= \Delta\Big(A(x)W(x)\big(\phi_m(x)\nabla\phi_n(x) - \phi_n(x)\nabla\phi_m(x)\big)\Big),$$

we obtain

$$\Delta\Big(A(x)W(x)\,(\phi_m(x)\nabla\phi_n(x) - \phi_n(x)\nabla\phi_m(x))\Big)$$
$$+ (\lambda_n - \lambda_m)\,C(x)W(x)\phi_n(x)\phi_m(x) + \frac{(-1)^m - (-1)^n}{2}E(x)W(x)\phi_n(x)\phi_m(x) = 0. \tag{2.52}$$

Now, a simple but important idea can appear here: "The sum of any odd summand on a symmetric counter set is equal to zero." This set can be one of the four introduced sets S_1, S_2, S_3, S_4. Consequently, summation on both sides of (2.52) gives

$$\sum_{x=\alpha}^{\beta-1}\Delta\Big(A(x)W(x)\big(\phi_m(x)\nabla\phi_n(x) - \phi_n(x)\nabla\phi_m(x)\big)\Big)$$
$$+(\lambda_n - \lambda_m)\sum_{x=\alpha}^{\beta-1}C(x)W(x)\phi_n(x)\phi_m(x)+\frac{(-1)^m - (-1)^n}{2}\sum_{x=\alpha}^{\beta-1}E(x)W(x)\phi_n(x)\phi_m(x) = 0, \tag{2.53}$$

in which $[\alpha, \beta - 1] \in \{S_1, S_2, S_3, S_4\}$ is assumed to be the main support.

On the other hand, $W(x)$ is a symmetric solution for the Pearson equation (2.49), and therefore the condition

$$W(x + 1)A(x + 1) - W(x)A(-x) = 0$$

is valid for all $x \in [\alpha, \beta - 1]$.

Hence without loss of generality, we can assume $[\alpha, \beta - 1] = [-\theta, \theta]$, i.e., $\alpha = -\theta$ and $\beta = \theta + 1$.

By noting the identity

$$\phi_m(x)\nabla\phi_n(x) - \phi_n(x)\nabla\phi_m(x) = \phi_n(x)\phi_m(x - 1) - \phi_m(x)\phi_n(x - 1),$$

the first sum of (2.53) is simplified as

$$\sum_{x=-\theta}^{\theta} \Delta \left(A(x)W(x) \left(\phi_m(x)\nabla\phi_n(x) - \phi_n(x)\nabla\phi_m(x) \right) \right)$$

$$= A(x)W(x) \left(\phi_n(x)\phi_m(x-1) - \phi_m(x)\phi_n(x-1) \right)\Big|_{x=-\theta}^{x=\theta+1}$$

$$= A(\theta+1)W(\theta+1) \left(\phi_n(\theta+1)\phi_m(\theta) - \phi_m(\theta+1)\phi_n(\theta) \right)$$

$$- A(-\theta)W(-\theta) \left(\phi_n(-\theta)\phi_m(-\theta-1) - \phi_m(-\theta)\phi_n(-\theta-1) \right). \quad (2.54)$$

Since $W(-x) = W(x)$, the solutions are symmetric, and the Pearson equation (2.49) is valid for $x = \theta$, i.e.,

$$A(\theta+1)W(\theta+1) = A(-\theta)W(\theta),$$

relation (2.54) changes to

$$\sum_{x=-\theta}^{\theta} \Delta \left(A(x)W(x) \left(\phi_m(x)\nabla\phi_n(x) - \phi_n(x)\nabla\phi_m(x) \right) \right)$$

$$= A(-\theta)W(\theta) \left(1 + (-1)^{n+m} \right) \left(\phi_m(\theta)\phi_n(\theta+1) - \phi_n(\theta)\phi_m(\theta+1) \right). \quad (2.55)$$

By noting that

$$\phi_m(\theta)\phi_n(\theta+1) - \phi_n(\theta)\phi_m(\theta+1) \neq 0,$$

two cases can occur for relation (2.55):

1. If $n + m$ is odd, then $1 + (-1)^{n+m} = 0$, and the right-hand side of equality (2.55) is automatically zero. This case is clear, since the sum of any odd summand on a symmetric counter set is equal to zero.
2. If $A(-\theta)W(\theta) = 0$, then

$$W(\theta) = 0 \quad \text{or} \quad A(-\theta) = 0, \quad \text{i.e.,} \quad A(\alpha) = A(1-\beta) = 0.$$

In this case, the right-hand side of equality (2.55) is again equal to zero.

Now, in order to prove the orthogonality property, it remains to show that

$$F(m,n) = \frac{(-1)^m - (-1)^n}{2} \sum_{x=\alpha}^{\beta-1} E(x)W(x)\phi_n(x)\phi_m(x) = 0.$$

For this purpose, four cases should be considered for values m, n respectively as follows:

1. If both m and n are even (or odd), then it is obvious that $F(n, m) = 0$.
2. If m is even and n is odd (or conversely), then we have

$$F(2i, 2j + 1) = \sum_{x=\alpha}^{\beta-1} E(x)W(x)\phi_{2j+1}(x)\phi_{2i}(x).$$

Since $E(x)$, $W(x)$, and $\phi_{2i}(x)$ are assumed to be even functions and $\phi_{2j+1}(x)$ is odd, we are computing a sum of an odd function on a symmetric set of points, which means that $F(2i, 2j + 1) = 0$. $\qquad\square$

2.4.1 Some Illustrative Examples

Although there are many examples of difference equation (2.45) whose solutions are symmetric functions, symmetric solutions of polynomial type are of special importance in physics and engineering. In this section, we first consider an example for the infinite case of Theorem 2.3 whose solution is a sequence of hypergeometric functions and then consider five examples of the extended Eq. (2.45) whose solutions are symmetric orthogonal polynomials. In each example, we obtain the corresponding orthogonality relation whose support for the first example is the infinite set

$$x \in \{\ldots, -2, -1, 0, 1, 2, \ldots\},$$

and for the five further examples is

$$x \in \{-N, \ldots, 0, 1, \ldots, N\}.$$

Example 2.1 Consider the difference equation

$$(x + b)\Delta\nabla\phi_n(x) - 2x\Delta\phi_n(x) + (2n + e\sigma_n)\phi_n(x) = 0 \tag{2.56}$$

as a special case of (2.45) for

$$A(x) = x + b,$$
$$C(x) = 1,$$
$$D(x) = 0,$$
$$E(x) = e,$$
$$\lambda_n = 2n.$$

We introduce the symmetric basis

$$\vartheta_n(x) = (-1)^{[n/2]} x^{\sigma_n} (\sigma_n - x)_{[n/2]} (\sigma_n + x)_{[n/2]} = (-1)^n \vartheta_n(-x), \tag{2.57}$$

satisfying the following properties:

$$\Delta \vartheta_n(x) = n \vartheta_{n-1}(x) + \frac{n(n-1)}{2} \vartheta_{n-2}(x) + \sigma_{n+1} \frac{n(n-1)(n-2)}{4} \vartheta_{n-3}(x), \tag{2.58}$$

$$x \vartheta_n(x) = \vartheta_{n+1}(x) + \sigma_{n+1} \left(\frac{n}{2}\right)^2 \vartheta_{n-1}(x), \tag{2.59}$$

and

$$\Delta \nabla \vartheta_n(x) = n(n-1) \vartheta_{n-2}(x). \tag{2.60}$$

If in (2.56),

$$\phi_n(x) = \sum_{j=0}^{\infty} a_j(n) \vartheta_j(x),$$

then the recurrence relation of the coefficients will be derived as

$$a_{j+2}(n) = \frac{-e\sigma_n + 2j - 2n - 4}{(j-1)j \left(\left[\frac{1-j}{2}\right] + b\right)} a_{j-2}(n), \qquad j \geq 0,$$

with two initial values $a_0(n)$ and $a_1(n)$.

Therefore, we have

$$a_{2j}(n) = \frac{(-4)^j a_0(n) \left(-\frac{e}{4} - \frac{n}{2}\right)_j}{(2j)!(b-j+1)_j}, \qquad j = 1, 2, \ldots,$$

and

$$a_{2j+1}(n) = \frac{(-4)^j a_1(n) \left(\frac{1}{4}(-e - 2n + 2)\right)_j}{(2j+1)!(b-j)_j}, \qquad j = 1, 2, \ldots,$$

where

$$(a)_j = \frac{\Gamma(a+j)}{\Gamma(a)}.$$

The above two relations imply that for odd n we get

$$\phi_n(x) = a_0(n) \; _3F_2 \left(\begin{array}{cc} -(e+2n)/4, \, -x, \, x \\ 1/2, \, -b \end{array} \middle| \, 1 \right)$$

$$+ x \, a_1(n) \; _3F_2 \left(\begin{array}{cc} -(e+2n-2)/4, \, 1-x, \, 1+x \\ 3/2, \, 1-b \end{array} \middle| \, 1 \right). \qquad (2.61)$$

But since $\phi_n(-x) = (-1)^n \phi_n(x)$, it follows that $\phi_{2m+1}(0) = 0$, and consequently $a_0(n) = 0$ in (2.61).

Similarly, for even n we obtain

$$\phi_n(x) = a_0(n) \; _3F_2 \left(\begin{array}{cc} -n/2, \, -x, \, x \\ 1/2, \, -b \end{array} \middle| \, 1 \right)$$

$$+ x a_1(n) \; _3F_2 \left(\begin{array}{cc} (1-n)/2, \, 1-x, \, 1+x \\ 3/2, \, 1-b \end{array} \middle| \, 1 \right). \qquad (2.62)$$

By combining the two formulas (2.61) and (2.62), one can conclude that

$$\phi_n(x; b, e) = x^{\sigma_n} \; _3F_2 \left(\begin{array}{cc} \frac{1}{4}((2-e)\sigma_n - 2n), \, \sigma_n - x, \, \sigma_n + x \\ \sigma_n + 1/2, \, \sigma_n - b \end{array} \middle| \, 1 \right)$$

$$= (-1)^n \phi_n(-x; b, e) \qquad (2.63)$$

is a symmetric solution for Eq. (2.56).

In this sense, if $e = 4k \in \mathbb{N}$ in (2.63), the sequence will be reduced to a finite polynomial sequence.

From Theorem 2.3, it is known that the sequence $\{\phi_n(x; b, e)\}$ satisfies an orthogonality relation of the form

$$\sum_{x=\alpha}^{\beta-1} W_1^*(x)\phi_n(x; b, e)\phi_m(x; b, e) = \left(\sum_{x=\alpha}^{\beta-1} W_1^*(x)\phi_n^2(x; b, e) \right) \delta_{n,m},$$

in which

$$W_1^*(x) = C(x)W_1(x) = W_1(x)$$

is the main weight function and $W_1(x)$ satisfies the equation

$$\frac{W_1(x+1)}{W_1(x)} = \frac{A(-x)}{A(x+1)} = \frac{b-x}{b+x+1}. \tag{2.64}$$

Up to a periodic function of period 1, a symmetric solution of Eq. (2.64) is

$$W_1(x) = \frac{1}{\Gamma(b+1-x)\Gamma(b+1+x)}. \tag{2.65}$$

It is clear that the convergence of the hypergeometric series (2.63) depends on the value of x and the parameters b and e. In order to have a full convergence, we can impose some restrictions on b and e to reduce the hypergeometric series (2.63) to a polynomial function that is convergent everywhere (except infinity). For this purpose, as we pointed out, if $e = 4k \in \mathbb{N}$, then

$$\bar{\phi}_n(x; b, 4k)$$

$$= (\sigma_n + 1/2)_{[n/2]} \, (\sigma_n - b)_{[n/2]} x^{\sigma_n} \, {}_3F_2 \left(\begin{array}{c} -[n/2] - k\sigma_n, \sigma_n - x, \sigma_n + x \\ \sigma_n + 1/2, \sigma_n - b \end{array} \middle| 1 \right) \tag{2.66}$$

is a monic finite sequence of symmetric polynomials orthogonal with respect to the weight function (2.65) with support $\mathbb{Z}$. For instance, for $n = 0, 1, 2, 3, 4,$ and 5, (2.66) takes the forms

$$\bar{\phi}_0(x; b, 4k) = 1,$$

$$\bar{\phi}_1(x; b, 4k) = x \, {}_3F_2 \left(\begin{array}{c} -k, 1 - x, 1 + x \\ 3/2, 1 - b \end{array} \middle| 1 \right),$$

$$\bar{\phi}_2(x; b, 4k) = x^2 - \frac{b}{2},$$

$$\bar{\phi}_3(x; b, 4k) = \frac{3(1-b)}{2} x \, {}_3F_2 \left(\begin{array}{c} -k - 1, 1 - x, 1 + x \\ 3/2, 1 - b \end{array} \middle| 1 \right),$$

$$\bar{\phi}_4(x; b, 4k) = x^4 + (2 - 3b)x^2 + \frac{3}{4}(b-1)b,$$

$$\bar{\phi}_5(x; b, 4k) = \frac{15}{4}(1-b)(2-b) x \, {}_3F_2 \left(\begin{array}{c} -k - 2, 1 - x, 1 + x \\ 3/2, 1 - b \end{array} \middle| 1 \right).$$

To compute the norm square value of the polynomials (2.66), for $n = 2m$ we obtain

$$\sum_{x=-\infty}^{\infty} \frac{(\bar{\phi}_{2m}(x; b, 4k))^2}{\Gamma(b+1-x)\,\Gamma(b+1+x)} = \frac{2^{2b-4m}(2m)!}{\Gamma(2b+1)} \frac{(-4b-1)_{2m}}{((1/2)_m)^2\,((-b)_m)^2} = N^*_{2m}, \tag{2.67}$$

where $b \notin \mathbb{Z}^+$, $b > -1/2$, and for $n = 2m + 1$ we have

$$\sum_{x=-\infty}^{\infty} \frac{(\bar{\phi}_{2m+1}(x; b, 4k))^2}{\Gamma(b+1-x)\,\Gamma(b+1+x)} = \frac{(m+k)!(b-(m+k+1))\Gamma(m+k+5/2)}{\sqrt{\pi}\,\Gamma(2b+1)}$$

$$\times \frac{2^{2b+1}(5/2-b)_{m+k}}{((3/2)_{m+k})^2\,(1-b)_{m+k}} = N_{2m+1}^* \qquad (b \notin \mathbb{Z}^+, b > -1/2).$$

(2.68)

By combining the two relations (2.67) and (2.68), the final form of the orthogonality relation is derived as

$$\sum_{x=-\infty}^{\infty} \frac{\bar{\phi}_n(x; b, 4k)\,\bar{\phi}_m(x; b, 4k)}{\Gamma(b+1-x)\,\Gamma(b+1+x)} = N_n^*\,\delta_{n,m} \qquad (b \notin \mathbb{Z}^+, b > -1/2).$$

Example 2.2 Consider the difference equation

$$(2x-7)(2x+1)(5x-9)(x+N)\Delta\nabla\phi_n(x) - 2x\left(4(5N-24)x^2 + 73N + 63\right)\Delta\phi_n(x)$$

$$+ \left(n\left(1-4x^2\right)(5n-10N+43) + 78(2N+1)\sigma_n\right)\phi_n(x) = 0, \qquad (2.69)$$

which is a special case of Eq. (2.45) for

$$A(x) = (2x-7)(2x+1)(5x-9)(x+N),$$

$$C(x) = 1 - 4x^2,$$

$$D(x) = 0,$$

$$E(x) = 78(2N+1),$$

$$\lambda_n = n(5n-10N+43).$$

We are interested in computing the symmetric polynomial solution of Eq. (2.69). Hence, by taking

$$\phi_n(x) = x^n + \delta_n x^{n-2} + \cdots$$

and replacing it in (2.69), the solution satisfies a three-term recurrence relation as

$$\phi_{n+1}(x) = x\phi_n(x) - \gamma_n\phi_{n-1}(x) \quad \text{with} \quad \phi_0(x) = 1 \text{ and } \phi_1(x) = x, \qquad (2.70)$$

where

$$\delta_n = \frac{5n^4 + (66 - 20N)n^3 + (1 - 228N)n^2 - (220N + 306)n + 234(2N + 1)\sigma_n}{24(10n - 10N + 33)}.$$

Since in general,

$$\gamma_n = \delta_n - \delta_{n+1}, \tag{2.71}$$

the explicit form of γ_n (for even and odd n) is derived as

$$\gamma_{2n} = -\frac{n(10n + 43)(5n - 5N + 4)(2n - 2N + 5)}{(20n - 10N + 33)(20n - 10N + 43)} \tag{2.72}$$

and

$$\gamma_{2n+1} = -\frac{(2n + 7)(5n + 9)(10n - 10N + 43)(n - N)}{(20n - 10N + 43)(20n - 10N + 53)}. \tag{2.73}$$

Some particular solutions of Eq. (2.69) are

$$\phi_2(x) = x^2 + \frac{63N}{10N - 53},$$

$$\phi_3(x) = x^3 + \frac{63 - 116N}{63 - 10N}x,$$

$$\phi_4(x) = x^4 + \frac{179 - 242N}{73 - 10N}x^2 + \frac{7938(N - 1)N}{(10N - 73)(10N - 63)}.$$

By using the well-known Navima algorithm, a solvable recurrence relation can be obtained for the coefficients of the polynomial solution of Eq. (2.69), which eventually leads to a hypergeometric series as

$$\phi_n(x) = \frac{(\sigma_n + 9/5)_{[n/2]}\,(\sigma_n + 7/2)_{[n/2]}\,(\sigma_n - N)_{[n/2]}}{(-N + [n/2] + \sigma_n + 43/10)_{[n/2]}}$$

$$\times\, x^{\sigma_n}\, {}_4F_3\left(\begin{array}{c} -[n/2],\, \sigma_n - x,\, \sigma_n + x,\, -N + [n/2] + \sigma_n + 43/10 \\ \sigma_n + 9/5,\, \sigma_n + 7/2,\, \sigma_n - N \end{array}\bigg|\, 1\right). \tag{2.74}$$

Since the support of orthogonality is $[\alpha, \beta] = [-N, N + 1]$, the recurrence relation (2.70) helps us directly compute the norm square value of the polynomials (2.74) as

$$\sum_{x=-N}^{N} W_2^*(x)\phi_n(x)\phi_m(x) = \left(\left(\prod_{k=1}^{n}\gamma_k\right)\sum_{x=-N}^{N} W_2^*(x)\right)\delta_{n,m}, \tag{2.75}$$

in which

$$W_2^*(x) = (1/4 - x^2)W_2(x)$$

is the main weight function and $W_2(x)$ satisfies the equation

$$\frac{W_2(x+1)}{W_2(x)} = \frac{A(-x)}{A(x+1)} = \frac{(2x-1)(2x+7)(5x+9)(x-N)}{(2x-5)(2x+3)(5x-4)(N+x+1)}. \tag{2.76}$$

Up to a periodic function of period 1, a symmetric solution of Eq. (2.76) is

$$W_2(x) = \left(\Gamma\left(-\frac{5}{2}-x\right)\Gamma\left(\frac{5}{2}+x\right)\Gamma\left(-\frac{4}{5}-x\right)\Gamma\left(-\frac{4}{5}+x\right) \right.$$
$$\left. \times \Gamma\left(\frac{3}{2}-x\right)\Gamma\left(\frac{3}{2}+x\right)\Gamma(N+1-x)\Gamma(N+1+x) \right)^{-1}.$$

Therefore,

$$W_2^*(x) = \left(\Gamma\left(-\frac{5}{2}-x\right)\Gamma\left(\frac{5}{2}+x\right)\Gamma\left(-\frac{4}{5}-x\right)\Gamma\left(-\frac{4}{5}+x\right) \right.$$
$$\left. \times \Gamma\left(\frac{1}{2}-x\right)\Gamma\left(\frac{1}{2}+x\right)\Gamma(N+1-x)\Gamma(N+1+x) \right)^{-1} \tag{2.77}$$

and

$$\sum_{x=-N}^{N} W_2^*(x) = 2\sum_{x=0}^{N} W_2^*(x) - W_2^*(0)$$
$$= \frac{225\left(2\,_4F_3\left(1, \frac{9}{5}, \frac{7}{2}, -N; -\frac{5}{2}, -\frac{4}{5}, N+1; 1\right) - 1\right)}{64\pi^2\Gamma^2\left(-\frac{4}{5}\right)\Gamma^2(N+1)} = S_2^*. \tag{2.78}$$

By substituting the results (2.72)–(2.74), (2.77), and (2.78) into (2.75), the final form of the orthogonality relation is derived as

$$\sum_{x=-N}^{N} W_2^*(x)\phi_n(x)\phi_m(x) = S_2^*(-1)^{\sigma_n} n!\, \delta_{n,m}$$
$$\times \frac{(-N)_{n+\sigma_n}(9/5)_{n+\sigma_n}(7/2)_{n+\sigma_n}(9/5-N)_n(7/2-N)_n(53/10)_n}{(53/10-N)_{2n+\sigma_n}(n-N+53/10)_n}.$$

Example 2.3 Consider the difference equation

$$(11 - 2x)^2(2x + 1)(x + N)\Delta\nabla\phi_n(x) - 2x(4(2N - 21)x^2 + 198N + 121)\Delta\phi_n(x)$$

$$+ \left(2n(n - 2N + 20)\left(1 - 4x^2\right) + 200(2N + 1)\sigma_n\right)\phi_n(x) = 0, \qquad (2.79)$$

which is a special case of (2.45) for

$$A(x) = (11 - 2x)^2(2x + 1)(x + N),$$

$$C(x) = 1 - 4x^2,$$

$$D(x) = 0,$$

$$E(x) = 200(2N + 1),$$

$$\lambda_n = 2n(n - 2N + 20).$$

Similar to the previous example, the polynomial solution of Eq. (2.79) can be written as

$$\phi_n(x) = x^n + \delta_n x^{n-2} + \cdots,$$

satisfying a recurrence relation of the form (2.70), in which

$$\delta_n = \frac{n^4 - 4(N - 9)n^3 + (185 - 114N)n^2 + (482N + 522)n + 300(2N + 1)\sigma_n}{48(n - N + 9)},$$

and from (2.71) the explicit form of γ_n (for even and odd n) is obtained as

$$\gamma_{2n} = -\frac{n(n + 10)(2n - 2N + 9)^2}{4(2n - N + 9)(2n - N + 10)} \qquad (2.80)$$

and

$$\gamma_{2n+1} = -\frac{(2n + 11)^2(n - N)(n - N + 10)}{4(2n - N + 10)(2n - N + 11)}. \qquad (2.81)$$

Some particular solutions of Eq. (2.79) are

$$\phi_2(x) = x^2 + \frac{121N}{4(N - 11)},$$

$$\phi_3(x) = x^3 + \frac{11(15N - 11)}{4(N - 12)}x,$$

$$\phi_4(x) = x^4 + \frac{167N - 143}{2(N - 13)}x^2 + \frac{20449(N - 1)N}{16(N - 13)(N - 12)}.$$

Using the symmetric basis (2.57), considering the properties (2.58)–(2.60), and using the Navima algorithm, we can again obtain a solvable recurrence relation for the coefficients of the polynomial solution of Eq. (2.79), which eventually leads to the hypergeometric representation

$$
\phi_n(x) = \frac{\left((\sigma_n + 11/2)_{[n/2]}\right)^2 (\sigma_n - N)_{[n/2]}}{(-N + [n/2] + \sigma_n + 10)_{[n/2]}}
$$

$$
\times x^{\sigma_n} \, {}_4F_3 \left(\begin{matrix} -[n/2], \sigma_n - x, x + \sigma_n, -N + [n/2] + \sigma_n + 10 \\ \sigma_n + 11/2, \sigma_n + 11/2, \sigma_n - N \end{matrix} \middle| 1 \right). \tag{2.82}
$$

For these polynomials,

$$
W_3^*(x) = (1/4 - x^2) W_3(x)
$$

is the main weight function, and $W_3(x)$ satisfies the equation

$$
\frac{W_3(x+1)}{W_3(x)} = \frac{A(-x)}{A(x+1)} = \frac{(2x-1)(2x+11)^2(x-N)}{(9-2x)^2(2x+3)(N+x+1)}. \tag{2.83}
$$

Up to a periodic function of period 1, a symmetric solution of Eq. (2.83) is given by

$$
W_3(x) = \frac{\Gamma\left(\frac{11}{2} - x\right) \Gamma\left(\frac{11}{2} + x\right)}{\Gamma\left(-\frac{9}{2} - x\right) \Gamma\left(-\frac{9}{2} + x\right) \Gamma\left(\frac{3}{2} - x\right) \Gamma\left(\frac{3}{2} + x\right) \Gamma(N+1-x) \Gamma(N+1+x)}. \tag{2.84}
$$

Therefore

$$
\sum_{x=-N}^{N} W_3^*(x) = \frac{797493650625 \left({}_2 4F_3 \left(1, \frac{11}{2}, \frac{11}{2}, -N; -\frac{9}{2}, -\frac{9}{2}, N+1; 1 \right) - 1 \right)}{1048576\pi \, \Gamma^2(N+1)} = S_3^*. \tag{2.85}
$$

By substituting the results (2.80)–(2.82), (2.84), and (2.85) into (2.75), the final form of the orthogonality relation is derived as

$$
\sum_{x=-N}^{N} W_3^*(x) \phi_n(x) \phi_m(x) = \frac{4S_3^*(-1)^{\sigma_{n+1}} n! \, N(n-N)^{\sigma_n}(n-N+10)^{\sigma_n}}{12658629375 \, \pi} \delta_{n,m}
$$

$$
\times \frac{\Gamma(n+11) \Gamma^2(11-N) \Gamma^2(n+\sigma_n+11/2)}{\Gamma(2n-N+11) \Gamma(2n-N+10+2\sigma_n)} (1-N)_{n-1}(11-N)_{n-1} \left((11/2 - N)_n\right)^2.
$$

Example 2.4 Consider the difference equation

$$(2x - 13)(2x + 1)(x + N)^2 \Delta \nabla \phi_n(x) - 2x \left(-2N(13 + 12N) + 8(N - 3)x^2 \right) \Delta \phi_n(x)$$

$$+ \left(n(n - 4N + 11) \left(1 - 4x^2 \right) - 12(2N + 1)^2 \sigma_n \right) \phi_n(x) = 0, \qquad (2.86)$$

which is a special case of (2.45) for

$$A(x) = (2x - 13)(2x + 1)(x + N)^2,$$

$$C(x) = 1 - 4x^2,$$

$$D(x) = 0,$$

$$E(x) = -12(2N + 1)^2,$$

$$\lambda_n = n(n - 4N + 11).$$

By noting the two previous examples, we respectively obtain

$$\gamma_{2n} = -\frac{n(n - 2N - 1)(2n - 2N + 11)^2}{(4n - 4N + 9)(4n - 4N + 11)} \qquad (2.87)$$

and

$$\gamma_{2n+1} = -\frac{(2n + 13)(2n - 4N + 11)(n - N)^2}{(4n - 4N + 11)(4n - 4N + 13)}. \qquad (2.88)$$

Some particular solutions of Eq. (2.86) are

$$\phi_2(x) = x^2 + \frac{13N^2}{13 - 4N},$$

$$\phi_3(x) = x^3 + \frac{(26 - 15N)N}{4N - 15}x,$$

$$\phi_4(x) = x^4 + \frac{30N^2 - 56N + 13}{17 - 4N}x^2 + \frac{195(N - 1)^2 N^2}{(4N - 17)(4N - 15)},$$

which are indeed particular cases of the hypergeometric polynomial

$$\phi_n(x) = \frac{(\sigma_n + 13/2)_{[n/2]} \left((\sigma_n - N)_{[n/2]} \right)^2}{(-2N + [n/2] + \sigma_n + 11/2)_{[n/2]}}$$

$$\times x^{\sigma_n} \, _4F_3 \left(\begin{array}{c} -[n/2], \sigma_n - x, x + \sigma_n, -2N + [n/2] + \sigma_n + 11/2 \\ \sigma_n + 13/2, \sigma_n - N, \sigma_n - N \end{array} \middle| 1 \right). \qquad (2.89)$$

Once again,

$$W_4^*(x) = (1/4 - x^2)W_4(x)$$

is the main weight function, and $W_4(x)$ satisfies the equation

$$\frac{W_4(x+1)}{W_4(x)} = \frac{A(-x)}{A(x+1)} = \frac{(2x-1)(2x+13)(N-x)^2}{(2x-11)(2x+3)(N+x+1)^2}. \qquad (2.90)$$

Up to a periodic function of period 1, a symmetric solution of Eq. (2.90) is given by

$$W_4(x) = \frac{\Gamma\left(-\frac{1}{2}-x\right)\Gamma\left(-\frac{1}{2}+x\right)\Gamma\left(\frac{13}{2}-x\right)\Gamma\left(\frac{13}{2}+x\right)}{\Gamma^2(N+1-x)\Gamma^2(N+1+x)}. \qquad (2.91)$$

Therefore

$$\sum_{x=-N}^{N} W_4^*(x) =$$

$$\frac{108056025\pi^2\left(2\,_4F_3\left(1, \frac{13}{2}, -N, -N; -\frac{11}{2}, N+1, N+1; 1\right) - 1\right)}{4096\Gamma^4(N+1)} = S_4^*. \qquad (2.92)$$

By substituting the results (2.87)–(2.89), (2.91), and (2.92) into (2.75), the final form of the orthogonality relation is derived as

$$\sum_{x=-N}^{N} W_4^*(x)\phi_n(x)\phi_m(x)$$

$$= \frac{64S_4^*(-1)^{\sigma_n}n!\,N^2(n-N)^{2\sigma_n}}{10395\sqrt{\pi}}(-2N)_n\left((1-N)_{n-1}\right)^2 \delta_{n,m}$$

$$\times \frac{\Gamma(13/2-2N)\Gamma^2(13/2-N+n)\Gamma(n+\sigma_n+13/2)\Gamma(n+\sigma_n+11/2-2N)}{\Gamma(2n-2N+13/2)\Gamma^2(13/2-N)\Gamma(2n-2N+11/2+2\sigma_n)}.$$

Example 2.5 Consider the difference equation

$$(2x+1)(2x+7)(N+x)\Delta\nabla\phi_n(x) - 2x\left(16N+4x^2+7\right)\Delta\phi_n(x)$$

$$+ \left(-2n\left(1-4x^2\right) + 16(2N+1)\sigma_n\right)\phi_n(x) = 0, \qquad (2.93)$$

which is a special case of (2.45) for

$$A(x) = (2x + 1)(2x + 7)(N + x),$$

$$C(x) = 1 - 4x^2,$$

$$D(x) = 0,$$

$$E(x) = 16(1 + 2N),$$

$$\lambda_n = -2n.$$

By referring to the previous examples, we can eventually obtain that

$$\gamma_n = \frac{1}{4}(n(-n + 2N + 9) - 8(2N + 1)\sigma_n). \tag{2.94}$$

Some particular solutions of Eq. (2.93) are

$$\phi_2(x) = x^2 + \frac{7N}{2},$$

$$\phi_3(x) = x^3 + \frac{1}{2}(5N - 7)x,$$

$$\phi_4(x) = x^4 + (5N - 6)x^2 + \frac{35}{4}(N - 1)N,$$

which are particular cases of the hypergeometric polynomial

$$\phi_n(x) = \frac{\Gamma\left(\frac{1}{2}(2[n/2] + 2\sigma_n - 7)\right)\Gamma(-N + [n/2] + \sigma_n)}{\Gamma\left(\frac{1}{2}(2\sigma n) - 7)\right)\Gamma(\sigma_n - N)}$$

$$\times x^{\sigma_n} \, {}_3F_2\left(\begin{array}{c} -[n/2], \sigma_n - x, x + \sigma_n \\ \sigma_n - 7/2, \sigma_n - N \end{array} \bigg| 1\right). \tag{2.95}$$

The function

$$W_5^*(x) = (1/4 - x^2)W_5(x)$$

is the main weight function for this example, where $W_5(x)$ satisfies the equation

$$\frac{W_5(x + 1)}{W_5(x)} = \frac{A(-x)}{A(x + 1)} = \frac{(2x - 7)(2x - 1)(N - x)}{(2x + 3)(2x + 9)(N + x + 1)}.$$

By solving the above equation, we finally obtain

$$
W_5^*(x) = \frac{1}{\Gamma\left(\frac{1}{2}-x\right)\Gamma\left(\frac{1}{2}+x\right)\Gamma\left(\frac{9}{2}-x\right)\Gamma\left(\frac{9}{2}+x\right)\Gamma(N+1-x)\Gamma(N+1+x)}.
\tag{2.96}
$$

Therefore we have

$$
\sum_{x=-N}^{N} W_5^*(x) = \frac{2^{2N+9}(N+1)}{105\pi^2\left(4N^2+24N+35\right)\Gamma(2N+4)}.
\tag{2.97}
$$

By substituting the results (2.94)–(2.97) into (2.75), the final form of the orthogonality relation is derived as

$$
\sum_{x=-N}^{N} W_5^*(x)\phi_n(x)\phi_m(x)
$$
$$
= \frac{(-1)^{\sigma_{n+1}}n!\,2^{2N+5}N(N+1)(1-N)_{n-\sigma_{n+1}}(-N-7/2)_n\,\Gamma(n+\sigma_n-7/2)}{\pi^2\sqrt{\pi}(4N^2+24N+35)\,\Gamma(2N+4)}\delta_{n,m}.
$$

Example 2.6 Consider the difference equation

$$
(2x+1)(N+x)^2\Delta\nabla\phi_n(x) - 4x\left(N^2+N+x^2\right)\Delta\phi_n(x)
$$
$$
+\left(-n\left(1-4x^2\right)+(2N+1)^2\sigma_n\right)\phi_n(x) = 0,
$$

which is a special case of (2.45) for

$$
A(x) = (2x+1)(N+x)^2,
$$
$$
C(x) = 1-4x^2,
$$
$$
D(x) = 0,
$$
$$
E(x) = (2N+1)^2,
$$
$$
\lambda_n = -n.
$$

After some computations we get

$$
\gamma_n = \frac{1}{4}\left(n(-n+4N+2)-(2N+1)^2\sigma_n\right)
\tag{2.98}
$$

and

$$\phi_2(x) = x^2 + N^2,$$

$$\phi_3(x) = x^3 + N(N-2)x,$$

$$\phi_4(x) = x^4 + (2(N-2)N+1)x^2 + (N-1)^2 N^2,$$

which are particular cases of the hypergeometric polynomial

$$\phi_n(x) = \frac{\Gamma^2\left(-N + [n/2] + \sigma_n\right)}{\Gamma^2(\sigma_n - N)} x^{\sigma_n} \, {}_3F_2\left(\begin{array}{c} -[n/2], \sigma_n - x, x + \sigma_n \\ \sigma_n - N, \sigma_n - N \end{array} \bigg| 1 \right). \qquad (2.99)$$

Since

$$W_6^*(x) = (1/4 - x^2) W_6(x)$$

is the main weight function and $W_6(x)$ satisfies the equation

$$\frac{W_6(x+1)}{W_6(x)} = \frac{A(-x)}{A(x+1)} = -\frac{(2x-1)(N-x)^2}{(2x+3)(N+x+1)^2},$$

solving the above equation eventually yields

$$W_6^*(x) = \frac{\Gamma\left(\frac{1}{2} - x\right) \Gamma\left(\frac{1}{2} + x\right)}{\Gamma^2(N+1-x)\Gamma^2(N+1+x)}. \qquad (2.100)$$

Hence

$$\sum_{x=-N}^{N} W_6^*(x) = \frac{\pi \left(2 \, {}_3F_2(1, -N, -N; N+1, N+1; -1) - 1\right)}{\Gamma^4(N+1)} = S_6^*. \qquad (2.101)$$

By substituting the results (2.98)–(2.101) into (2.75), the final form of the orthogonality relation is derived as

$$\sum_{x=-N}^{N} W_6^*(x)\phi_n(x)\phi_m(x) = (-1)^{\sigma_n} n! \, N^2 S_6^* \left((1-N)_{n-\sigma_{n+1}}\right)^2 (-2N)_n \, \delta_{n,m}.$$

In the next section, we will introduce a main sequence of symmetric orthogonal polynomials of a discrete variable such that all examples introduced in this section are just particular cases of it.

2.5 A Basic Class of Symmetric Orthogonal Polynomials of a Discrete Variable

Using the extended Sturm–Liouville theorem in discrete spaces, we will introduce a basic class of symmetric orthogonal polynomials of a discrete variable with four free parameters that generalizes all classical symmetric orthogonal polynomials presented in Sect. 2.3. We then obtain standard properties of these polynomials such as a second-order difference equation, an explicit form for the polynomials, a three-term recurrence relation, and a generic orthogonality relation. We show that two infinite types and two finite types of hypergeometric orthogonal sequences with different weight functions can be extracted from this class, and moments corresponding to the derived weight functions can be explicitly computed. We also present a particular example containing all classical discrete symmetric orthogonal polynomials described in Sect. 2.3.

Let us recall the differential equation (1.100) as

$$x^2(p^*x^2 + q^*)\Phi_n''(x) + x(r^*x^2 + s^*)\Phi_n'(x)$$
$$- (n(r^* + (n-1)p^*)x^2 + \sigma_n s^*)\Phi_n(x) = 0. \quad (2.102)$$

We observe in (2.102) that $A(x) = x^2(p^*x^2 + q^*)$ is a polynomial of degree at most 4, $C(x) = x^2$ is a symmetric quadratic polynomial, $D(x) = 0$, and $E(x) = s^*$. Since discrete orthogonal polynomials have a direct relationship with continuous polynomials, motivated by (2.102), we suppose in the main Eq. (2.45) that

$$A(x) = \sum_{i=0}^{4} a_i x^i,$$

$$C(x) = c_2 x^2 + c_0,$$

$$D(x) = 0, \quad (2.103)$$

and

$$E(x) = e_0.$$

We are interested in obtaining a symmetric orthogonal polynomial solution as

$$\phi_n(x) = x^n + \delta_n x^{n-2} + \cdots, \quad (2.104)$$

which satisfies the three-term recurrence relation

$$\phi_{n+1}(x) = x\phi_n(x) - \gamma_n \phi_{n-1}(x) \quad \text{with } \phi_0(x) = 1 \text{ and } \phi_1(x) = x. \quad (2.105)$$

From (2.105), (2.104), and (2.103), equating the coefficient in x^{n+2} gives the eigenvalues as

$$\lambda_n = \frac{n\left(2a_3 - a_4(n-1)\right)}{c_2}, \tag{2.106}$$

provided that $c_2 \neq 0$ and $|a_3| + |a_4| \neq 0$.

By considering λ_n in (2.106) and equating the coefficient in x^n, we obtain

$$\delta_n = \left\{6c_2\left(4a_1 n - 2a_2(n-1)n + e_0\left((-1)^n - 1\right)\right)\right.$$

$$+a_4(n-1)n\left(12c_0 - c_2(n-3)(n-2)\right)$$

$$\left.+4a_3 n\left(c_2(n-2)(n-1) - 6c_0\right)\right\} / \left\{24c_2\left(a_4(3-2n) + 2a_3\right)\right\}.$$

Also from (2.104) and (2.105) we have

$$x^{n+1} + \delta_{n+1} x^{n-1} + \cdots = x\left(x^n + \delta_n x^{n-2} + \cdots\right) - \gamma_n\left(x^{n-1} + \delta_{n-1} x^{n-3} + \cdots\right),$$

which implies

$$\gamma_n = \delta_n - \delta_{n+1}. \tag{2.107}$$

Therefore, in order that $\phi_n(x)$ be a solution of Eq. (2.103), the following extra conditions must be considered for the initial values of n:

$$e_0 = 2a_1 - \frac{2a_3 c_0}{c_2},$$

$$a_0 = \frac{c_0\left((2a_3 - a_4)c_0 + (a_2 - 2a_1)c_2\right)}{c_2^2},$$

and

$$c_2 = -4c_0.$$

Summarizing the data, we get

$$A(x) = a_4 x^4 + a_3 x^3 + a_2\left(x^2 - \frac{1}{4}\right) + a_1\left(x + \frac{1}{2}\right) + \frac{a_3}{8} - \frac{a_4}{16},$$

$$B(x) = A(-x) - A(x) = -2x\left(a_3 x^2 + a_1\right),$$

$$C(x) = c_0\left(1 - 4x^2\right),$$

$$D(x) = 0, \tag{2.108}$$

$$E(x) = \frac{1}{2}\left(4a_1 + a_3\right),$$

$$\lambda_n = \frac{n\left(a_4(n-1) - 2a_3\right)}{4c_0},$$

and

$$\delta_n = \{12a_1\left(2n + (-1)^n - 1\right) + 3a_3\left((-1)^n - 1\right)$$
$$+n\left(-12a_2(n-1) + 2a_3(2(n-3)n+7)\right)$$
$$-a_4(n-1)((n-5)n+9))\}/\{24a_4(3-2n) + 48a_3\}. \qquad (2.109)$$

For simplicity, if we set

$$a_4 = 2a, \quad a_3 = a + 2b, \quad a_2 = b + 2c, \quad a_1 = c + 2d,$$

then (2.108) changes to

$$A(x) = (2x+1)(ax^3 + bx^2 + cx + d),$$

$$B(x) = -2x\left((a+2b)x^2 + c + 2d\right),$$

$$C(x) = \frac{1}{4} - x^2,$$

$$D(x) = 0,$$

$$E(x) = \frac{1}{2}(a + 2b + 4c + 8d),$$

$$\lambda_n = 2n(an - 2(a+b)) \quad \text{for} \quad |a| + |b| \neq 0.$$

Consequently, the following basic difference equation appears:

$$(2x+1)(ax^3 + bx^2 + cx + d)\Delta\nabla\phi_n(x) - 2x\left(x^2(a+2b) + c + 2d\right)\Delta\phi_n(x)$$
$$+ \left(2n(an - 2(a+b))\left(\frac{1}{4} - x^2\right) + \sigma_n\left(\frac{a}{2} + b + 2c + 4d\right)\right)\phi_n(x) = 0. \qquad (2.110)$$

Since the polynomial solution of Eq. (2.110) is symmetric, we use the notation

$$\phi_n(x) = S_n^*\left(\begin{array}{cc|c} a & b & \\ c & d & x \end{array}\right)$$

for mathematical display formulas and $S_n^*(x\,;\,a,b,c,d)$ in the text. This means that we are dealing with just one characteristic vector $\mathbf{V} = (a,b,c,d)$ for any given subcase.

If $S_n^*(x\,;\,a,b,c,d)$ satisfies a three-term recurrence relation of type (2.105), then by referring to (2.107) and (2.109), we obtain

$$\gamma_n = \gamma_n \begin{pmatrix} a & b \\ c & d \end{pmatrix} = \frac{\sum_{i=0}^{4} K_i(a,b,c,d)n^i}{32(b-a(n-2))(b-a(n-1))}, \tag{2.111}$$

where

$$K_4(a,b,c,d) = -2a^2,$$

$$K_3(a,b,c,d) = 4a(3a+2b),$$

$$K_2(a,b,c,d) = -8\left(3a^2 + a(4b+c) + b^2\right),$$

$$K_1(a,b,c,d) = 2(3a+2b)(3a+4(b+c)) - 2a(-1)^n(a+2b+4c+8d),$$

$$K_0(a,b,c,d) = \left((-1)^n - 1\right)(3a+2b)(a+2b+4c+8d).$$

For $n = 2m$ and $n = 2m+1$, γ_n in (2.111) is decomposed as

$$\gamma_{2m} = \frac{m\left(-a^2(m-1)^3 + a\left(2b(m-1)^2 + c(1-m) - d\right) + b(b(1-m)+c)\right)}{(b-2a(m-1))(b-a(2m-1))}$$

and

$$\gamma_{2m+1} = \frac{(a(m-1)-b)\left(-am^3 + bm^2 - cm + d\right)}{(2am - (a+b))(2am - b)}.$$

Theorem 2.4 *The explicit form of the polynomial $S_n^*(x\,;\,a,b,c,d)$ is denoted by*

$$S_n^*\begin{pmatrix} a & b \\ c & d \end{pmatrix} x\end{pmatrix} = x^{\sigma_n} \sum_{j=0}^{[n/2]} (-1)^j \binom{[n/2]}{j}$$

$$\times \left(\prod_{i=j}^{[n/2]-1} \frac{a(i+\sigma_n)^3 - b(i+\sigma_n)^2 + c(i+\sigma_n) - d}{a(i+[n/2]-\sigma_{n+1}) - b} \right) (\sigma_n - x)_j (\sigma_n + x)_j,$$

$$\tag{2.112}$$

where $[x]$ denotes the integer part of x and $\prod_{i=0}^{-1}(.) = 1$.

Moreover, since

$$(-x)_j(x)_j = (-1)^j \prod_{k=0}^{j-1}(x^2 - k^2),$$

for n = 2m and n = 2m + 1, (2.112) respectively changes to

$$S_{2m}^* \left(\begin{matrix} a & b \\ c & d \end{matrix} \middle| x \right) = \sum_{j=0}^{m} \binom{m}{j} \left(\prod_{i=j}^{m-1} \frac{ai^3 - bi^2 + ci - d}{a(i+m-1) - b} \right) \prod_{k=0}^{j-1} (x^2 - k^2),$$

and

$$S_{2m+1}^* \left(\begin{matrix} a & b \\ c & d \end{matrix} \middle| x \right) =$$

$$x \sum_{j=0}^{m} \binom{m}{j} \left(\prod_{i=j}^{m-1} \frac{a(i+1)^3 - b(i+1)^2 + c(i+1) - d}{a(i+m) - b} \right) \prod_{k=1}^{j} (x^2 - k^2).$$

Proof Once again, we consider the symmetric basis

$$\vartheta_n(x) = (-1)^{[n/2]} x^{\sigma_n} (\sigma_n - x)_{[n/2]} (\sigma_n + x)_{[n/2]} = (-1)^n \vartheta_n(-x), \qquad (2.113)$$

together with the following straightforward properties:

$$\Delta \vartheta_n(x) = n\vartheta_{n-1}(x) + \frac{n(n-1)}{2} \vartheta_{n-2}(x) + \sigma_{n+1} \frac{n(n-1)(n-2)}{4} \vartheta_{n-3}(x), \qquad (2.114)$$

$$x\vartheta_n(x) = \vartheta_{n+1}(x) + \sigma_{n+1} \left(\frac{n}{2} \right)^2 \vartheta_{n-1}(x), \qquad (2.115)$$

$$\Delta \nabla \vartheta_n(x) = n(n-1)\vartheta_{n-2}(x). \qquad (2.116)$$

If $S_n^*(x; a, b, c, d)$ is expanded as

$$S_n^*(x; a, b, c, d) = \sum_{j=0}^{[n/2]} c_j(n) \vartheta_{2j+\sigma_n}(x)$$

$$= \begin{cases} \sum_{j=0}^{m} (-1)^j c_j(2m)(-x)_j(x)_j, & n = 2m, \\ \sum_{j=0}^{m} (-1)^j c_j(2m+1)x(1-x)_j(1+x)_j, & n = 2m+1, \end{cases} \qquad (2.117)$$

then using the Navima algorithm, we reach a solvable recurrence relation for the connection coefficients $c_j(n)$. So, using the properties (2.114)–(2.116) and substituting (2.117) into (2.110), we obtain

$$0 = \sum_{j=0}^{[n/2]} c_j(n) \left(E_j(n) + \vartheta_{2j+\sigma_n+2} + F_j(n)\vartheta_{2j+\sigma_n} + G_j(n)\vartheta_{2j+\sigma_n-2} \right),$$

where

$$E_j(n) = 2(2j + \sigma_n - n)(a(2j + n - 2) + a\sigma_n - 2b),$$

$$F_j(n) = 4a(j-1)^2 j(2j-1)\sigma_{n-1}$$

$$+ \frac{1}{2}\sigma_{n+2}(2j + \sigma_n + 1)^2(2j + \sigma_n - n)(a(2j + n - 2) + a\sigma_n - 2b)$$

$$+ 2j\sigma_{n+1}\left(a\left(4j^3 - 6j^2 - j(n-3)(n+1) - 1\right) + 2b(j(-4j + n + 3) - 1)\right)$$

$$+ \frac{1}{2}(\sigma_n(\sigma_n(\sigma_n(\sigma_n(8aj + a\sigma_n - 4a - 2b) + a(24(j-1)j + 5) + 4b(1 - 3j))$$

$$+ 2a(2j-1)(8(j-1)j + 1) + 8b(2-3j)j + 4c) + a\left(4(j-1)j(1-2j)^2 + 1\right)$$

$$+ 4j(b(1 - 4(j-1)j) + 4c) - 4c) - 2n(a+b) + an^2$$

$$- 4j(b + 4(c+d)) + 8j^2(b + 2c)\Big),$$

and

$$G_j(n) = (j-1)j(2j-1)\left(2(j-1)\sigma_{n-1}\left(2a(j-1)^2\sigma_{n-1} + b + 2c\right)\right.$$

$$\left. - 2\sigma_{n+1}\left((j-1)^2(a+2b)\sigma_{n-1} + c + 2d\right)\right)$$

$$+ \frac{1}{8}(2j + \sigma_n - 1)(2j + \sigma_n)(\sigma_n(2j + \sigma_n - 1)(\sigma_n(2((j-1)(-4j(a+b)$$

$$\left. + 4aj^2 + a\right) + 2c) + \sigma_n(\sigma_n(6aj + a\sigma_n - 4a - 2b)$$

$$+ (2j-1)(a(6j-5) - 4b))) + b(4j-2) - 8(c+d) + 8cj) + 8d).$$

But since $\vartheta_n(x)$ is linearly independent, the coefficients $c_j(n)$ satisfy the relation

$$E_{j-1}(n)c_{j-1}(n) + F_j(n)c_j(n) + G_{j+1}(n)c_{j+1}(n) = 0,$$

which is explicitly solvable with the initial conditions $c_m(n) = 0$ for $m > [n/2]$ and $c_{[n/2]}(n) = 1$, providing (2.112).   □

Since the recurrence relation (2.105) is now explicitly known, the complete form of the orthogonality relation can be designed as

$$
\sum_{x=-\theta}^{\theta} W^* \begin{pmatrix} a & b \\ c & d \end{pmatrix} x \Big| S_n^* \begin{pmatrix} a & b \\ c & d \end{pmatrix} x \Big| S_m^* \begin{pmatrix} a & b \\ c & d \end{pmatrix} x \Big|
$$

$$
= \prod_{k=1}^{n} \gamma_k \begin{pmatrix} a & b \\ c & d \end{pmatrix} \left(\sum_{x=-\theta}^{\theta} W^* \begin{pmatrix} a & b \\ c & d \end{pmatrix} x \Big| \right) \delta_{n,m}, \qquad (2.118)
$$

where

$$
W^* \begin{pmatrix} a & b \\ c & d \end{pmatrix} x \Big| = \left(\frac{1}{4} - x^2 \right) W^*(x)
$$

is the original weight function and $W^*(x)$ satisfies the difference equation

$$
\frac{W^*(x+1)}{W^*(x)} = \frac{(1/2) - x}{(3/2) + x} \frac{-ax^3 + bx^2 - cx + d}{a(x+1)^3 + b(x+1)^2 + c(c+1) + d}. \qquad (2.119)
$$

By noting that

$$
A(x) = (2x + 1)(ax^3 + bx^2 + cx + d) \quad \text{for} \quad |a| + |b| \neq 0,
$$

we see that two cases can generally occur for the parameter a in (2.119), i.e., when $a \neq 0$ and b arbitrary or $a = 0$ and $b \neq 0$.

In the first case, since every polynomial of degree 3 has at least one real root, say $x = p \in \mathbb{R}$, the aforementioned $A(x)$ can be decomposed in three different forms, i.e.,

$$
A(x) = (2x + 1)(x - p)\left(a x^2 + u x + v\right)
$$

$$
= \begin{cases} (2x + 1)(x - p)\left(a x^2 + u x + v\right), & (u^2 < 4av), \\ (2x + 1) a(x - p)(x - q)(x - r), & (u^2 > 4av), \\ (2x + 1) a(x - p)(x - q)^2, & (u^2 = 4av). \end{cases} \qquad (2.120)
$$

Similarly, in the second case, when $a = 0$ and $b \neq 0$, $A(x)$ can be decomposed as

$$
A(x) = (2x + 1)\left(b x^2 + c x + d\right) = \begin{cases} (2x + 1)\left(b x^2 + c x + d\right), & (c^2 < 4bd), \\ (2x + 1) b(x - p)(x - q), & (c^2 > 4bd), \\ (2x + 1) b(x - p)^2, & (c^2 = 4bd). \end{cases}
$$

$$
(2.121)
$$

For the two subcases

$$A(x) = (2x + 1)\,(x - p)\left(a\,x^2 + u\,x + v\right)$$

in (2.120) and

$$A(x) = (2x + 1)\left(b\,x^2 + c\,x + d\right)$$

in (2.121), the difference equations corresponding to (2.119) respectively take the forms

$$\frac{W^*(x+1)}{W^*(x)} = \frac{(1/2) - x}{(3/2) + x}\,\frac{p + x}{p - x - 1}\,\frac{a\,x^2 - u\,x + v}{a\,(x+1)^2 + u\,(x+1) + v} \qquad (u^2 < 4av)$$

$$\tag{2.122}$$

and

$$\frac{W^*(x+1)}{W^*(x)} = \frac{(1/2) - x}{(3/2) + x}\,\frac{b\,x^2 - c\,x + d}{b\,(x+1)^2 + c\,(x+1) + d} \qquad (c^2 < 4bd). \tag{2.123}$$

Since the denominators of the two fractions (2.122) and (2.123) are not decomposable in the real line, we should separately consider them in finite types of symmetric orthogonal polynomials in the next section. Note that the cases analyzed in this part allow us to recover all classical symmetric orthogonal polynomials of a discrete variable in Sect. 2.5.3.

For the second subcase of (2.120), without loss of generality take $a = 1$ to get

$$A(x) = (2x + 1)(x - p)(x - q)(x - r),$$

where $p, q, r \in \mathbb{R}$, which indeed covers the third subcase of (2.120) too.

For the second subcase of (2.121) we can similarly consider

$$A(x) = (2x + 1)(x - p)(x - q),$$

where $p, q \in \mathbb{R}$.

This means that there exist only two infinite types of orthogonal sequences of $S_n^*(x\,;\,a, b, c, d)$ when $A(x)$ is decomposable. In other words, when $A(x)$ is a polynomial of degree four, a three-parameter family appears, and when $A(x)$ is of degree three, a two-parameter family will appear. Finally, when $A(x)$ is of degree 2, the well-known classical symmetric discrete families appear.

The question is now how to determine restrictions on the parameter θ in the orthogonality support $[-\theta, \theta]$ in (2.118). To answer, we should reconsider the main difference equation (2.110) on $[\alpha, \beta - 1] = [-\theta, \theta]$ and write it in a self-adjoint form to obtain

$$\sum_{x=-\theta}^{\theta} \Delta\Big(A(x)W^*(x)\,(\phi_m(x)\nabla\phi_n(x) - \phi_n(x)\nabla\phi_m(x))\Big)$$

$$+ (\lambda_n - \lambda_m) \sum_{x=-\theta}^{\theta} \Big(\frac{1}{4} - x^2\Big) W^*(x)\phi_n(x)\phi_m(x)$$

$$+ \frac{(-1)^m - (-1)^n}{2}\Big(\frac{a}{2} + b + 2c + 4d\Big) \sum_{x=-\theta}^{\theta} W^*(x)\phi_n(x)\phi_m(x) = 0. \qquad (2.124)$$

On the other hand, the identity

$$\phi_m(x)\nabla\phi_n(x) - \phi_n(x)\nabla\phi_m(x) = \phi_n(x)\phi_m(x-1) - \phi_m(x)\phi_n(x-1)$$

simplifies the first sum of (2.124) as

$$\sum_{x=-\theta}^{\theta} \Delta\Big(A(x)W^*(x)\,(\phi_m(x)\nabla\phi_n(x) - \phi_n(x)\nabla\phi_m(x))\Big)$$

$$= A(x)W^*(x)\Big(\phi_n(x)\phi_m(x-1) - \phi_m(x)\phi_n(x-1)\Big)\Big|_{x=-\theta}^{|x=\theta+1}$$

$$= A(\theta+1)W^*(\theta+1)\,(\phi_n(\theta+1)\phi_m(\theta) - \phi_m(\theta+1)\phi_n(\theta))$$

$$- A(-\theta)W^*(-\theta)\,(\phi_n(-\theta)\phi_m(-\theta-1) - \phi_m(-\theta)\phi_n(-\theta-1)). \qquad (2.125)$$

By taking into account that all weight functions are even, i.e., $W^*(-x) = W^*(x)$, the polynomials are symmetric, i.e., $\phi_n(x) = (-1)^n\phi_n(-x)$, and the corresponding Pearson difference equation is also valid for $x = \theta$, i.e.,

$$A(\theta+1)W^*(\theta+1) = A(-\theta)W^*(\theta),$$

relation (2.125) may finally be simplified as

$$\sum_{x=-\theta}^{\theta} \Delta\Big(A(x)W^*(x)\,(\phi_m(x)\nabla\phi_n(x) - \phi_n(x)\nabla\phi_m(x))\Big)$$

$$= A(-\theta)W^*(\theta)\big(1 + (-1)^{n+m}\big)\,(\phi_m(\theta)\phi_n(\theta+1) - \phi_n(\theta)\phi_m(\theta+1)). \qquad (2.126)$$

Since

$$\phi_m(\theta)\phi_n(\theta+1) - \phi_n(\theta)\phi_m(\theta+1) \neq 0,$$

two cases can in general occur for the right-hand side of (2.126):

1. If $n+m$ is odd, then $1 + (-1)^{n+m} = 0$, and (2.126) is automatically zero.
2. If simultaneously $A(-\theta) = 0$ and $W^*(\theta) \neq 0$, then (2.126) is again equal to zero.

2.5.1 Two Infinite Types of $S_n^*(x\,;\,a,b,c,d)$

Here we introduce two infinite hypergeometric sequences of symmetric orthogonal polynomials that are particular cases of $S_n^*(x\,;\,a,b,c,d)$ and obtain all possible weight functions together with orthogonality supports for them.

First Sequence For $p,q,r \in \mathbb{R}$, if the characteristic vector

$$(a,b,c,d) = (1, -(p+q+r), pq+pr+qr, -pqr)$$

is replaced in Eq. (2.110), then

$$
(2x+1)(x-p)(x-q)(x-r)\Delta\nabla\phi_n(x)
$$
$$
- 2x\left(x^2(1-2p-2q-2r) + pq + pr + qr - 2pqr\right)\Delta\phi_n(x)
$$
$$
+ \left(2n(n+2(p+q+r-1))\left(\frac{1}{4}-x^2\right)\right.
$$
$$
\left. - \frac{\sigma_n}{2}(2p-1)(2q-1)(2r-1)\right)\phi_n(x) = 0
$$

has a basis hypergeometric-type solution as

$$
\phi_n(x) = S_n^*\left(\begin{matrix} 1 & -(p+q+r) \\ pq+pr+qr & -pqr \end{matrix}\,\middle|\, x\right)
$$
$$
= \frac{(p+\sigma_n)_{[n/2]}(q+\sigma_n)_{[n/2]}(r+\sigma_n)_{[n/2]}}{([n/2]+p+q+r-1+\sigma_n)_{[n/2]}}
$$
$$
\times x^{\sigma_n} \,_4F_3\left(\begin{matrix} -[n/2],\ [n/2]+p+q+r-1+\sigma_n,\ \sigma_n-x,\ \sigma_n+x \\ p+\sigma_n,\ q+\sigma_n,\ r+\sigma_n \end{matrix}\,\middle|\, 1\right), \qquad (2.127)
$$

which satisfies the recurrence relation

$$\phi_{n+1}(x) = x\phi_n(x) - \gamma_n \begin{pmatrix} 1 & -(p+q+r) \\ pq + pr + qr & -pqr \end{pmatrix} \phi_{n-1}(x),$$

$$(\phi_0(x) = 1, \quad \phi_1(x) = x), \qquad (2.128)$$

where

$$\gamma_n \begin{pmatrix} 1 & -(p+q+r) \\ pq + pr + qr & -pqr \end{pmatrix} = \{-2n^4 - 4(-3 + 2p + 2q + 2r)n^3$$

$$-8\left(p^2 + p(3q + 3r - 4) + 3qr + (q-4)q + r^2 - 4r + 3\right)n^2$$

$$+ \left(2(-1)^n(-1+2p)(-1+2q)(-1+2r)\right.$$

$$-2(-3 + 2p + 2q + 2r)(3 + 4q(-1+r) - 4r + 4p(-1+q+r)))n$$

$$+(-1 + (-1)^n)(-1+2p)(-1+2q)(-1+2r)(-3+2p+2q+2r)\}$$

$$/\{32(n + p + q + r - 2)(n + p + q + r - 1)\}, \qquad (2.129)$$

dividing to

$$\gamma_{2n} = -\frac{n(n+p+q-1)(n+p+r-1)(n+q+r-1)}{(2n+p+q+r-2)(2n+p+q+r-1)}$$

and

$$\gamma_{2n+1} = -\frac{(n+p)(n+q)(n+r)(n+p+q+r-1)}{(2n+p+q+r-1)(2n+p+q+r)}.$$

The latter representations will allow us to analyze the sign of γ_n in terms of the values of p, q, and r.

By noting the relations (2.128) and (2.129), the orthogonality relation of the first sequence can be expressed as

$$\sum_{x=-\theta}^{\theta} \left(W^* \begin{pmatrix} 1 & -(p+q+r) \\ pq+pr+qr & -pqr \end{pmatrix} x \right) S_n^{(1)}(x) S_m^{(1)}(x) \right)$$

$$= \prod_{k=1}^{n} \gamma_k \begin{pmatrix} 1 & -(p+q+r) \\ pq+pr+qr & -pqr \end{pmatrix}$$

$$\times \left(\sum_{x=-\theta}^{\theta} W^* \begin{pmatrix} 1 & -(p+q+r) \\ pq+pr+qr & -pqr \end{pmatrix} x \right) \delta_{n,m},$$

in which

$$S_n^{(1)}(x) = S_n^* \left(\begin{array}{cc|c} 1 & -(p+q+r) & \\ pq+pr+qr & -pqr & \end{array} x \right)$$

and

$$W^* \left(\begin{array}{cc|c} 1 & -(p+q+r) & \\ pq+pr+qr & -pqr & \end{array} x \right) = \left(\frac{1}{4} - x^2 \right) W^*(x)$$

denotes the original weight function corresponding to the hypergeometric polynomial (2.127), and finally $W^*(x)$ satisfies a particular case of the difference equation (2.119) in the form

$$\frac{W^*(x+1)}{W^*(x)} = \frac{-x+(1/2)}{x+(3/2)} \frac{-x-p}{x+1-p} \frac{-x-q}{x+1-q} \frac{-x-r}{x+1-r}. \tag{2.130}$$

It is interesting that there exist 16 symmetric solutions for Eq. (2.130), as follows:

$$W_1(x; p, q, r) = (\Gamma(1-p+x)\Gamma(1-p-x)\Gamma(1-q+x)\Gamma(1-q-x)$$

$$\times \Gamma(1-r+x)\Gamma(1-r-x)\Gamma(3/2+x)\Gamma(3/2-x))^{-1},$$

$W_{2,1}(x; p, q, r)$

$$= \frac{\Gamma(p+x)\Gamma(p-x)}{\Gamma(1-q+x)\Gamma(1-q-x)\Gamma(1-r+x)\Gamma(1-r-x)\Gamma(3/2+x)\Gamma(3/2-x)},$$

$W_{2,2}(x; p, q, r) = W_{2,1}(x; q, p, r)$

$$= \frac{\Gamma(q+x)\Gamma(q-x)}{\Gamma(1-p+x)\Gamma(1-p-x)\Gamma(1-r+x)\Gamma(1-r-x)\Gamma(3/2+x)\Gamma(3/2-x)},$$

$W_{2,3}(x; p, q, r) = W_{2,1}(x; r, q, p)$

$$= \frac{\Gamma(r+x)\Gamma(r-x)}{\Gamma(1-p+x)\Gamma(1-p-x)\Gamma(1-q+x)\Gamma(1-q-x)\Gamma(3/2+x)\Gamma(3/2-x)},$$

$$W_{3,1}(x; p, q, r) = \frac{\Gamma(p+x)\Gamma(p-x)\Gamma(q+x)\Gamma(q-x)}{\Gamma(1-r+x)\Gamma(1-r-x)\Gamma(3/2+x)\Gamma(3/2-x)},$$

$$W_{3,2}(x; p, q, r) = W_{3,1}(x; p, r, q) = \frac{\Gamma(p+x)\Gamma(p-x)\Gamma(r+x)\Gamma(r-x)}{\Gamma(1-q+x)\Gamma(1-q-x)\Gamma(3/2+x)\Gamma(3/2-x)},$$

$$W_{3,3}(x; p, q, r) = W_{3,2}(x; r, q, p) = \frac{\Gamma(q+x)\Gamma(q-x)\Gamma(r+x)\Gamma(r-x)}{\Gamma(1-p+x)\Gamma(1-p-x)\Gamma(3/2+x)\Gamma(3/2-x)},$$

$$W_{4,1}(x; p, q, r) = \frac{\Gamma(p+x)\Gamma(p-x)\Gamma(-1/2+x)\Gamma(-1/2-x)}{\Gamma(1-q+x)\Gamma(1-q-x)\Gamma(1-r+x)\Gamma(1-r-x)},$$

$$W_{4,2}(x; p, q, r) = W_{4,1}(x; q, p, r)$$

$$= \frac{\Gamma(q+x)\Gamma(q-x)\Gamma(-1/2+x)\Gamma(-1/2-x)}{\Gamma(1-p+x)\Gamma(1-p-x)\Gamma(1-r+x)\Gamma(1-r-x)},$$

$$W_{4,3}(x; p, q, r) = W_{4,1}(x; r, q, p)$$

$$= \frac{\Gamma(r+x)\Gamma(r-x)\Gamma(-1/2+x)\Gamma(-1/2-x)}{\Gamma(1-p+x)\Gamma(1-p-x)\Gamma(1-q+x)\Gamma(1-q-x)},$$

$$W_{5,1}(x; p, q, r) = \frac{\Gamma(p+x)\Gamma(p-x)\Gamma(q+x)\Gamma(q-x)\Gamma(-1/2+x)\Gamma(-1/2-x)}{\Gamma(1-r+x)\Gamma(1-r-x)},$$

$$W_{5,2}(x; p, q, r) = W_{5,1}(x; p, r, q)$$

$$= \frac{\Gamma(p+x)\Gamma(p-x)\Gamma(r+x)\Gamma(r-x)\Gamma(-1/2+x)\Gamma(-1/2-x)}{\Gamma(1-q+x)\Gamma(1-q-x)},$$

$$W_{5,3}(x; p, q, r) = W_{5,1}(x; r, q, p)$$

$$= \frac{\Gamma(q+x)\Gamma(q-x)\Gamma(r+x)\Gamma(r-x)\Gamma(-1/2+x)\Gamma(-1/2-x)}{\Gamma(1-p+x)\Gamma(1-p-x)},$$

$$W_6(x; p, q, r)$$

$$= \frac{\Gamma(-1/2+x)\Gamma(-1/2-x)}{\Gamma(1-p+x)\Gamma(1-p-x)\Gamma(1-q+x)\Gamma(1-q-x)\Gamma(1-r+x)\Gamma(1-r-x)},$$

$$W_7(x; p, q, r) = \frac{\Gamma(p+x)\Gamma(p-x)\Gamma(q+x)\Gamma(q-x)\Gamma(r+x)\Gamma(r-x)}{\Gamma(3/2+x)\Gamma(3/2-x)},$$

$$W_8(x; p, q, r) = \Gamma(p+x)\Gamma(p-x)\Gamma(q+x)\Gamma(q-x)$$

$$\times \Gamma(r+x)\Gamma(r-x)\Gamma(-1/2+x)\Gamma(-1/2-x).$$

Since the original weight functions corresponding to all 16 cases are

$$\left(\frac{1}{4} - x^2\right) W_k(x),$$

the two following identities are remarkable in this direction:

$$(1/4 - x^2)\Gamma(-1/2 + x)\Gamma(-1/2 - x) = \Gamma(1/2 + x)\Gamma(1/2 - x),$$

and

$$\frac{1/4 - x^2}{\Gamma(3/2 + x)\Gamma(3/2 - x)} = \frac{1}{\Gamma(1/2 + x)\Gamma(1/2 - x)}.$$

For example, for the last given case, the original weight function becomes

$$W_8^*\left(\begin{matrix} 1 & -(p+q+r) \\ pq + pr + qr & -pqr \end{matrix}\middle| x\right) = \left(\frac{1}{4} - x^2\right) W_8(x; p, q, r)$$

$$= \Gamma(p+x)\Gamma(p-x)\Gamma(q+x)\Gamma(q-x)\Gamma(r+x)\Gamma(r-x)\Gamma\left(\frac{1}{2}+x\right)\Gamma\left(\frac{1}{2}-x\right).$$

Table 2.2 shows the orthogonality supports of each weight function together with their parameter restrictions in which $\mathbf{Z}^- = \{0, -1, -2, \ldots\}$.

Table 2.2 Orthogonality supports for the first hypergeometric sequence

$W_k(x)$	Support	Parameter restrictions
$W_1(x; p, q, r)$	$[-p, p]$	$p \in \mathbf{Z}^-, 1 - q \pm p \notin \mathbf{Z}^-, 1 - r \pm p \notin \mathbf{Z}^-$.
	$[-q, q]$	$q \in \mathbf{Z}^-, 1 - p \pm q \notin \mathbf{Z}^-, 1 - r \pm q \notin \mathbf{Z}^-$.
	$[-r, r]$	$r \in \mathbf{Z}^-, 1 - q \pm r \notin \mathbf{Z}^-, 1 - p \pm r \notin \mathbf{Z}^-$.
$W_{2,1}(x; p, q, r)$	$[-q, q]$	$q \in \mathbf{Z}^-, p \pm q \notin \mathbf{Z}^-, 1 - r \pm q \notin \mathbf{Z}^-$.
	$[-r, r]$	$r \in \mathbf{Z}^-, p \pm r \notin \mathbf{Z}^-, 1 - q \pm r \notin \mathbf{Z}^-$.
$W_{3,1}(x; p, q, r)$	$[-r, r]$	$r \in \mathbf{Z}^-, p \pm r \notin \mathbf{Z}^-, q \pm r \notin \mathbf{Z}^-$.
$W_{4,1}(x; p, q, r)$	$[-q, q]$	$q \in \mathbf{Z}^-, p \pm q \notin \mathbf{Z}^-, 1 - r \pm q \notin \mathbf{Z}^-$.
	$[-r, r]$	$r \in \mathbf{Z}^-, p \pm r \notin \mathbf{Z}^-, 1 - q \pm r \notin \mathbf{Z}^-$.
$W_{5,1}(x; p, q, r)$	$[-r, r]$	$r \in \mathbf{Z}^-, p \pm r \notin \mathbf{Z}^-, q \pm r \notin \mathbf{Z}^-$.
$W_6(x; p, q, r)$	$[-p, p]$	$p \in \mathbf{Z}^-, 1 - q \pm p \notin \mathbf{Z}^-, 1 - r \pm q \notin \mathbf{Z}^-$.
	$[-q, q]$	$q \in \mathbf{Z}^-, 1 - r \pm q \notin \mathbf{Z}^-, 1 - p \pm q \notin \mathbf{Z}^-$.
	$[-r, r]$	$r \in \mathbf{Z}^-, 1 - q \pm r \notin \mathbf{Z}^-, 1 - p \pm r \notin \mathbf{Z}^-$.
$W_7(x; p, q, r)$	–	–
$W_8(x; p, q, r)$	–	–

Note that since some weight functions are symmetric with respect to the parameters p, q, and r, e.g.,

$$W_{4,2}(x; p, q, r) = W_{4,1}(x; q, p, r) \quad \text{and} \quad W_{4,3}(x; p, q, r) = W_{4,1}(x; r, q, p),$$

their orthogonality supports and parameter restrictions can be directly derived via the rows of Table 2.2 by just interchanging the parameters. Also $W_7(x; p, q, r)$ and $W_8(x; p, q, r)$ have no valid orthogonality support. Therefore, there are 24 eligible orthogonality supports altogether for the first sequence.

- **A numerical example for the first infinite sequence**

  Let us replace $p = -9$, $q = -10$, and $r = -11$ in (2.127) to get

  $$S_n^* \left(\begin{array}{cc} 1 & 30 \\ 299 & 990 \end{array} \bigg| x \right) = \frac{(-9 + \sigma_n)_{[n/2]}(-10 + \sigma_n)_{[n/2]}(-11 + \sigma_n)_{[n/2]}}{(-31 + [n/2] + \sigma_n)_{[n/2]}}$$

  $$\times x^{\sigma_n} {}_4F_3 \left(\begin{array}{c} -[n/2], \ -31 + [n/2] + \sigma_n, \ \sigma_n - x, \ \sigma_n + x \\ -9 + \sigma_n, \ -10 + \sigma_n, \ -11 + \sigma_n \end{array} \bigg| 1 \right). \qquad (2.131)$$

  Since $p \in \mathbf{Z}^-$, $1 - q \pm p \notin \mathbf{Z}^-$, and $1 - r \pm p \notin \mathbf{Z}^-$, referring to Table 2.2 shows that there are two possible weight functions for the selected parameters that are orthogonal with respect to the sequence $S_n^*(x ; 1, 30, 299, 990)$ on the support $\{-9, -8, \ldots, 8, 9\}$, i.e.,

  $$W_1^* \left(\begin{array}{cc} 1 & 30 \\ 299 & 990 \end{array} \bigg| x \right) = \left(\frac{1}{4} - x^2 \right) W_1(x; -9, -10, -11)$$

  $$= (\Gamma(10 + x)\Gamma(10 - x)\Gamma(11 + x)\Gamma(11 - x)$$

  $$\times \ \Gamma(12 + x)\Gamma(12 - x)\Gamma(1/2 + x)\Gamma(1/2 - x))^{-1}$$

  and

  $$W_6^* \left(\begin{array}{cc} 1 & 30 \\ 299 & 990 \end{array} \bigg| x \right) = \left(\frac{1}{4} - x^2 \right) W_6(x; -9, -10, -11)$$

  $$= \frac{\Gamma(1/2 + x)\Gamma(1/2 - x)}{\Gamma(10 + x)\Gamma(10 - x)\Gamma(11 + x)\Gamma(11 - x)\Gamma(12 + x)\Gamma(12 - x)}.$$

Hence there are two orthogonality relations corresponding to the polynomials (2.131) respectively as follows:

$$\sum_{x=-9}^{9} W_i^* \left(\begin{array}{cc|} 1 & 30 \\ 299 & 990 \end{array} x \right) S_n^*(x \; ; \; 1, 30, 299, 990) S_m^*(x \; ; \; 1, 30, 299, 990)$$

$$= \alpha_i \frac{36183421612800000(-1)^n(-29)_{\lfloor \frac{n-1}{2} \rfloor - 1}(-19)_{\lfloor \frac{n}{2} \rfloor - 2}(-18)_{\lfloor \frac{n}{2} \rfloor - 2}(-17)_{\lfloor \frac{n}{2} \rfloor - 2}}{(-31)_n(-30)_n}$$

$$\times (-9)_{\lfloor \frac{n-1}{2} \rfloor - 1}(-8)_{\lfloor \frac{n-1}{2} \rfloor - 1}(-7)_{\lfloor \frac{n-1}{2} \rfloor - 1} \Gamma\left(\left\lfloor \frac{n}{2} \right\rfloor + 1 \right) \delta_{n,m},$$

where $\lfloor x \rfloor$ denotes the floor function, and

$$\alpha_i = \sum_{x=-9}^{9} W_i^* \left(\begin{array}{cc|} 1 & 30 \\ 299 & 990 \end{array} x \right) = \begin{cases} \beta/\pi, & i = 1, \\ \beta\pi, & i = 6, \end{cases}$$

in which $\beta = \dfrac{667}{19985300949284669299866050679181148160000000000}$.

Second Sequence   For $p, q \in \mathbb{R}$, if the characteristic vector

$$(a, b, c, d) = (0, 1, -p - q, pq)$$

is replaced in Eq. (2.110), then

$$(2x + 1)(x - p)(x - q)\Delta\nabla\phi_n(x) - 2x\left(2x^2 + p(2q - 1) - q\right)\Delta\phi_n(x)$$

$$+ \left(-4n\left(\frac{1}{4} - x^2\right) + \frac{1 - (-1)^n}{2}(2p - 1)(2q - 1)\right)\phi_n(x) = 0$$

has a basis solution as

$$\phi_n(x) = S_n^* \left(\begin{array}{cc|} 0 & 1 \\ -p - q & pq \end{array} x \right) = (p + \sigma_n)_{[n/2]}(q + \sigma_n)_{[n/2]}$$

$$\times x^{\sigma_n} {}_3F_2 \left(\begin{array}{c} -[n/2], \; \sigma_n - x, \; \sigma_n + x \\ p + \sigma_n, \; q + \sigma_n \end{array} \Bigg| 1 \right), \qquad (2.132)$$

satisfying a recurrence relation of type (2.105) with

$$\gamma_n \left(\begin{array}{cc|} 0 & 1 \\ -p - q & pq \end{array} \right) = \frac{1}{8}\left(-2n^2 - 4n(p + q - 1) + \left((-1)^n - 1\right)(2p - 1)(2q - 1)\right),$$

which implies

$$\gamma_{2n} = -n(-1 + n + p + q)$$

and

$$\gamma_{2n+1} = -(n + p)(n + q).$$

Hence the orthogonality relation corresponding to this sequence takes the form

$$\sum_{x=-\theta}^{\theta} \left(W^* \left(\begin{matrix} 0 & 1 \\ -p - q & pq \end{matrix} \middle| x \right) S_n^* \left(\begin{matrix} 0 & 1 \\ -p - q & pq \end{matrix} \middle| x \right) S_m^* \left(\begin{matrix} 0 & 1 \\ -p - q & pq \end{matrix} \middle| x \right) \right)$$

$$= \prod_{k=1}^{n} \gamma_k \left(\begin{matrix} 0 & 1 \\ -p - q & pq \end{matrix} \right) \left(\sum_{x=-\theta}^{\theta} W^* \left(\begin{matrix} 0 & 1 \\ -p - q & pq \end{matrix} \middle| x \right) \right) \delta_{n,m},$$

where

$$W^* \left(\begin{matrix} 0 & 1 \\ -p - q & pq \end{matrix} \middle| x \right) = \left(\frac{1}{4} - x^2 \right) W^*(x)$$

is the original weight function and $W^*(x)$ satisfies a particular case of the difference equation (2.119) as

$$\frac{W^*(x + 1)}{W^*(x)} = \frac{1/2 - x}{3/2 + x} \frac{-x - p}{x + 1 - p} \frac{-x - q}{x + 1 - q}. \qquad (2.133)$$

There exist eight symmetric solutions for Eq. (2.133) respectively as follows:

$W_9(x; p, q)$

$$= \frac{1}{\Gamma(1 - p + x)\Gamma(1 - p - x)\Gamma(1 - q + x)\Gamma(1 - q - x)\Gamma(3/2 + x)\Gamma(3/2 - x)},$$

$$W_{10,1}(x; p, q) = \frac{\Gamma(p + x)\Gamma(p - x)}{\Gamma(1 - q + x)\Gamma(1 - q - x)\Gamma(3/2 + x)\Gamma(3/2 - x)},$$

$$W_{10,2}(x; p, q) = \frac{\Gamma(q + x)\Gamma(q - x)}{\Gamma(1 - p + x)\Gamma(1 - p - x)\Gamma(3/2 + x)\Gamma(3/2 - x)},$$

$$W_{11}(x; p, q) = \frac{\Gamma(p + x)\Gamma(p - x)\Gamma(q + x)\Gamma(q - x)}{\Gamma(3/2 + x)\Gamma(3/2 - x)},$$

$$W_{12}(x; p, q) = \frac{\Gamma(-1/2 + x)\Gamma(-1/2 - x)}{\Gamma(1 - p + x)\Gamma(1 - p - x)\Gamma(1 - q + x)\Gamma(1 - q - x)},$$

$$W_{13,1}(x; p, q) = \frac{\Gamma(p + x)\Gamma(p - x)\Gamma(-1/2 + x)\Gamma(-1/2 - x)}{\Gamma(1 - q + x)\Gamma(1 - q - x)},$$

$$W_{13,2}(x; p, q) = \frac{\Gamma(q + x)\Gamma(q - x)\Gamma(-1/2 + x)\Gamma(-1/2 - x)}{\Gamma(1 - p + x)\Gamma(1 - p - x)},$$

$$W_{14}(x; p, q) = \Gamma(p + x)\Gamma(p - x)\Gamma(q + x)\Gamma(q - x)\Gamma(-1/2 + x)\Gamma(-1/2 - x).$$

Table 2.3 shows the orthogonality supports of each of the above-mentioned weights and their parameter restrictions.

As the main orthogonality relation (2.118) shows, an important part that we have to compute in norm square value is

$$\sum_{x=-\theta}^{\theta} W^*(x; a, b, c, d),$$

where $[-\theta, \theta]$ is the same orthogonality support as determined in Tables 2.2 and 2.3. Since $\{W_k(x; a, b, c, d)\}_{k=1}$ are all even functions, the aforementioned sums can be simplified on their orthogonality supports using the two identities

$$\Gamma(p + x) = \Gamma(p)(p)_x \quad \text{and} \quad \Gamma(p - x) = \frac{\Gamma(p)(-1)^x}{(-p)_x}$$

and the fact that

$$\sum_{x=-\theta}^{\theta} W^*(x; a, b, c, d) = 2\sum_{x=0}^{\theta} W^*(x; a, b, c, d) - W^*(0; a, b, c, d).$$

Table 2.3 Orthogonality supports for the second hypergeometric sequence

$W_k(x)$	Support	Parameter restrictions
$W_9(x; p, q)$	$[-p, p]$	$p \in \mathbf{Z}^-, 1 - q \pm p \notin \mathbf{Z}^-$.
	$[-q, q]$	$q \in \mathbf{Z}^-, 1 - p \pm q \notin \mathbf{Z}^-$.
$W_{10,1}(x; p, q)$	$[-q, q]$	$q \in \mathbf{Z}^-, p \pm q \notin \mathbf{Z}^-$.
$W_{11}(x; p, q)$	–	–
$W_{12}(x; p, q)$	$[-p, p]$	$p \in \mathbf{Z}^-, 1 - q \pm p \notin \mathbf{Z}^-$.
	$[-q, q]$	$q \in \mathbf{Z}^-, 1 - p \pm q \notin \mathbf{Z}^-$.
$W_{13,1}(x; p, q)$	$[-q, q]$	$q \in \mathbf{Z}^-, p \pm q \notin \mathbf{Z}^-$.
$W_{14}(x; p, q)$	–	–

For example, for the first given weight function $W_1(x; p, q, r)$ on, e.g., $[-p, p]$ we have

$$\sum_{x=-p}^{p} \left(\frac{1}{4} - x^2\right) W_1(x; p, q, r)$$

$$= \frac{1}{\pi \Gamma^2(1-p)\Gamma^2(1-q)\Gamma^2(1-r)} \left(2 \sum_{x=0}^{p} \frac{(p)_x (q)_x (r)_x}{(1-p)_x (1-q)_x (1-r)_x} - 1\right).$$

- **A numerical example for the second infinite sequence**

Let us replace $p = 14$ and $q = -10$ in (2.132) to get

$$S_n^* \left(\begin{array}{cc} 0 & 1 \\ -4 & -140 \end{array} \middle| x\right) = (14 + \sigma_n)_{[n/2]}(-10 + \sigma_n)_{[n/2]}$$

$$\times x^{\sigma_n} {}_3F_2 \left(\begin{array}{c} -[n/2], \sigma_n - x, \sigma_n + x \\ 14 + \sigma_n, -10 + \sigma_n \end{array} \middle| 1\right). \qquad (2.134)$$

Since $q \in \mathbf{Z}^-$ and $p \pm q \notin \mathbf{Z}^-$, referring to Table 2.3 shows that there are two possible weight functions for the selected parameters that are orthogonal with respect to the sequence $S_n^*(x; 0, 1, -4, -140)$ on the support $\{-10, -9, \ldots, 9, 10\}$, i.e.,

$$W_{10,1}^* \left(\begin{array}{cc} 0 & 1 \\ -4 & -140 \end{array} \middle| x\right) = \left(\frac{1}{4} - x^2\right) W_{10,1}(x; 14, -10)$$

$$= \frac{(13-x)(12-x)(11-x)(x+11)(x+12)(x+13)\cos(\pi x)}{\pi}$$

and

$$W_{13,1}^* \left(\begin{array}{cc} 0 & 1 \\ -4 & -140 \end{array} \middle| x\right) = \left(\frac{1}{4} - x^2\right) W_{13,1}(x; 14, -10)$$

$$= \pi(13-x)(12-x)(11-x)(x+11)(x+12)(x+13)\sec(\pi x).$$

Hence there are two orthogonality relations corresponding to the polynomials (2.134) respectively as follows:

$$\sum_{x=-10}^{10} W_{j,1}^* \left(\begin{array}{cc} 0 & 1 \\ -4 & -140 \end{array} \middle| x\right) S_n^*(x; 0, 1, -4, -140) S_m^*(x; 0, 1, -4, -140)$$

$$= \tilde{\alpha}_j \frac{(-1)^{n+1}(-9)_{\lfloor \frac{n-1}{2} \rfloor} \Gamma\left(\lfloor \frac{n-1}{2} \rfloor + 15\right) \Gamma\left(\lfloor \frac{n}{2} \rfloor + 1\right) \Gamma\left(\lfloor \frac{n}{2} \rfloor + 4\right)}{3736212480},$$

where

$$\tilde{\alpha}_j = \sum_{x=-10}^{10} W_{j,1}^* \begin{pmatrix} 0 & 1 \\ -4 & -140 \end{pmatrix} x = \begin{cases} 10296/\pi, & j = 10, \\ 10296\pi, & j = 13. \end{cases}$$

2.5.2 Moments of the Two Introduced Infinite Sequences

To compute the moments of a continuous distribution, different bases are usually considered. For example, for the normal distribution, the classical basis $\{x^j\}_j$ is used to get

$$\int_{-\infty}^{\infty} x^n \frac{e^{-x^2/2}}{\sqrt{2\pi}} dx = \begin{cases} 0, & n = 2m + 1, \\ \dfrac{2^{m+1/2}}{\sqrt{2\pi}} \Gamma\left(m + \dfrac{1}{2}\right), & n = 2m, \end{cases}$$

while for the Jacobi weight function $(1 - x)^\alpha (1 + x)^\beta$ as the shifted beta distribution on $[-1, 1]$, using one of the two bases $\{(1 - x)^j\}_j$ or $\{(1 + x)^j\}_j$ is appropriate for this purpose.

This matter similarly holds for the moments of discrete orthogonal polynomials. For instance, in the negative hypergeometric distribution corresponding to Hahn polynomials, it is more convenient to use the Pochhammer basis $\{(-x)_n\}_n$, instead of the classical basis, to get

$$\sum_{x=0}^{N-1} \frac{\Gamma(N)\Gamma(\alpha + \beta + 2)\Gamma(\alpha + N - x)\Gamma(\beta + x + 1)}{\Gamma(\alpha + 1)\Gamma(\beta + 1)\Gamma(\alpha + \beta + N + 1)\Gamma(N - x)\Gamma(x + 1)} (-x)_n$$

$$= (-1)^n \frac{(1 - N)_n (\beta + 1)_n}{(\alpha + \beta + 2)_n}.$$

Following this approach, for the weight functions $\varrho_i(x)$ appearing in the two introduced hypergeometric sequences, we can compute the moments of the form

$$(\varrho_i)_n = \sum_{x=-\theta}^{\theta} \vartheta_n(x)\varrho_i(x) = \sum_{x=-\theta}^{\theta} \vartheta_n(x)\left(\frac{1}{4} - x^2\right) W_i(x),$$

where the basis $\vartheta_n(x)$ is defined in (2.113).

Since $\vartheta_{2n+1}(x)$ are odd polynomials, clearly all odd moments with respect to this basis are zero. Moreover, from definitions (2.127) and (2.132) and using their orthogonality

property, it can be proved by induction that the even moments corresponding to the first
and second sequences respectively satisfy the following recurrence relations:

$$(\varrho_i)_{2n} = -\frac{(p+n-1)(q+n-1)(r+n-1)}{p+q+r+n-1}(\varrho_i)_{2n-2}$$

and

$$(\varrho_j)_{2n} = -(n+p-1)(n+q-1)(\varrho_j)_{2n-2}.$$

Therefore, if these weight functions are normalized with the first moment equal to one, we
eventually obtain

$$\sum_{x=-\theta}^{\theta}\left(\frac{1}{4}-x^2\right)W_i(x)\vartheta_n(x)=\begin{cases}0, & n=2m+1,\\[2mm]\dfrac{(-1)^m(p)_m(q)_m(r)_m}{(p+q+r)_m}, & n=2m,\end{cases}$$

for the weight functions of the first sequence, and

$$\sum_{x=-\theta}^{\theta}\left(\frac{1}{4}-x^2\right)W_j(x)\vartheta_n(x)=\begin{cases}0, & n=2m+1,\\[2mm](-1)^m(p)_m(q)_m, & n=2m,\end{cases}$$

for the weight functions of the second sequence.

2.5.3   A Special Case of $S_n^*(x\,;\,a,b,c,d)$ Generating All Classical Symmetric Orthogonal Polynomials of a Discrete Variable

In this part, we introduce a particular example of $S_n^*(x\,;\,a,b,c,d)$ that generates all the
classical symmetric orthogonal polynomials of a discrete variable studied in Sect. 2.3 and
is different from the two introduced infinite hypergeometric sequences.

If $a = -2(b + 2c + 4d)$ is replaced in the main difference equation (2.110), after
simplification of a common factor we get

$$\left(x^2(b+2c+4d)+x(c+2d)+d\right)\Delta\nabla\phi_n(x)-2x(c+2d)\Delta\phi_n(x)$$

$$+n(-(b(n-1)+2(n-2)(c+2d)))\phi_n(x)=0,$$

which is the same equation as analyzed in Sect. 2.3, i.e.,

$$(\hat{a}x^2+\hat{b}x+\hat{c})\Delta\nabla y_n(x)-2\hat{b}x\Delta y(x)+n(\hat{a}(1-n)+2\hat{b})y(x)=0,$$

for

$$d = \hat{c}, \quad c = \hat{b} - 2\hat{c}, \quad b = \hat{a} - 2\hat{b}, \quad \text{and} \quad a = -2\hat{a}.$$

This means that the particular polynomial

$$y_n(x) = S_n^* \left(\begin{matrix} -2\hat{a} & \hat{a} - 2\hat{b} \\ \hat{b} - 2\hat{c} & \hat{c} \end{matrix} \middle| x \right) = x^{\sigma_n} \sum_{j=0}^{[n/2]} (-1)^j \binom{[n/2]}{j}$$

$$\times \left(\prod_{i=j}^{[n/2]-1} \frac{(2i + 2\sigma_n + 1)\left(-\hat{a}(i + \sigma_n)^2 + \hat{b}(i + \sigma_n) - \hat{c}\right)}{2\left(\hat{b} + \hat{a}(\sigma_{n+1} - i - [n/2])\right) - \hat{a}} \right) (\sigma_n - x)_j (\sigma_n + x)_j$$

(2.135)

generates all cases of classical symmetric orthogonal polynomials of a discrete variable as follows:

Case 1 If $\hat{a} = 0$, $\hat{b} = 1$, and $\hat{c}$ is free in (2.135), then $\sigma(x) = \hat{a}x^2 + \hat{b}x + \hat{c} = x + \hat{c}$ has one real root, and the symmetric Kravchuk polynomials are derived as

$$S_n^* \left(\begin{matrix} 0 & -2 \\ 1 - 2\hat{c} & \hat{c} \end{matrix} \middle| x \right) = 2^{-n}(-2\hat{c})_n \, {}_2F_1 \left(\begin{matrix} -n, -\hat{c} - x \\ -2\hat{c} \end{matrix} \middle| 2 \right) = k_n^{(1/2)}(x + \hat{c}; 2\hat{c}),$$

which are orthogonal with respect to the weight function

$$\varrho(x) = \frac{1}{\Gamma(\hat{c} - x + 1)\Gamma(\hat{c} + x + 1)} \quad \text{for } x \in \{-\hat{c}, -\hat{c} + 1, \ldots, \hat{c} - 1, \hat{c}\} \text{ when } 2\hat{c} \in \mathbb{N}.$$

Case 2 If $\hat{a} = 1$, $\hat{b}$ is free, and $\hat{c} = \hat{b}^2/4$ in (2.135), then $\sigma(x) = \hat{a}x^2 + \hat{b}x + \hat{c} = (x + \hat{b}/2)^2$ has a double real root and the symmetric Hahn–Eberlein polynomials

$$S_n^* \left(\begin{matrix} -2 & 1 - 2\hat{b} \\ \hat{b}(2 - \hat{b})/2 & \hat{b}^2/4 \end{matrix} \middle| x \right) = \frac{((-\hat{b})_n)^2}{(n - 2\hat{b} - 1)_n}$$

$$\times \, {}_3F_2 \left(\begin{matrix} -n, n - 2\hat{b} - 1, -x - \hat{b}/2 \\ -\hat{b}, -\hat{b} \end{matrix} \middle| 1 \right) = \tilde{h}_n^{(0,0)}(x + \hat{b}/2, \hat{b} + 1)$$

are orthogonal with respect to the weight function

$$\varrho(x) = \frac{1}{\Gamma^2(x + 1 + \hat{b}/2)\Gamma^2(-x + 1 + \hat{b}/2)},$$

for $x \in \{-\hat{b}/2, -\hat{b}/2 + 1, \ldots, \hat{b}/2 - 1, \hat{b}/2\}$ when $b \in \mathbb{N}$.

Case 3 If $\hat{a} = 1$, $\hat{b} = -\delta_1 - \delta_2$, and $\hat{c} = \delta_1\delta_2$ in (2.135), then $\sigma(x)$ has two different real roots δ_1 and δ_2, and depending on the values δ_1 and δ_2, the symmetric Hahn–Eberlein polynomials are derived as

$$S_n^* \left(\begin{array}{cc} -2 & 1 + 2(\delta_1 + \delta_2) \\ -\delta_2 - \delta_1(1 + 2\delta_2) & \delta_1\delta_2 \end{array} \middle| x \right)$$

$$= \frac{(\delta_1 + \delta_2)_n \, (2\delta_2)_n}{(n + 2(\delta_1 + \delta_2) - 1)_n} \, {}_3F_2 \left(\begin{array}{c} -n, n + 2(\delta_1 + \delta_2) - 1, \delta_2 - x \\ \delta_1 + \delta_2, 2\delta_2 \end{array} \middle| 1 \right)$$

$$= \tilde{h}_n^{(\delta_2 - \delta_1, \delta_2 - \delta_1)}(x - \delta_2; 1 - 2\delta_2),$$

when $-2\delta_2 \in \mathbb{N}$, while the symmetric Hahn polynomials

$$S_n^* \left(\begin{array}{cc} -2 & 1 + 2(\delta_1 + \delta_2) \\ -\delta_2 - \delta_1(1 + 2\delta_2) & \delta_1\delta_2 \end{array} \middle| x \right)$$

$$= \frac{(\delta_1 + \delta_2)_n \, (2\delta_1)_n}{(n + 2(\delta_1 + \delta_2) - 1)_n} \, {}_3F_2 \left(\begin{array}{c} -n, n + 2(\delta_1 + \delta_2) - 1, \delta_1 - x \\ \delta_1 + \delta_2, 2\delta_1 \end{array} \middle| 1 \right)$$

$$= h_n^{(\delta_1 + \delta_2 - 1, \delta_1 + \delta_1 - 1)}(x - \delta_1; 1 - 2\delta_1),$$

are derived when $-2\delta_1 \in \mathbb{N}$.

2.5.4   Two Finite Types of $S_n^*(x \, ; \, a, b, c, d)$

In this section, we introduce two finite sequences of hypergeometric symmetric orthogonal polynomials of a discrete variable and obtain their basic properties.

To define such polynomials, we need to recall that the limit definition

$$\Gamma(z) = \lim_{n \to \infty} \frac{n! \, n^z}{\prod\limits_{k=0}^{n} z + k}$$

implies that

$$\Gamma(p + iq) \, \Gamma(p - iq) = \Gamma^2(p) \prod_{k=0}^{\infty} \frac{(p + k)^2}{(p + k)^2 + q^2} \tag{2.136}$$

is always a positive real value for every $p > 0$ and $q \in \mathbb{R}$.

First Finite Sequence For $p, q, r \in \mathbb{R}$, consider the equation

$$(2x + 1)(x - p)(x^2 - 2qx + q^2 + r^2)\Delta\nabla\phi_n(x)$$
$$- 2x\left((1 - 2p - 4q)x^2 + 2pq + (q^2 + r^2)(1 - 2p)\right)\Delta\phi_n(x)$$
$$+ \left(2n(n + 2(p + 2q - 1))(\frac{1}{4} - x^2) - \frac{\sigma_n}{2}(2p - 1)(4r^2 + (2q - 1)^2)\right)\phi_n(x) = 0,$$

having a monic polynomial solution

$$S_n^*\left(\begin{matrix} 1, & -(p + 2q) \\ 2pq + q^2 + r^2, & -p(q^2 + r^2) \end{matrix}\middle| x\right) = \bar{\phi}_n(x; p, q, r)$$

$$= \frac{(p + \sigma_n)_{[n/2]}(q + ir + \sigma_n)_{[n/2]}(q - ir + \sigma_n)_{[n/2]}}{([n/2] + p + 2q - 1 + \sigma_n)_{[n/2]}}$$

$$\times x^{\sigma_n}\,_4F_3\left(\begin{matrix} -[n/2], \ [n/2] + p + 2q - 1 + \sigma_n, \ \sigma_n - x, \ \sigma_n + x \\ p + \sigma_n, \ q + ir + \sigma_n, \ q - ir + \sigma_n \end{matrix}\middle| 1\right) \qquad (2.137)$$

that satisfies the recurrence relation

$$\bar{\phi}_{n+1}(x; p, q, r) = x\bar{\phi}_n(x; p, q, r) - \gamma_n(p, q, r)\,\bar{\phi}_{n-1}(x; p, q, r), \qquad (2.138)$$

in which

$$32(n + p + 2q - 2)(n + p + 2q - 1)\gamma_n(p, q, r) = \qquad\qquad\qquad (2.139)$$
$$- 2n^4 + 4(3 - 2p - 4q)n^3 - 8(p^2 + 5q^2 + r^2 + 6pq - 4p - 8q + 3)n^2$$
$$+ 2\left((3 - 2p - 4q)(3 - 4p - 8q(1 - p) + 4q^2 + 4r^2) + (-1)^n(2p - 1)(4r^2 + (2q - 1)^2)\right)n$$
$$- (1 - (-1)^n)(2p - 1)(2p + 4q - 3)(4r^2 + (2q - 1)^2).$$

Relation (2.139) can be decomposed as

$$\gamma_n(p, q, r) = -\frac{1}{16(n + p + 2q - 2)(n + p + 2q - 1)}$$
$$\times \left(n + (2p - 1)\sigma_n\right)\left(n + 2q + 2ri - 1 + (2p - 1)(1 - \sigma_n)\right)$$
$$\times \left(n + 2q - 2ri - 1 + (2p - 1)(1 - \sigma_n)\right)\left(n + 4q - 2 + (2p - 1)\sigma_n\right). \qquad (2.140)$$

By noting the relations (2.137)–(2.140), the orthogonality relation of the first finite sequence now takes the general form

$$\sum_{x=-\theta}^{\theta} W^* \left(\begin{matrix} 1 & -(p+2q) \\ 2pq+q^2+r^2 & -p(q^2+r^2) \end{matrix} \middle| x \right) \bar{\phi}_n(x; p,q,r) \, \bar{\phi}_m(x; p,q,r)$$

$$= \prod_{k=1}^{n} \gamma_k(p,q,r) \left(\sum_{x=-\theta}^{\theta} W^* \left(\begin{matrix} 1 & -(p+2q) \\ 2pq+q^2+r^2 & -p(q^2+r^2) \end{matrix} \middle| x \right) \right) \delta_{n,m},$$

in which

$$W^* \left(\begin{matrix} 1 & -(p+2q) \\ 2pq+q^2+r^2 & -p(q^2+r^2) \end{matrix} \middle| x \right) = \left(\frac{1}{4} - x^2 \right) W^*(x)$$

denotes the original weight function and $W^*(x)$ satisfies the difference equation

$$\frac{W^*(x+1)}{W^*(x)} = \frac{(1/2)-x}{(3/2)+x} \frac{-x-p}{x+1-p} \frac{-x-q-ir}{x+1-q-ir} \frac{-x-q+ir}{x+1-q+ir}. \tag{2.141}$$

There are 16 symmetric solutions for Eq. (2.141) as follows:

$$W_1(x) = \Big(\Gamma(1-p+x)\Gamma(1-p-x)\Gamma(1-q-ir+x)\Gamma(1-q-ir-x)$$

$$\times \Gamma(1-q+ir+x)\Gamma(1-q+ir-x)\Gamma(3/2+x)\Gamma(3/2-x) \Big)^{-1},$$

$$W_2(x) =$$

$$\frac{\Gamma(p+x)\Gamma(p-x)}{\Gamma(1-q-ir+x)\Gamma(1-q-ir-x)\Gamma(1-q+ir+x)\Gamma(1-q+ir-x)\Gamma(3/2+x)\Gamma(3/2-x)},$$

$$W_3(x) =$$

$$\frac{\Gamma(q+ir+x)\Gamma(q+ir-x)}{\Gamma(1-p+x)\Gamma(1-p-x)\Gamma(1-q+ir+x)\Gamma(1-q+ir-x)\Gamma(3/2+x)\Gamma(3/2-x)},$$

$$W_4(x) =$$

$$\frac{\Gamma(q-ir+x)\Gamma(q-ir-x)}{\Gamma(1-p+x)\Gamma(1-p-x)\Gamma(1-q-ir+x)\Gamma(1-q-ir-x)\Gamma(3/2+x)\Gamma(3/2-x)},$$

$$W_5(x) = \frac{\Gamma(p+x)\Gamma(p-x)\Gamma(q+ir+x)\Gamma(q+ir-x)}{\Gamma(1-q+ir+x)\Gamma(1-q+ir-x)\Gamma(3/2+x)\Gamma(3/2-x)},$$

$$W_6(x) = \frac{\Gamma(p+x)\Gamma(p-x)\Gamma(q-ir+x)\Gamma(q-ir-x)}{\Gamma(1-q-ir+x)\Gamma(1-q-ir-x)\Gamma(3/2+x)\Gamma(3/2-x)},$$

$$W_7(x) = \frac{\Gamma(q+ir+x)\Gamma(q+ir-x)\Gamma(q-ir+x)\Gamma(q-ir-x)}{\Gamma(1-p+x)\Gamma(1-p-x)\Gamma(3/2+x)\Gamma(3/2-x)},$$

$$W_8(x) = \frac{\Gamma(p+x)\Gamma(p-x)\Gamma(-1/2+x)\Gamma(-1/2-x)}{\Gamma(1-q-ir+x)\Gamma(1-q-ir-x)\Gamma(1-q+ir+x)\Gamma(1-q+ir-x)},$$

$$W_9(x) = \frac{\Gamma(q+ir+x)\Gamma(q+ir-x)\Gamma(-1/2+x)\Gamma(-1/2-x)}{\Gamma(1-p+x)\Gamma(1-p-x)\Gamma(1-q+ir+x)\Gamma(1-q+ir-x)},$$

$$W_{10}(x) = \frac{\Gamma(q-ir+x)\Gamma(q-ir-x)\Gamma(-1/2+x)\Gamma(-1/2-x)}{\Gamma(1-p+x)\Gamma(1-p-x)\Gamma(1-q-ir+x)\Gamma(1-q-ir-x)},$$

$$W_{11}(x) = \frac{\Gamma(p+x)\Gamma(p-x)\Gamma(q+ir+x)\Gamma(q+ir-x)\Gamma(-1/2+x)\Gamma(-1/2-x)}{\Gamma(1-q+ir+x)\Gamma(1-q+ir-x)},$$

$$W_{12}(x) = \frac{\Gamma(p+x)\Gamma(p-x)\Gamma(q-ir+x)\Gamma(q-ir-x)\Gamma(-1/2+x)\Gamma(-1/2-x)}{\Gamma(1-q-ir+x)\Gamma(1-q-ir-x)},$$

$$W_{13}(x) =$$
$$\frac{\Gamma(q+ir+x)\Gamma(q+ir-x)\Gamma(q-ir+x)\Gamma(q-ir-x)\Gamma(-1/2+x)\Gamma(-1/2-x)}{\Gamma(1-p+x)\Gamma(1-p-x)},$$

$$W_{14}(x) =$$
$$\frac{\Gamma(-1/2+x)\Gamma(-1/2-x)}{\Gamma(1-p+x)\Gamma(1-p-x)\Gamma(1-q-ir+x)\Gamma(1-q-ir-x)\Gamma(1-q+ir+x)\Gamma(1-q+ir-x)},$$

$$W_{15}(x) = \frac{\Gamma(p+x)\Gamma(p-x)\Gamma(q+ir+x)\Gamma(q+ir-x)\Gamma(q-ir+x)\Gamma(q-ir-x)}{\Gamma(3/2+x)\Gamma(3/2-x)},$$

and

$$W_{16}(x) = \Gamma(p+x)\Gamma(p-x)\Gamma(q+ir+x)\Gamma(q+ir-x)$$
$$\times \Gamma(q-ir+x)\Gamma(q-ir-x)\Gamma(-1/2+x)\Gamma(-1/2-x).$$

Although $\{W_k(x)\}_{k=1}^{16}$ are all symmetric, only the three weight functions W_1, W_2, and W_{14} are applicable, since they are real-valued functions according to (2.136). On the other hand, using the two identities

$$\Gamma(p+x) = \Gamma(p)(p)_x \quad \text{and} \quad \Gamma(p-x) = \frac{\Gamma(p)(-1)^x}{(1-p)_x},$$

and the fact that

$$\sum_{x=-\theta}^{\theta} \left(\frac{1}{4} - x^2\right)W^*(x) = 2\sum_{x=0}^{\theta} \left(\frac{1}{4} - x^2\right)W^*(x) - \frac{1}{4}W^*(0),$$

if we assume

$$B(p,q,r) = 2\ _4F_3 \left(\begin{array}{c} q+ir, \ q-ir, \ p, \ 1 \\ 1-q+ir, \ 1-q-ir, \ 1-p \end{array} \middle| 1 \right) - 1,$$

then after some computations we obtain

$$\sum_{x=-\infty}^{\infty} \left(\frac{1}{4} - x^2\right)W_1(x) = \frac{B(p,q,r)}{\pi\, \Gamma^2(1-p)\Gamma^2(1-q-ir)\Gamma^2(1-q+ir)},$$

$$\sum_{x=-\infty}^{\infty} \left(\frac{1}{4} - x^2\right)W_2(x) = \frac{\Gamma^2(p)}{\pi\, \Gamma^2(1-q-ir)\,\Gamma^2(1-q+ir)}\, B(p,q,r),$$

and

$$\sum_{x=-\infty}^{\infty} \left(\frac{1}{4} - x^2\right)W_{14}(x) = \frac{\pi\, B(p,q,r)}{\Gamma^2(1-p)\Gamma^2(1-q-ir)\Gamma^2(1-q+ir)}.$$

The above computations show that there exists a unique representation for the original weight function as

$$W(x; p,q,r) = \frac{(p)_x\,(q+ir)_x\,(q-ir)_x}{(1-p)_x\,(1-q+ir)_x\,(1-q-ir)_x}. \tag{2.142}$$

Also, by noting the identity

$$(p)_{-x} = \frac{(-1)^x}{(1-p)_x},$$

(2.143)

we see that (2.142) is symmetric on the support $(-\infty, \infty)$, i.e., using (2.143) we can prove that

$$W(-x; p, q, r) = W(x; p, q, r).$$

Now, to prove the finite orthogonality, since in the first sequence

$$A(x) = (2x + 1)(x - p)(x^2 - 2qx + q^2 + r^2)$$

and

$$W^*(x) = \frac{W(x; p, q, r)}{(1/4) - x^2} = \frac{K}{1 - 4x^2} \frac{\Gamma(p + x)\, \Gamma(q + ir + x)\, \Gamma(q - ir + x)}{\Gamma(1 - p + x)\, \Gamma(1 - q + ir + x)\, \Gamma(1 - q - ir + x)},$$

where

$$K = 4 \frac{\Gamma(1 - p)\, \Gamma(1 - q + ir)\, \Gamma(1 - q - ir)}{\Gamma(p)\, \Gamma(q + ir)\, \Gamma(q - ir)},$$

the key condition

$$\lim_{\theta \to \infty} A(-\theta) W^*(\theta)\, \theta^{n+m} = 0$$

(2.144)

implies that we have

$$\lim_{\theta \to \infty} \frac{(2\theta - 1)(\theta + p)(\theta^2 + 2q\theta + q^2 + r^2)\Gamma(p + \theta)\Gamma(q + ir + \theta)\Gamma(q - ir + \theta)}{(1 - 4\theta^2)\Gamma(1 - p + \theta)\Gamma(1 - q + ir + \theta)\Gamma(1 - q - ir + \theta)}\, \theta^{n+m} = 0.$$

(2.145)

On the other hand, since

$$\lim_{\theta \to \infty} \frac{(2\theta - 1)(\theta + p)}{1 - 4\theta^2} = -\frac{1}{2}$$

and

$$\lim_{\theta \to \infty} \frac{1}{\ln \theta} \ln \frac{\Gamma(p + \theta)\Gamma(q + ir + \theta)\Gamma(q - ir + \theta)}{\Gamma(1 - p + \theta)\Gamma(1 - q + ir + \theta)\Gamma(1 - q - ir + \theta)} = 2p + 4q - 3,$$

relation (2.145) is equivalent to

$$\lim_{\theta \to \infty} \theta^{2+n+m+2p+4q-3} = 0. \tag{2.146}$$

If in (2.146) we assume that $\max\{m, n\} = N$, then it yields

$$2N + 2p + 4q - 1 < 0 \quad \text{or} \quad N < \frac{1}{2} - p - 2q.$$

Theorem 2.5 *If*

$$\phi_n(x; p, q, r) = x^{\sigma_n} \, {}_4F_3 \left(\begin{matrix} -[n/2], \ [n/2] + p + 2q - 1 + \sigma_n, \ \sigma_n - x, \ \sigma_n + x \\ p + \sigma_n, \ q + ir + \sigma_n, \ q - ir + \sigma_n \end{matrix} \,\middle|\, 1 \right),$$

then the polynomial set $\{\phi_n(x; p, q, r)\}_{n=0}^{N}$, where $N < \frac{1}{2} - p - 2q$, is finitely orthogonal with respect to the weight function (2.142) on the support $(-\infty, \infty)$, so that we have

$$\sum_{x=-\infty}^{\infty} \frac{(p)_x \, (q + ir)_x \, (q - ir)_x}{(1 - p)_x \, (1 - q + ir)_x \, (1 - q - ir)_x} \, \phi_n(x; p, q, r) \, \phi_m(x; p, q, r) =$$

$$\frac{[n/2]!(p + 2q - 1 + \sigma_n)_{[n/2]}\left((p + 2q - 1 + \sigma_n + [n/2])_{[n/2]}\right)^2 (2q)_{[n/2]}(p + q + ir)_{[n/2]}(p + q - ir)_{[n/2]}}{4^{n-\sigma_n}(p + \sigma_n)_{[n/2]}(p/2 + q + \sigma_n)_{[n/2]}((p-1)/2 + q)_{[n/2]}(p/2 + q)_{[n/2]}((p+1)/2 + q)_{[n/2]}(q + \sigma_n + ir)_{[n/2]}(q + \sigma_n - ir)_{[n/2]}}$$

$$\times \left({}_2{}_4F_3 \left(\begin{matrix} q + ir \ q - ir \ p \ 1 \\ 1 - q + ir \ 1 - q - ir \ 1 - p \end{matrix} \,\middle|\, 1 \right) - 1 \right) \delta_{n,m}. \tag{2.147}$$

Moreover, the orthogonality (2.147) is valid on $(-\infty, \infty)$ only if

$$2q, \, p + 2q \notin \mathbb{Z}^-, \ p + 2q < 1, \ p \in (0, 1), \ \text{and} \ r \in \mathbb{R}.$$

For instance, the finite set $\{\phi_n(x; 1/3, -31/3, 2)\}_{n=0}^{N=20}$ is orthogonal with respect to the weight function

$$\frac{(1/3)_x((-31/3) + 2i)_x((-31/3) - 2i)_x}{(2/3)_x((34/3) + 2i)_x((34/3) - 2i)_x}$$

on the support $(-\infty, \infty)$.

Second Finite Sequence   For $p, q \in \mathbb{R}$, consider the equation

$$(2x + 1)(x^2 - 2px + p^2 + q^2)\Delta\nabla\phi_n^*(x) - 4x\left(x^2 + p^2 + q^2 - p\right)\Delta\phi_n^*(x)$$

$$+ \left(-4n\left(\frac{1}{4} - x^2\right) + \sigma_n\left((2p - 1)^2 + 4q^2\right) \right)\phi_n^*(x) = 0,$$

having a monic polynomial solution

$$S_n^* \left(\begin{matrix} 0 & 1 \\ -2p & p^2 + q^2 \end{matrix} \middle| x \right) = \bar{\phi}_n^*(x; p, q)$$

$$= (p + iq + \sigma_n)_{[n/2]} (p - iq + \sigma_n)_{[n/2]} x^{\sigma_n} \, {}_3F_2 \left(\begin{matrix} -[n/2], \ \sigma_n - x, \ \sigma_n + x \\ p + iq + \sigma_n, \ p - iq + \sigma_n \end{matrix} \middle| 1 \right)$$

that satisfies a recurrence relation of type (2.105) and (2.111) with

$$\gamma_n^*(p, q) = -\frac{1}{4} \left(n^2 + 2n(2p - 1) + \sigma_n \left(4q^2 + (2p - 1)^2 \right) \right)$$

$$= -\frac{1}{4} \left(n + (2p - 1 - 2qi)\sigma_n \right) \left(n + 3p + qi - \frac{3}{2} + (-1)^n \left(p - \frac{1}{2} - qi \right) \right).$$

Hence the orthogonality relation corresponding to the second sequence takes the general form

$$\sum_{x=-\theta}^{\theta} W^* \left(\begin{matrix} 0 & 1 \\ -2p & p^2 + q^2 \end{matrix} \middle| x \right) \bar{\phi}_n^*(x; p, q) \, \bar{\phi}_m^*(x; p, q)$$

$$= \prod_{k=1}^{n} \gamma_k^*(p, q) \left(\sum_{x=-\theta}^{\theta} W^* \left(\begin{matrix} 0 & 1 \\ -2p & p^2 + q^2 \end{matrix} \middle| x \right) \right) \delta_{n,m},$$

where

$$W^* \left(\begin{matrix} 0 & 1 \\ -2p & p^2 + q^2 \end{matrix} \middle| x \right) = \left(\frac{1}{4} - x^2 \right) W^*(x)$$

is the original weight function and $W^*(x)$ satisfies the difference equation

$$\frac{W^*(x + 1)}{W^*(x)} = \frac{(1/2) - x}{(3/2) + x} \frac{-x - p - iq}{x + 1 - p - iq} \frac{-x - p + iq}{x + 1 - p + iq}. \tag{2.148}$$

There are eight symmetric solutions for Eq. (2.148) as follows:

$$W_{17}(x) =$$

$$\frac{1}{\Gamma(1 - p - iq + x)\Gamma(1 - p - iq - x)\Gamma(1 - p + iq + x)\Gamma(1 - p + iq - x)\Gamma(3/2 + x)\Gamma(3/2 - x)},$$

$$W_{18}(x) = \frac{\Gamma(p+iq+x)\Gamma(p+iq-x)}{\Gamma(1-p+iq+x)\Gamma(1-p+iq-x)\Gamma(3/2+x)\Gamma(3/2-x)},$$

$$W_{19}(x) = \frac{\Gamma(p-iq+x)\Gamma(p-iq-x)}{\Gamma(1-p-iq+x)\Gamma(1-p-iq-x)\Gamma(3/2+x)\Gamma(3/2-x)},$$

$$W_{20}(x) = \frac{\Gamma(p+iq+x)\Gamma(p+iq-x)\Gamma(p-iq+x)\Gamma(p-iq-x)}{\Gamma(3/2+x)\Gamma(3/2-x)},$$

$$W_{21}(x) = \frac{\Gamma(-1/2+x)\Gamma(-1/2-x)}{\Gamma(1-p-iq+x)\Gamma(1-p-iq-x)\Gamma(1-p+iq+x)\Gamma(1-p+iq-x)},$$

$$W_{22}(x) = \frac{\Gamma(p+iq+x)\Gamma(p+iq-x)\Gamma(-1/2+x)\Gamma(-1/2-x)}{\Gamma(1-p+iq+x)\Gamma(1-p+iq-x)},$$

$$W_{23}(x) = \frac{\Gamma(p-iq+x)\Gamma(p-iq-x)\Gamma(-1/2+x)\Gamma(-1/2-x)}{\Gamma(1-p-iq+x)\Gamma(1-p-iq-x)},$$

and

$$W_{24}(x) = \Gamma(p+iq+x)\Gamma(p+iq-x)\Gamma(p-iq+x)\Gamma(p-iq-x)\Gamma(-1/2+x)\Gamma(-1/2-x).$$

Once again, although $\{W_k\}_{k=17}^{24}$ are all symmetric, only two weight functions W_{17} and W_{21} are real-valued functions according to (2.136). Hence, similarly to the previous case, if we assume

$$C(p,q) = 2 \, {}_3F_2 \left(\begin{matrix} p+iq, \ p-iq, \ 1 \\ 1-p+iq, \ 1-p-iq \end{matrix} \middle| 1 \right) - 1,$$

then we obtain

$$\sum_{x=-\infty}^{\infty} \left(\frac{1}{4} - x^2 \right) W_{17}(x) = \frac{C(p,q)}{\pi \, \Gamma^2(1-p-iq) \, \Gamma^2(1-p+iq)}$$

and

$$\sum_{x=-\infty}^{\infty} \left(\frac{1}{4} - x^2 \right) W_{21}(x) = \frac{\pi \, C(p,q)}{\Gamma^2(1-p-iq) \, \Gamma^2(1-p+iq)}.$$

These computations show that there exists a unique representation for the original weight function as

$$W(x; p, q) = \frac{(p+iq)_x \, (p-iq)_x}{(1-p+iq)_x \, (1-p-iq)_x} = W(-x; p, q), \tag{2.149}$$

which can be summed by Dougall's bilateral sum

$$\sum_{n=-\infty}^{\infty} \frac{\Gamma(a+n)\Gamma(b+n)}{\Gamma(c+n)\Gamma(d+n)} = \frac{\Gamma(a)\Gamma(1-a)\Gamma(b)\Gamma(1-b)\Gamma(c+d-a-b-1)}{\Gamma(c-a)\Gamma(c-b)\Gamma(d-a)\Gamma(d-b)}$$

as

$$C(p,q) = \sum_{x=-\infty}^{\infty} \frac{(p+iq)_x\,(p-iq)_x}{(1-p+iq)_x\,(1-p-iq)_x}$$

$$= \frac{\Gamma^2(1-p+iq)\,\Gamma^2(1-p-iq)\,\Gamma(1-4p)}{\Gamma(1-2p+2iq)\,\Gamma(1-2p-2iq)\,\Gamma^2(1-2p)},$$

where $p < \frac{1}{4}$ and $q \in \mathbb{R}$.

Now we can apply the key condition (2.144) for

$$A(x) = (2x+1)(x^2 - 2px + p^2 + q^2)$$

and

$$W^*(x) = \frac{W(x;p,q)}{(1/4) - x^2} = \frac{K^*}{1-4x^2}\,\frac{\Gamma(p+iq+x)\,\Gamma(p-iq+x)}{\Gamma(1-p+iq+x)\,\Gamma(1-p-iq+x)},$$

where

$$K^* = 4\,\frac{\Gamma(1-p+iq)\Gamma(1-p-iq)}{\Gamma(p+iq)\Gamma(p-iq)}.$$

So, noting that

$$\lim_{\theta \to \infty} \frac{1}{\ln\theta}\,\ln \frac{\Gamma(p+iq+\theta)\Gamma(p-iq+\theta)}{\Gamma(1-p+iq+\theta)\Gamma(1-p-iq+\theta)} = 4p-2,$$

we obtain the following theorem.

Theorem 2.6 *If*

$$\phi_n^*(x;p,q) = x^{\sigma_n}\,{}_3F_2\left(\begin{array}{c} -[n/2],\ \sigma_n - x,\ \sigma_n + x \\ p+iq+\sigma_n,\ p-iq+\sigma_n \end{array} \middle|\, 1 \right),$$

then the polynomial set $\{\phi_n^*(x; p, q)\}_{n=0}^N$, where $N < \frac{1}{2} - 2p$, is finitely orthogonal with respect to the weight function (2.149) on the support $(-\infty, \infty)$, so that we have

$$\sum_{x=-\infty}^{\infty} \frac{(p+iq)_x (p-iq)_x}{(1-p+iq)_x (1-p-iq)_x} \phi_n^*(x; p, q) \phi_m^*(x; p, q) =$$

$$\frac{[n/2]!(2p)_{[n/2]}}{(p+\sigma_n+iq)_{[n/2]}(p+\sigma_n-iq)_{[n/2]}} \frac{\Gamma^2(1-p+iq)\,\Gamma^2(1-p-iq)\,\Gamma(1-4p)}{\Gamma(1-2p+2iq)\,\Gamma(1-2p-2iq)\,\Gamma^2(1-2p)} \delta_{n,m},$$

provided that $p \notin \mathbb{Z}^-$, $p < \frac{1}{4}$, and $q \in \mathbb{R}$.

For example, the finite set $\{\phi_n^*(x; -11/2, 1)\}_{n=0}^{N=11}$ is orthogonal with respect to the weight function

$$\frac{((-11/2)+i)_x((-11/2)-i)_x}{((13/2)+i)_x((13/2)-i)_x}$$

on the support $(-\infty, \infty)$.

References

1. I. Area, E. Godoy, A. Ronveaux, A. Zarzo, Classical symmetric orthogonal polynomials of a discrete variable. Integr. Transf. Spec. Funct. **15**(1), 1–12 (2004)
2. P.A. Lesky. Vervollständigung der klassischen Orthogonalpolynome durch Ergänzungen zum Askey–Schema der hypergeometrischen orthogonalen Polynome. Sitzungsber. Abt. II Österr. Akad. Wiss. Math.-Naturwiss. Kl. **204**, 151–166 (1995)
3. A. Ronveaux, Semi classical weights $(-\infty, +\infty)$: semi-Hermite orthogonal polynomials, in *On Orthogonal Polynomials and Their Applications, 2nd International Symposium*, Segovia 1986. Monografias de la Real Academia de Ciencias Exactas, Fisicas, Quimicas y Naturales de Zaragoza (1988), pp. 129–137

Further Reading

- I. Area, E. Godoy, A. Ronveaux, A. Zarzo, Classical discrete orthogonal polynomials, Lah numbers, and involutory matrices. Appl. Math. Lett. **16**(3), 383–387 (2003)
- I. Area, E. Godoy, A. Ronveaux, A. Zarzo, Corrigendum: "Minimal recurrence relations for connection coefficients between classical orthogonal polynomials: discrete case". J. Comput. Appl. Math. **89**, 309–325 (1998)
- N.M. Atakishiyev, K.B. Wolf, Approximation on a finite set of points through Kravchuk functions. Rev. Mexicana Fís. **40**(3), 366–377 (1994)

- J. Baik, T. Kriecherbauer, K.T.R. McLaughlin, P.D. Miller, *Discrete Orthogonal Polynomials*. Annals of Mathematics Studies, vol. 164 (Princeton University Press, Princeton, 2007)
- G. Bangerezako, Discrete Darboux transformation for discrete polynomials of hypergeometric type. J. Phys. A **31**, 2191–2196 (1998)
- G. Bangerezako, M.N. Hounkonnou, The transformation of polynomial eigenfunctions of linear second-order difference operators: a special case of Meixner polynomials. J. Phys. A **34** 5653–5666 (2001)
- M. Cámara, J. Fàbrega, M.A. Fiol, E. Garriga, Some families of orthogonal polynomials of a discrete variable and their applications to graphs and codes. Electron. J. Combin. **16**(1), 83 (2009)
- A. Gelb, R.B. Platte, W.S. Rosenthal, The discrete orthogonal polynomial least squares method for approximation and solving partial differential equations. Commun. Comput. Phys. **3**, 734–758 (2008)
- T. Hakioğlu, K.B. Wolf, The canonical Kravchuk basis for discrete quantum mechanics. J. Phys. A **33**, 3313–3323 (2000)
- W. Harle, Symmetriche Orthogonalpolynome als Lösungen Sturm–Liouvillescher Differenzengleichungen. Ph.D. Thesis, Universität Stuttgart (1986)
- A. Jirari, *Second-Order Sturm–Liouville Difference Equations and Orthogonal Polynomials*. Memoirs of the American Mathematical Society, vol. 542 (American Mathematical Society, Providence, 1995)
- A.K. Khanmamedov, The inverse scattering problem for the discrete Sturm–Liouville equation on the line. Mat. Sb. **202**, 147–160 (2011)
- K. Kristinsson, G.A. Dumont, Cross-directional control on paper machines using Gram polynomials. Automatica J. IFAC **32**, 533–548 (1996)
- O.E. Lancaster, Orthogonal polynomials defined by difference equations. Am. J. Math. **63**(1), 185–207 (1941)
- P. Maroni, M. Mejri, The symmetric D_ω-semi-classical orthogonal polynomials of class one. Numer. Algorithms **49**(1–4), 251–282 (2008)
- M. Masjed-Jamei, I. Area, A symmetric generalization of Sturm–Liouville problems in discrete spaces. J. Difference. Equ. Appl. **19**, 1544–1562 (2013)
- M. Masjed-Jamei, I. Area, A basic class of symmetric orthogonal polynomials of a discrete variable. J. Math. Anal. Appl. **399**, 291–305 (2013)
- A.F. Nikiforov, S.K. Suslov, V.B. Uvarov, *Classical Orthogonal Polynomials of a Discrete Variable*. Springer Series in Computational Physics (Springer, Berlin, 1991)
- E.M. Nikishin, The discrete Sturm–Liouville operator and some problems of function theory. Trudy Sem. Petrovsk. **237**, 3–77 (1984)
- M. Petkovšek, Hypergeometric solutions of linear recurrences with polynomial coefficients. J. Symb. Comput. **14**, 243–264 (1992)
- A. Ronveaux, S. Belmehdi, E. Godoy, A. Zarzo, Recurrence relation approach for connection coefficients. Applications to classical discrete orthogonal polynomials, in *Symmetries and Integrability of Difference Equations*, Estérel, PQ, 1994. CRM

Proceedings and Lecture Notes, vol. 9 (American Mathematical Society, Providence, 1996)

- S.K. Suslov, On the theory of difference analogues of special functions of hypergeometric type. Uspekhi Mat. Nauk **44**(2:266), 185–226 (1989)
- M. van Hoeij, Finite singularities and hypergeometric solutions of linear recurrence equations. J. Pure Appl. Algebra **139**, 109–131 (1999)

Special Functions Generated by Generalized Sturm–Liouville Problems in q-Spaces

3.1 Introduction

In many applications of mathematical physics, q-orthogonal polynomials appear in a natural way. Our aim in this chapter is to introduce new classes of q-orthogonal polynomials that are generated by some q-Sturm–Liouville problems.

A q-Sturm–Liouville problem is a boundary value problem of the form

$$L[y](x; q) + \lambda_q w(x; q) y(qx; q) = 0, \tag{3.1}$$

where

$$L[y](x; q) = \left(D_q \big(r(x; q) \, D_q y \big) \right)(x; q) \quad \text{for} \quad r(x; q) > 0 \quad \text{and} \quad w(x; q) > 0, \tag{3.2}$$

with the boundary conditions

$$\alpha_1 y(a; q) + \beta_1 D_q y(a; q) = 0, \tag{3.3}$$
$$\alpha_2 y(b; q) + \beta_2 D_q y(b; q) = 0,$$

in which D_q is the q-difference operator defined by

$$D_q f(x) = \frac{f(qx) - f(x)}{(q - 1)x} \quad (x \neq 0, \ q \neq 1),$$

with $D_q f(0) := f'(0)$ (provided $f'(0)$ exists).

© Springer Nature Switzerland AG 2020
M. Masjed-Jamei, *Special Functions and Generalized Sturm–Liouville Problems*,
Frontiers in Mathematics, https://doi.org/10.1007/978-3-030-32820-7_3

Two eigenfunctions of the q-Sturm–Liouville problem (3.1)–(3.3) are orthogonal with respect to the positive weight function $w(x; q)$ on (a, b). This result is a consequence of the following fact, which is a q-analogue of integration by parts. Let us consider the identity

$$\int_{x_1}^{x_2} (fLg - gLf)(x; q)d_q x = r(x; q)\left((D_q g)(q^{-1}x; q)f(x; q) - (D_q f)(q^{-1}x; q)g(x; q)\right)\Big|_{x_1}^{x_2},$$

(3.4)

where the q-integral operator is defined by

$$\int_0^x f(t)d_q t = (1 - q)x \sum_{j=0}^{\infty} q^j f(q^j x),$$

and for two nonnegative numbers x_1 and x_2 with $x_1 < x_2$ we have

$$\int_{x_1}^{x_2} f(x)d_q x = \int_0^{x_2} f(x)d_q x - \int_0^{x_1} f(x)d_q x.$$

Let $y_m(x; q)$ and $y_n(x; q)$ be solutions of Eq. (3.1) corresponding to two eigenvalues $\lambda_{m,q}$ and $\lambda_{n,q}$. If $x_1 = a$, $x_2 = b$, $f(x; q) = y_m(x; q)$, and $g(x; q) = y_n(x; q)$ in (3.4), then

$$\int_a^b (y_m L[y_n] - y_n L[y_m])(x; q)d_q x = 0.$$

This means that $y_n(x; q)$ and $y_m(x; q)$ as eigenfunctions of Eq. (3.1) are orthogonal with respect to the weight function $w(x; q)$, and we have

$$\int_a^b y_m(x; q) y_n(x; q) w(x; q)d_q x = 0, \quad (\lambda_m \neq \lambda_n).$$

Hence, one would be able to obtain some q-orthogonal polynomials through the solutions of a q-Sturm–Liouville problem. For instance, classical q-orthogonal polynomials are discrete versions of continuous classical orthogonal polynomials. In 1884, Markov found a q-analogue of Chebyshev polynomials that was indeed the first example of q-orthogonal polynomial families. In 1949, Hahn studied a problem similar to the case considered by Bochner, i.e., finding the sequences of q-orthogonal polynomials that are solutions of a linear second-order q-difference equation.

Today they are known as the Hahn class of q-orthogonal polynomials. He also obtained the most general q-orthogonal polynomial on the exponential lattice, i.e., big q-Jacobi polynomials.

Afterward, many authors considered q-polynomials with different aspects and found various applications in theoretical and mathematical physics such as continued fractions, Eulerian series, algebras and quantum groups, discrete mathematics, algebraic combinatorics (coding theory, design theory, various theories of group representation), the q-Schrödinger equation and q-harmonic oscillators. The theory of q-special functions has received a new impulse with the introduction of quantum algebras and groups. Indeed, real interpretations of these functions have been derived from representation theory of quantum algebras in similar ways to how Lie algebras are used to describe classical special functions.

The q-classical polynomials satisfy a q-difference equation of hypergeometric type as

$$\sigma(x)(D_q^2 y_n)(x; q) + \tau(x)(D_q y_n)y(x; q) + \lambda_{n,q} y_n(qx; q) = 0, \tag{3.5}$$

where $\sigma(x)$ and $\tau(x)$ are polynomials of degree at most 2 and 1, respectively, and $\lambda_{n,q}$ is computed as

$$\lambda_{n,q} = -\frac{[n]_q}{q^n}(\sigma''[n-1]_q + \tau'), \qquad n = 0, 1, \ldots.$$

To prove the orthogonality relation of q-polynomials satisfied by Eq. (3.5), we can first assume in Eq. (3.1) that $r(x; q) = \sigma(x)w(x; q)$ and then write Eq. (3.5) in the self-adjoint form

$$\left(D_q\left(\sigma w D_q y_n\right)\right)(x; q) + w(x; q)\lambda_{n,q} y_n(qx; q) = 0$$

to conclude that they are orthogonal with respect to the weight function w as

$$\int_a^b P_m(x; q)P_n(x; q)w(x; q)d_q x = d_n^2 \delta_m,$$

where d_n shows the norm square value and w is a solution of the q-Pearson equation

$$D_q\left(\sigma(q^{-1}x)w(x; q)\right) = w(qx; q)\tau(x), \tag{3.6}$$

such that the following boundary conditions always holds

$$\sigma(x)w(x; q)\Big|_a^b = 0. \tag{3.7}$$

Thus, similar to the continuous and discrete cases, if some extera constraints are required for satisfying the boundary conditions (3.7), then finite classes of q-orthogonal polynomials will appear.

In this chapter, we introduce some finite classes of q-orthogonal polynomials via introducing some q-Sturm–Liouville problems. We also concentrate on their general properties and compare the results with the corresponding continuous cases. Next, using a symmetric generalization of Sturm–Liouville problems in q-difference spaces, we introduce some classes of symmetric q-orthogonal polynomials and obtain their standard properties.

For this purpose, we should first state some basic definitions and necessary preliminaries related to q-calculus that are less well known.

3.2   Some Preliminaries and Definitions

3.2.1   The q-Shifted Factorial

One of the basic concepts of the q-calculus is the limit relation

$$\lim_{q \to 1} \frac{1 - q^\alpha}{1 - q} = \alpha.$$

Based on this limit,

$$[n]_q = \frac{1 - q^n}{1 - q} = \sum_{k=0}^{n-1} q^k \quad \text{with} \quad [0]_q = 1 \tag{3.8}$$

is called a q-number.

The q-shifted factorial is defined by

$$(a; q)_k = (1 - a)(1 - aq) \cdots (1 - aq^{k-1}) \quad \text{with} \quad (a; q)_0 = 1.$$

So it is clear that

$$\lim_{q \to 1} \frac{(q^\alpha; q)_k}{(1 - q)^k} = (\alpha)_k.$$

For negative indices, the q-shifted factorial is defined by

$$(a; q)_{-k} = \frac{1}{\prod_{i=1}^{k}(1 - aq^{-i})}, \quad \text{where} \quad a \neq q, q^2, q^3, \ldots, q^k \quad \text{and} \quad k = 1, 2, 3, \ldots.$$

Also, as an extension we have

$$(a; q)_\infty = \prod_{k=0}^{\infty}(1 - aq^k) \quad \text{for} \quad 0 < |q| < 1.$$

This implies that for every complex number λ,

$$(a; q)_\lambda = \frac{(a; q)_\infty}{(aq^\lambda; q)_\infty}, \qquad 0 < |q| < 1,$$

where the principal value of q^λ is taken.

The following relations hold for q-shifted factorials:

$$(a; q)_\infty = (a; q^2)_\infty (aq; q^2)_\infty, \qquad 0 < |q| < 1,$$

and

$$(a^2; q^2)_\infty = (a; q)_\infty (-a; q)_\infty, \qquad 0 < |q| < 1.$$

3.2.2 q-Hypergeometric Series

Since a usual hypergeometric series is defined as

$$_r F_s \left(\begin{array}{c} a_1, \ldots, a_r \\ b_1, \ldots, b_s \end{array} \middle| z \right) := \sum_{k=0}^{\infty} \frac{(a_1, \ldots, a_r)_k}{(b_1, \ldots, b_s)_k} \frac{z^k}{k!},$$

where

$$(a_1, \ldots, a_r)_k := (a_1)_k \cdots (a_r)_k,$$

a basic or q-hypergeometric series is defined as

$$_r \phi_s \left(\begin{array}{c} a_1, \ldots, a_r \\ b_1, \ldots, b_s \end{array} \middle| q; z \right) := \sum_{k=0}^{\infty} \frac{(a_1, \ldots, a_r; q)_k}{(b_1, \ldots, b_s; q)_k} \frac{z^k}{(q; q)_k} \left((-1)^k q^{\frac{k(k-1)}{2}} \right)^{1+s-r},$$

where

$$(a_1, \ldots, a_r; q)_k := (a_1; q)_k \cdots (a_r; q)_k,$$

and $a_1, a_2, \ldots, a_r, b_1, b_2, \ldots, b_s, z \in \mathbb{C}$.

In this case, we must assume that

$$b_1, b_2, \ldots, b_s \neq q^{-k} \ (k = 0, 1, \ldots)$$

in order to have a well-defined series. By noting these assumptions, we see that the following limit relation holds:

$$\lim_{q \to 1} {}_r\phi_s \left(\begin{matrix} q^{a_1}, \ldots, q^{a_r} \\ q^{b_1}, \ldots, q^{b_s} \end{matrix} \middle| q; (q-1)^{1+s-r} z \right) = {}_rF_s \left(\begin{matrix} a_1, \ldots, a_r \\ b_1, \ldots, b_s \end{matrix} \middle| z \right).$$

3.2.3 q-Binomial Coefficients and the q-Binomial Theorem

A q-binomial coefficient is defined as

$$\begin{bmatrix} n \\ k \end{bmatrix}_q = \begin{bmatrix} n \\ n-k \end{bmatrix}_q = \frac{(q; q)_n}{(q; q)_k (q; q)_{n-k}} \quad \text{for} \quad k = 0, 1, 2, \ldots, n,$$

where n is a nonnegative integer. Moreover, for all complex α and β and $0 < |q| < 1$, we have

$$\begin{bmatrix} \alpha \\ \beta \end{bmatrix}_q = \frac{\Gamma_q(\alpha+1)}{\Gamma_q(\beta+1)\Gamma_q(\alpha-\beta+1)} = \frac{(q^{\beta+1}; q)_\infty (q^{\alpha-\beta+1}; q)_\infty}{(q; q)_\infty (q^{\alpha+1}; q)_\infty}.$$

Note that

$$\lim_{q \to 1} \begin{bmatrix} \alpha \\ \beta \end{bmatrix}_q = \binom{\alpha}{\beta} = \frac{\alpha!}{\beta!(\alpha-\beta)!}.$$

The q-binomial theorem,

$$ {}_1\phi_0 \left(\begin{matrix} a \\ - \end{matrix} \middle| q; z \right) = \sum_{n=0}^\infty \frac{(a; q)_n}{(q; q)_n} z^n = \frac{(az; q)_\infty}{(z; q)_\infty} \quad \text{for} \quad 0 < |q| < 1 \quad \text{and} \quad |z| < 1,$$

$$(3.9)$$

is indeed a q-analogue of the usual binomial theorem

$$ {}_1F_0 \left(\begin{matrix} a \\ - \end{matrix} \middle| z \right) = \sum_{n=0}^\infty \frac{(a)_n}{n!} z^n = (1-z)^{-a} \quad \text{for} \quad |z| < 1.$$

If in (3.9) we set $a = q^{-n}$, then we obtain

$$ {}_1\phi_0 \left(\begin{matrix} q^{-n} \\ - \end{matrix} \middle| q; z \right) = (zq^{-n}; q)_n \quad \text{for all} \quad n = 0, 1, 2, \ldots.$$

3.2.4 q-Gamma and q-Beta Functions

The q-gamma function defined by

$$\Gamma_q(x) = \frac{(q; q)_\infty}{(q^x; q)_\infty}(1 - q)^{1-x} \tag{3.10}$$

holds for $0 < |q| < 1$ if the principal values of q^x and $(1 - q)^{1-x}$ are considered.
   For $q > 1$, it is defined as

$$\Gamma_q(x) = \frac{(q^{-1}; q^{-1})_\infty}{(q^{-x}; q^{-1})_\infty} q^{\binom{x}{2}}(q - 1)^{1-x}.$$

Naturally we have

$$\lim_{q \to 1} \Gamma_q(x) = \Gamma(x).$$

Also, there exists a q-extension of the functional equation of the ordinary gamma function as

$$\Gamma_q(z + 1) = \frac{1 - q^z}{1 - q}\Gamma_q(z) \quad \text{with} \quad \Gamma_q(1) = 1.$$

The q-beta function defined by

$$B_q(t, s) = \frac{\Gamma_q(s)\Gamma_q(t)}{\Gamma_q(s + t)}$$

satisfies the following q-integral representation, which is a q-analogue of Euler's formula:

$$B_q(t, s) = \int_0^1 x^{t-1}(qx; q)_{s-1} d_q x \qquad \text{for all} \quad t, s > 0.$$

3.2.5 q-Analogues of Some Special Functions

The exponential function has two different q-extensions as follows:

$$e_q(z) = \sum_{n=0}^{\infty} \frac{(1 - q)^n}{(q; q)_n} z^n = \frac{1}{((1 - q)z; q)_\infty} \quad \text{for} \quad |z| < 1 \tag{3.11}$$

and

$$E_q(z) = \sum_{n=0}^{\infty} \frac{q^{\binom{n}{2}}(1-q)^n}{(q;q)_n} z^n = (-(1-q)z; q)_\infty,$$

which are related to each other by

$$e_q(z)E_q(z) = 1.$$

In the limit case we have

$$\lim_{q \to 1} e_q(z) = \lim_{q \to 1} E_q(z) = e^z.$$

By assuming $|z| < 1$, we can now define

$$\sin_q(z) = \frac{e_q(iz) - e_q(-iz)}{2i} = \sum_{n=0}^{\infty} \frac{(-1)^n (1-q)^{2n+1} z^{2n+1}}{(q;q)_{2n+1}}$$

and

$$\cos_q(z) = \frac{e_q(iz) + e_q(-iz)}{2} = \sum_{n=0}^{\infty} \frac{(-1)^n (1-q)^{2n} z^{2n}}{(q;q)_{2n}}.$$

Another q-analogue of the trigonometric functions may be defined as

$$\mathrm{Sin}_q(z) = \frac{E_q(iz) - E_q(-iz)}{2i}$$

and

$$\mathrm{Cos}_q(z) = \frac{E_q(iz) + E_q(-iz)}{2},$$

where

$$e_q(iz) = \cos_q(z) + i \sin_q(z) \quad \text{and} \quad E_q(iz) = \mathrm{Cos}_q(z) + i \ \mathrm{Sin}_q(z).$$

3.2.6 q-Difference Operators

A linear q-difference operator D_q defined by

$$(D_q f)(x) = \frac{f(qx) - f(x)}{(q-1)x}, \quad q \in \mathbb{R} \setminus \{-1, 0, 1\}, \quad x \in \mathbb{C} \setminus \{0\},$$

is valid for arbitrary complex-valued functions f whose domain contains $qx \in \mathbb{C}$, $x \in \mathbb{C}$, and satisfies the following relations:

$$D_q D_{q^{-1}} = q^{-1} D_{q^{-1}} D_q,$$

$$D_q = D_{q^{-1}} + (q-1) x D_q D_{q^{-1}},$$

$$D_q f(q^{-1} x) = D_{q^{-1}} f(x),$$

and

$$D_q D_{q^{-1}} f(x) = q^{-1} D_q^2 f(q^{-1} x).$$

Moreover, for two functions f_1 and f_2 we have

$$(D_q(f_1 f_2))(x) = (D_q f_1)(x) f_2(x) + f_1(qx)(D_q f_2)(x).$$

It is important to note that if we apply the operator D_q to the function $f(\cdot + c)$, where $c \in \mathbb{R}$ is a constant, then we have

$$\left(D_q f(\cdot + c) \right)(x) = \frac{f(qx + c) - f(x + c)}{(q-1)x},$$

while if we first apply the operator D_q to the function f and then replace the argument x by $x + c$, we have

$$\left(D_q f(\cdot) \right)(x + c) = \frac{f(qx + c) - f(x + c)}{(q-1)(x + c)}.$$

For simplicity, if we take

$$(D_q f)(x) = D_q \ f(x) = \frac{f(qx) - f(x)}{(q-1)x} = p(x),$$

then we get

$$(D_q^2 f)(x) = (D_q p)(x) = \frac{p(qx) - p(x)}{(q-1)x} = \frac{f(q^2 x) - (1+q)f(qx) + qf(x)}{q(q-1)^2 x^2},$$

$$\tag{3.12}$$

as well as

$$(D_{q^{-1}} D_q f)(x) = \frac{f(qx) - (1+q)f(x) + qf(q^{-1}x)}{(q-1)^2 x^2}.$$

For example, we have

$$D_q(x^n) = \frac{(qx)^n - x^n}{(q-1)x} = \sum_{k=0}^{n-1}(qx)^{n-1-k}x^k$$

$$= x^{n-1}\sum_{k=0}^{n-1}q^k + r(x) = [n]_q x^{n-1} + r(x) \qquad \text{for all} \quad n = 2, 3, 4, \ldots,$$

where r is a polynomial of degree at most $n - 2$.

If the operator $\mathcal{L}_q$ is defined as

$$(\mathcal{L}_q f)(x) = f(qx), \tag{3.13}$$

we can conclude that

$$\mathcal{L}_q^m f(x) = f(q^m x) \tag{3.14}$$

and

$$(\mathcal{L}_q^{-1} f)(x) = f\left(\frac{x}{q}\right).$$

3.2.7   q-Integral Operators

A q-integral, as the inverse of a q-difference operator, is defined as

$$\int_0^x f(t)d_q t = (1-q)x \sum_{j=0}^{\infty} q^j f(q^j x), \tag{3.15}$$

provided that the series converges.

For two nonnegative numbers a and b with $a < b$, the definition (3.15) yields

$$\int_a^b f(t)d_q t = \int_0^b f(t)d_q t - \int_0^a f(t)d_q t.$$

Furthermore, we have

$$\int_0^\infty f(t)d_q t = (1-q)\sum_{n=-\infty}^{\infty} q^n f(q^n)$$

and

$$\int_{-\infty}^{\infty} f(t)d_q t = (1-q) \sum_{n=-\infty}^{\infty} q^n \left(f(q^n) + f(-q^n) \right).$$

A function f that is defined on a q-geometric set A with $0 \in A$ is said to be q-regular at zero if

$$\lim_{n\to\infty} f(xq^n) = f(0) \quad \text{for every} \quad x \in A.$$

The rule of q-integration by parts is given by

$$\int_0^a g(x)D_q f(x)d_q x = (fg)(a) - \lim_{n\to\infty} (fg)(aq^n) - \int_0^a D_q g(x)f(qx)d_q x. \quad (3.16)$$

Note that if f, g are q-regular at zero, then $\lim_{n\to\infty}(fg)(aq^n)$ on the right-hand side of (3.16) can be replaced by $(fg)(0)$.

For $0 < R \le \infty$, let Ω_R denote the disk $\{z \in \mathbb{C} : |z| < R\}$. The q-analogue of the fundamental theorem of calculus says that if $f : \Omega_R \to \mathbb{C}$ is q-regular at zero and $\theta \in \Omega_R$ is fixed, then the function

$$F(x) = \int_\theta^x f(t)d_q t \quad (x \in \Omega_R) \quad (3.17)$$

is q-regular at zero, $D_q F(x)$ exists for all $x \in \Omega_R$, and $D_q F(x) = f(x)$.

Conversely, if $a, b \in \Omega_R$, we have

$$\int_a^b D_q f(t)d_q t = f(b) - f(a).$$

Relation (3.15) can be obtained via (3.17) and applying the properties (3.13) and (3.14), because the equality $(D_q F)(x) = f(x)$ implies that

$$\frac{F(qx) - F(x)}{(q-1)x} = \frac{(\mathfrak{L}_q - I)F(x)}{(q-1)x} = f(x).$$

Hence we have

$$F(x) = \frac{1}{I - \mathfrak{L}_q}\left((1-q)xf(x)\right) = \sum_{j=0}^{\infty} \mathfrak{L}_q^j((1-q)xf(x)) = (1-q)x\sum_{j=0}^{\infty} q^j f(q^j x).$$

Finally, if $u(x) = \alpha x^{\beta}$, the following q-integral relation holds:

$$\int_{u(a)}^{u(b)} f(u)d_q u = \int_a^b f(u(x)) D_{q^{\frac{1}{\beta}}} u(x) d_{q^{\frac{1}{\beta}}} x. \tag{3.18}$$

- **Some Definite q-Integrals**

  For $|q| < 1$, $|a| > |q|$, $|b| < 1$, and $|\frac{b}{a}| < |x| < 1$, Ramanujan's celebrated formula is given by

$$\Psi(a, b; q; x) = \sum_{n=-\infty}^{\infty} \frac{(a; q)_n}{(b; q)_n} x^n = \prod_{n=0}^{\infty} \frac{\left(1 - \frac{bq^n}{a}\right)(1 - q^{n+1})\left(1 - \frac{q^{n+1}}{ax}\right)(1 - axq^n)}{(1 - bq^n)\left(1 - \frac{q^{n+1}}{a}\right)\left(1 - \frac{bq^n}{ax}\right)(1 - xq^n)}. \tag{3.19}$$

Now consider the integrals

$$\int_0^{\infty/A} \frac{x^{t-1}}{(-x; q)_{t+s}} d_q x = \frac{B_q(t, s)}{K(A, t)} \tag{3.20}$$

and

$$\int_0^{\infty/(1-q)} x^{t-1} e_q^{-x} d_q x = \frac{\Gamma_q(t)}{K(1, t)}, \tag{3.21}$$

where

$$\int_0^{\infty/A} f(t)d_q t = (1 - q) \sum_{n=-\infty}^{\infty} \frac{q^n}{A} f\left(\frac{q^n}{A}\right)$$

and

$$K(x, t) = \frac{x^t}{1+x}\left(-\frac{1}{x}; q\right)_t (-x; q)_{1-t}.$$

For instance, the integral (3.21) can be directly derived by Ramanujan's formula as follows:

$$\int_0^{\frac{\infty}{1-q}} x^{t-1} e_q^{-x} d_q x = (1 - q) \sum_{n=-\infty}^{\infty} \frac{q^n}{1-q} \left(\frac{q^n}{1-q}\right)^{t-1} e_q^{-\frac{q^n}{1-q}}$$

$$= (1 - q)^{1-t} \sum_{n=-\infty}^{\infty} q^n \frac{q^{nt}}{q^n} e_q^{-\frac{q^n}{1-q}} = (1 - q)^{1-t} \sum_{n=-\infty}^{\infty} \frac{q^{nt}}{(-q^n; q)_{\infty}}$$

$$= \frac{(1-q)^{1-t}}{(-1;q)_\infty} \sum_{n=-\infty}^{\infty} \frac{q^{nt}(-1;q)_n}{(0;q)_n} = \frac{(1-q)^{1-t}}{(-1;q)_\infty} \Psi(-1,0;q;q^t)$$

$$= \frac{(1-q)^{1-t}}{(-1;q)_\infty} \prod_{n=0}^{\infty} \frac{(1-q^{n+1})(1+\frac{q^{n+1}}{q^t})(1+q^t q^n)}{(1+q^{n+1})(1-q^t q^n)}$$

$$= \frac{(1-q)^{1-t}}{(-1;q)_\infty} \frac{(q;q)_\infty(-q^{1-t};q)_\infty(-q^t;q)_\infty}{(-q;q)_\infty(q^t;q)_\infty} = \frac{\Gamma_q(t)}{K(1,t)}.$$

3.2.8   An Analytical Solution for the q-Pearson Difference Equation

If Eq. (3.6) is expanded, a general q-difference equation will appear as

$$\frac{W(qx)}{W(x)} = A(x;q). \tag{3.22}$$

Theorem 3.1 *The solution of the generic equation (3.22) can be represented as*

$$W(x) = \prod_{j=0}^{\infty} \frac{1}{A(q^j x; q)}.$$

Proof Taking the logarithm on both sides of (3.22) first yields

$$\frac{\ln W(qx) - \ln W(x)}{(q-1)x} = \frac{\ln A(x;q)}{(q-1)x}.$$

Now by referring to the definition

$$\int_0^x f(t)\,d_q t = (1-q)x \sum_{j=0}^{\infty} q^j f(xq^j), \tag{3.23}$$

replace

$$f(t) = \frac{\ln W(qt) - \ln W(t)}{(q-1)t}$$

in (3.23) to get

$$\int_0^x \frac{\ln W(qt) - \ln W(t)}{(q-1)t}\, d_q t = (1-q)x \sum_{j=0}^{\infty} q^j \frac{\ln W(q^{j+1}x) - \ln W(q^j x)}{(q-1)xq^j}$$

$$= \lim_{n\to\infty} \sum_{j=0}^{n} \ln \frac{W(q^j x)}{W(q^{j+1}x)} = \lim_{n\to\infty} \ln \prod_{j=0}^{n} \frac{W(q^j x)}{W(q^{j+1}x)}$$

$$= \lim_{n\to\infty} \ln \frac{W(x)}{W(q^{n+1}x)} = \ln \frac{W(x)}{W(0)}.$$

Therefore, we have

$$\int_0^x \frac{\ln A(t;q)}{(q-1)t}\, d_q t = \ln \frac{W(x)}{W(0)}.$$

Since $W(0) \neq 0$ has no effect on the initial q-difference equation, we finally obtain

$$W(x) = \exp\left(\int_0^x \frac{\ln A(t;q)}{(q-1)t}\, d_q t \right) = \exp\left((1-q)x \sum_{j=0}^{\infty} q^j \frac{\ln A(q^j x; q)}{(q-1)xq^j} \right)$$

$$= \exp\left(-\ln \prod_{j=0}^{\infty} A(q^j x; q) \right) = \prod_{j=0}^{\infty} \frac{1}{A(q^j x; q)}.$$

$\square$

One of the direct consequences of this theorem is that the q-difference equation

$$\frac{W(x)}{W(qx)} = B(x;q)$$

has a solution of the form

$$W(x) = \prod_{j=0}^{\infty} B(q^j x; q).$$

3.2.9 Difference Equations of q-Hypergeometric Series

Let

$$y(z) = {}_r\phi_s \left(\begin{matrix} a_1, \dots, a_r \\ b_1, \dots, b_s \end{matrix} \,\middle|\, q; z \right).$$

Then the above-mentioned series satisfies a general difference equation as

$$\Delta \Delta_{b_1/q} \Delta_{b_2/q} \cdots \Delta_{b_s/q} \, y(z) = z(\Delta_{a_1} \cdots \Delta_{a_r}) y(zq^{1+s-r}), \qquad (3.24)$$

where

$$\Delta_b f(x) = bf(qx) - f(x).$$

For instance, suppose in (3.24) that $r = 2$ and $s = 1$. Then the equation

$$\Delta(\Delta_{c/q} y(x)) - x \Delta_a (\Delta_b y(x)) = 0$$

is expanded as

$$q^{-1} cy(q^2 x) - y(qx) - q^{-1} cy(qx) + y(x)$$
$$- x\big(aby(q^2 x) - ay(qx) - by(qx) + y(x)\big) = 0.$$

This means that the explicit form of the q-difference equation of $_2\phi_1$ is

$$(cq^{-1} - abx)y(q^2 x) - (cq^{-1} + 1 - (a+b)x)y(qx) + (1-x)y(x) = 0,$$

which is equivalent to

$$(1-q)^2 x(c - abqx) D_q^2 y(x) + (1-q)\big(1 - c - (a+b-ab-abq)x\big) D_q y(x)$$
$$- (1-a)(1-b)y(x) = 0.$$

Similarly, by taking $r = s = 1$ in (3.24), the equation

$$\Delta(\Delta_{c/q} y(x)) - x(\Delta_a y(qx)) = 0$$

is expanded as

$$q^{-1} cy(q^2 x) - y(qx) - q^{-1} cy(qx) + y(x) - x\big(ay(q^2 x) - y(qx)\big) = 0.$$

Therefore, the explicit form of the q-difference equation of $_1\phi_1$ is

$$(cq^{-1} - ax)y(q^2 x) - (cq^{-1} + 1 - x)y(qx) + y(x) = 0,$$

which is equivalent to

$$q(1-q)^2 x(c - aqx) D_q^2 y(x) + (1-q)\big(q(1-c) - q(1-a-aq)x\big) D_q y(x)$$
$$+ q(1-a)y(x) = 0.$$

It is not difficult to verify that

$$\lim_{b \to \infty} {}_2\phi_1 \left(\begin{matrix} a, b \\ c \end{matrix} \bigg| q; \frac{x}{b} \right) = {}_1\phi_1 \left(\begin{matrix} a \\ c \end{matrix} \bigg| q; x \right).$$

3.2.10 *q*-Analogues of Jacobi Polynomials

There are two q-analogues of Jacobi orthogonal polynomials whose standard properties are represented as follows.

1. Big *q*-Jacobi Polynomials

- Basic hypergeometric representation:

$$P_n(x; a, b, c; q) = {}_3\phi_2 \left(\begin{matrix} q^{-n}, abq^{n+1}, x \\ aq, cq \end{matrix} \bigg| q; q \right). \tag{3.25}$$

- Orthogonality relation:
  For $0 < aq < 1, 0 \le bq < 1$, and $c < 0$ we have

$$\int_{cq}^{aq} \frac{(a^{-1}x, c^{-1}x; q)_\infty}{(x, bc^{-1}x; q)_\infty} P_m(x; a, b, c; q) P_n(x; a, b, c; q) d_q x$$

$$= aq(1-q) \frac{(q, abq^2, a^{-1}c, ac^{-1}q; q)_\infty}{(aq, bq, cq, abc^{-1}q; q)_\infty}$$

$$\times \frac{(1-abq)}{(1-abq^{2n+1})} \frac{(q, bq, abc^{-1}q; q)_n}{(aq, abq, cq; q)_n} (-acq^2)^n q^{\binom{n}{2}} \delta_{m,n}.$$

- Recurrence relation

$$(x-1) P_n(x; a, b, c; q)$$

$$= A_n P_{n+1}(x; a, b, c; q) - (A_n + C_n) P_n(x; a, b, c; q) + C_n P_{n-1}(x; a, b, c; q),$$

where

$$A_n = \frac{(1-aq^{n+1})(1-abq^{n+1})(1-cq^{n+1})}{(1-abq^{2n+1})(1-abq^{2n+2})}$$

and

$$C_n = -acq^{n+1} \frac{(1-q^n)(1-abc^{-1}q^n)(1-bq^n)}{(1-abq^{2n})(1-abq^{2n+1})}.$$

- q-Difference equation:

$$q^{-n}(1-q^n)(1-abq^{n+1})x^2 y(x) = B(x)y(qx) - \big(B(x)+D(x)\big)y(x) + D(x)y(q^{-1}x),$$

where $y(x) = P_n(x; a, b, c; q)$,

$$B(x) = aq(x-1)(bx-c), \quad \text{and} \quad D(x) = (x-aq)(x-cq).$$

- Limit relation:
  If we set $c = 0$, $a = q^\alpha$, and $b = q^\beta$ in (3.25) and then let $q \to 1$, we obtain the usual shifted Jacobi polynomials as

$$\lim_{q \to 1} P_n(x; q^\alpha, q^\beta, 0; q) = \frac{P_n^{(\alpha,\beta)}(2x-1)}{P_n^{(\alpha,\beta)}(1)},$$

and if $c = -q^\gamma$ for arbitrary real γ instead of $c = 0$, then we get

$$\lim_{q \to 1} P_n(x; q^\alpha, q^\beta, -q^\gamma; q) = \frac{P_n^{(\alpha,\beta)}(x)}{P_n^{(\alpha,\beta)}(1)}.$$

Note that the big q-Legendre polynomials are a particular case of the big q-Jacobi polynomials for $a = b = 1$.

2. **Little q-Jacobi Polynomials**

- Basic hypergeometric representation:

$$p_n(x; a, b; q) = {}_2\phi_1 \left(\begin{matrix} q^{-n}, abq^{n+1} \\ aq \end{matrix} \bigg| q; qx \right). \tag{3.26}$$

- Orthogonality relation:
  For $0 < aq < 1$ and $0 \le bq < 1$ we have

$$\sum_{k=0}^{\infty} \frac{(bq; q)_k}{(q; q)_k} (aq)^k p_m(q^k; a, b; q) P_n(q^k; a, b; q)$$

$$= \frac{(abq^2; q)_\infty}{(aq; q)_\infty} \frac{(1-abq)(aq)^n}{(1-abq^{2n+1})} \frac{(q, bq; q)_n}{(aq, abq; q)_n} \delta_{m,n}.$$

- Recurrence relation:

$$-xp_n(x; a, b; q) = A_n p_{n+1}(x; a, b; q) - (A_n + C_n)p_n(x; a, b; q) + C_n p_{n-1}(x; a, b; q),$$

where

$$A_n = q^n \frac{(1 - aq^{n+1})(1 - abq^{n+1})}{(1 - abq^{2n+1})(1 - abq^{2n+2})}$$

and

$$C_n = aq^n \frac{(1 - q^n)(1 - bq^n)}{(1 - abq^{2n})(1 - abq^{2n+1})}.$$

- q-Difference equation:

$$q^{-n}(1-q^n)(1-abq^{n+1})xy(x) = B(x)y(qx) - \big(B(x)+D(x)\big)y(x) + D(x)y(q^{-1}x),$$

where $y(x) = p_n(x; a, b; q)$,

$$B(x) = a(bqx - 1), \quad \text{and} \quad D(x) = x - 1.$$

- Limit relation:
  The little q-Jacobi polynomials tend to the usual continuous Laguerre polynomials, so that

$$\lim_{q \to 1} p_n\left(\frac{1}{2}(1 - q)x; q^\alpha, -q^\beta; q\right) = \frac{L_n^{(\alpha)}(x)}{L_n^{(\alpha)}(0)}.$$

Moreover, the little q-Legendre polynomials are a particular case of the little q-Jacobi polynomials for $a = b = 1$.

3.2.11 q-Analogues of Laguerre Polynomials

There are three q-analogues of the Laguerre orthogonal polynomials, whose standard properties are represented as follows.

1. Big q-Laguerre Polynomials

- Basic hypergeometric representation:

$$L_n(x; a, b; q) = {}_3\phi_2\left(\begin{matrix} q^{-n}, 0, x \\ aq, bq \end{matrix} \bigg| q; q\right) = \frac{1}{(b^{-1}q^{-n}; q)_n} {}_2\phi_1\left(\begin{matrix} q^{-n}, aqx^{-1} \\ aq \end{matrix} \bigg| q; \frac{x}{b}\right).$$

$$\tag{3.27}$$

- Orthogonality relation:
  For $0 < aq < 1$ and $b < 0$ we have

$$\int_{bq}^{aq} \frac{(a^{-1}x, b^{-1}x; q)_\infty}{(x; q)_\infty} L_m(x; a, b; q) L_n(x; a, b; q) d_q x$$

$$= aq(1-q) \frac{(q, a^{-1}b, ab^{-1}q; q)_\infty}{(aq, bq; q)_\infty} \frac{(q; q)_n}{(aq, bq; q)_n} (-abq^2)^n q^{\binom{n}{2}} \delta_{m,n}.$$

- Recurrence relation:

$$(x-1)L_n(x; a, b; q) = A_n L_{n+1}(x; a, b; q) - (A_n + C_n) L_n(x; a, b; q) + C_n L_{n-1}(x; a, b; q),$$

where

$$A_n = (1 - aq^{n+1})(1 - bq^{n+1})$$

and

$$C_n = -abq^{n+1}(1 - q^n).$$

- q-Difference equation

$$q^{-n}(1-q^n)x^2 y(x) = B(x)y(qx) - \big(B(x) + D(x)\big)y(x) + D(x)y(q^{-1}x),$$

where $y(x) = L_n(x; a, b; q)$,

$$B(x) = abq(1-x), \quad \text{and} \quad D(x) = (x - aq)(x - bq).$$

- Limit relations:
  If $b = 0$, $a = q^\alpha$, and $c = q^\beta$ are replaced in (3.25), the big q-Laguerre polynomials given by (3.27) are derived, i.e.,

$$P_n(x; a, 0, c; q) = L_n(x; a, c; q).$$

Also, the shifted Laguerre polynomials can be obtained from the limit relation

$$\lim_{q \to 1} L_n(x; q^\alpha, (1-q)^{-1} q^\beta; q) = \frac{L_n^{(\alpha)}(x - 1)}{L_n^{(\alpha)}(0)}.$$

2. Little q-Laguerre Polynomials

- Basic hypergeometric representation:

$$l_n(x; a; q) = {}_2\phi_1 \left(\begin{matrix} q^{-n}, 0 \\ aq \end{matrix} \,\middle|\, q; qx \right) = \frac{1}{(a^{-1}q^{-n}; q)_n} {}_2\phi_0 \left(\begin{matrix} q^{-n}, x^{-1} \\ - \end{matrix} \,\middle|\, q; \frac{x}{a} \right).$$

(3.28)

- Orthogonality relation:

$$\sum_{k=0}^{\infty} \frac{(aq)^k}{(q; q)_k} l_m(q^k; a; q) l_n(q^k; a; q) = \frac{(aq)^n}{(aq; q)_\infty} \frac{(q; q)_n}{(aq; q)_n} \delta_{m,n}, \quad 0 < aq < 1.$$

- Recurrence relation:

$$- x l_n(x; a; q) = A_n l_{n+1}(x; a; q) - (A_n + C_n) l_n(x; a; q) + C_n l_{n-1}(x; a; q),$$

where

$$A_n = q^n(1 - aq^{n+1}) \quad \text{and} \quad C_n = aq^n(1 - q^n).$$

- q-Difference equation:

$$q^{-n}(1-q^n)xy(x) = ay(qx) + (x-a-1)y(x) + (1-x)y(q^{-1}x) \quad \text{with} \quad y(x) = l_n(x; a; q).$$

- Limit relations:
  The little q-Laguerre polynomials (3.28) are a limit case of the big q-Laguerre polynomials, so that

$$\lim_{b \to -\infty} L_n(bqx; a, b; q) = l_n(x; a; q).$$

In this sense, note that they are also a particular case of the little q-Jacobi polynomials for $b = 0$.

Also, if $a = q^\alpha$ and $x \to (1 - q)x$ in (3.28), then

$$\lim_{q \to 1} l_n((1 - q)x; q^\alpha; q) = \frac{L_n^{(\alpha)}(x)}{L_n^{(\alpha)}(0)}.$$

3. q-Laguerre Polynomials

- Basic hypergeometric representation:

$$L_n^{(\alpha)}(x; q) = \frac{(q^\alpha + 1; q)_n}{(q; q)_n} \, {}_1\phi_1 \left(\begin{matrix} q^{-n} \\ q^{\alpha+1} \end{matrix} \Big| q; -q^{n+\alpha+1} x \right)$$

$$= \frac{1}{(q; q)_n} \, {}_2\phi_1 \left(\begin{matrix} q^{-n}, -x \\ 0 \end{matrix} \Big| q; q^{n+\alpha+1} \right).$$

- Orthogonality relation for $\alpha > -1$:

$$\int_0^\infty \frac{x^\alpha}{(-x; q)_\infty} L_m^{(\alpha)}(x; q) L_n^{(\alpha)}(x; q) dx = \frac{(q^{-\alpha}; q)_\infty}{(q; q)_\infty} \frac{(q^{\alpha+1}; q)_n}{(q; q)_n q^n} \Gamma(-\alpha) \Gamma(\alpha + 1) \delta_{m,n}.$$

- Recurrence relation:

$$- q^{2n+\alpha+1} x L_n^{(\alpha)}(x; q) =$$

$$(1-q^{n+1}) L_{n+1}^{(\alpha)}(x; q) - [(1-q^{n+1}) + q(1-q^{n+\alpha})] L_n^{(\alpha)}(x; q) + q(1-q^{n+\alpha}) L_{n-1}^{(\alpha)}(x; q).$$

- q-Difference equation for $y(x) = L_n^{(\alpha)}(x; q)$:

$$-q^\alpha(1 - q^n) x y(x) = q^\alpha(1 + x) y(qx) - [1 + q^\alpha(1 + x)] y(x) + y(q^{-1} x).$$

- Limit relations:
  If we substitute $a = q^\alpha$ and $x \to -b^{-1} q^{-1} x$ in (3.26) and then take $b \to -\infty$, we obtain the q-Laguerre polynomials as

$$\lim_{b \to -\infty} p_n(-b^{-1} q^{-1} x; q^\alpha, b; q) = \frac{(q; q)_n}{(q^{\alpha+1}; q)_n} L_n^{(\alpha)}(x; q).$$

3.2.12 q-Analogue of Hermite Polynomials

- Basic hypergeometric representation:

$$h_n(x; q) = q^{\binom{n}{2}} \, {}_2\phi_1 \left(\begin{matrix} q^{-n}, x^{-1} \\ 0 \end{matrix} \Big| q; -qx \right) = x^n \, {}_2\phi_0 \left(\begin{matrix} q^{-n}, q^{-n+1} \\ - \end{matrix} \Big| q^2; \frac{q^{2n-1}}{x^2} \right).$$

- Orthogonality relation:

$$\int_{-1}^{1} (qx, -qx; q)_\infty h_m(x; q) h_n(x; q) d_q x = (1-q)(q; q)_\infty (q, -1, -q; q; q)_\infty q^{\binom{n}{2}} \delta_{m,n}.$$

- Recurrence relation:

$$x h_n(x; q) = h_{n+1}(x; q) + q^{n-1}(1 - q^n) h_{n-1}(x; q).$$

- q-Difference equation:

$$-q^{-n+1} x^2 y(x) = y(qx) - (1+q) y(x) + q(1 - x^2) y(q^{-1}x) \quad \text{with} \quad y(x) = h_n(x; q).$$

- Limit relation:
  The usual continuous Hermite polynomials can be found via the following limit relation:

$$\lim_{q \to 1} \frac{h_n(x\sqrt{1 - q^2}; q)}{(1 - q^2)^{\frac{n}{2}}} = \frac{H_n(x)}{2^n}.$$

3.2.13 A Biorthogonal Exponential Sequence

Using the Fourier transform of the Stieltjes–Wigert polynomials, in this section we derive a sequence of exponential functions that are biorthogonal with respect to a complex weight function like

$$\exp\left(q_1(ix + p_1)^2 + q_2(ix + p_2)^2\right) \quad \text{on} \quad (-\infty, \infty).$$

Then we restrict these biorthogonal functions to a special case to obtain a real trigonometric sequence of orthogonal functions with respect to the weight function

$$\exp(-qx^2) \quad \text{on} \quad (-\infty, \infty).$$

In 1895, Stieltjes [6] proved that

$$\int_0^\infty x^n x^{-\ln x} \sin(2\pi \ln x) \, dx = 0, \quad n \in \mathbb{Z}.$$

Hence, independent of λ, we have

$$\frac{1}{\sqrt{\pi}} \int_0^\infty x^n x^{-\ln x}(1 + \lambda \, \sin(2\pi \, \ln x)) \, dx = \frac{1}{\sqrt{\pi}} \int_0^\infty x^n x^{-\ln x} dx = \exp\left(\frac{1}{4}(n + 1)^2\right),$$

which shows that for all $\lambda \in [-1, 1]$, the densities

$$W_\lambda(x) = \frac{1}{\sqrt{\pi}} x^{-\ln x}(1 + \lambda \, \sin(2\pi \, \ln x)), \quad x > 0, \tag{3.29}$$

have the same moments.

Using this result, one can conclude that there exist many different weight functions for orthogonal polynomials corresponding to positive measures (3.29), known later as Stieltjes–Wigert polynomials. In this sense, it was Wigert [7] who first succeeded in 1923 in finding the explicit form of the orthogonal polynomials corresponding to the log-normal weight function

$$\rho(x; q) = \frac{\gamma}{\sqrt{\pi}} \exp(-\gamma^2 \ln^2 x) = \frac{\gamma}{\sqrt{\pi}} x^{(-\gamma^2 \ln x)} \quad \text{for} \quad -\gamma^2 = \frac{1}{\ln q^2}, \tag{3.30}$$

which is a more general case of (3.29) for $\lambda = 0$.

The Wigert polynomials are defined as

$$W_n(x; q) = \sum_{k=0}^n (-1)^k \begin{bmatrix} n \\ k \end{bmatrix}_q q^{k(k+\frac{1}{2})} x^k \quad \text{for} \quad 0 < q < 1, \tag{3.31}$$

in which, as before,

$$\begin{bmatrix} n \\ i \end{bmatrix}_q = \frac{(q; q)_n}{(q; q)_i (q; q)_{n-i}} = \prod_{j=0}^{i-1} \frac{1 - q^{n-j}}{1 - q^{j+1}} \quad \text{and} \quad (a; q)_n = \prod_{i=0}^{n-1} 1 - aq^i.$$

According to [2], they satisfy the orthogonality relation

$$\int_0^\infty \rho(x; q) \, W_n(x; q) \, W_m(x; q) \, dx = \left(q^{-(n+1/2)}(q; q)_n\right) \delta_{n,m}. \tag{3.32}$$

Although Stieltjes considered the weight function (3.30) only for $\gamma = 1$ (or equivalently $\lambda = 0$ in (3.29)) as an example leading to an indeterminate Stieltjes moment problem [6], the author in [3] has considered such a problem associated with the Stieltjes–Wigert polynomials in a general case. Consequently, there should be various weight functions for polynomials (3.31). For instance, the shifted polynomials

$$S_n(x; q) = \frac{1}{(q; q)_n} W_n(q^{-\frac{1}{2}}x; q)$$

(3.33)

satisfy the orthogonality relation

$$\int_0^\infty \rho^*(x; q)\, S_n(x; q)\, S_m(x; q)\, dx = \left(-\frac{\ln q}{q^n} \frac{(q; q)_\infty}{(q; q)_n}\right) \delta_{n,m},$$

(3.34)

where [4]

$$\rho^*(x; q) = \frac{1}{(-x; q)_\infty(-qx^{-1}; q)_\infty} = \prod_{i=0}^\infty \frac{1}{(1 + xq^i)(1 + x^{-1}q^{i+1})}.$$

(3.35)

Now if (3.33) is replaced in (3.34), we get

$$\int_0^\infty \rho^*(q^{\frac{1}{2}}x; q)\, W_n(x; q)\, W_m(x; q)\, dx = \left(-\frac{\ln q}{q^{n+1/2}}(q; q)_n(q; q)_\infty\right) \delta_{n,m}.$$

Notice that the shape of $\rho^*(q^{1/2}x; q)$, by noting the definition (3.35), is completely different from the log-normal function, though both of them play the role of a weight function for the Stieltjes–Wigert polynomials.

Hence, let us define the shifted polynomials

$$V_n(x; p, q) = W_n(q^{p-\frac{1}{2}}x; q) = \sum_{k=0}^n (-1)^k \begin{bmatrix} n \\ k \end{bmatrix}_q q^{k(k+p)}x^k,$$

(3.36)

in which p is a free real parameter and $q \in (0, 1)$.

If (3.36) is replaced in (3.32), then

$$\int_0^\infty \rho(q^{p-\frac{1}{2}}x; q)\, V_n(x; p, q)\, V_m(x; p, q)\, dx = \left(q^{-(p+n)}(q; q)_n\right) \delta_{n,m}.$$

(3.37)

Since the weight function of (3.37) takes the form

$$\rho(q^{p-\frac{1}{2}}x; q) = \frac{1}{\sqrt{(-2\pi)\ln q}} \exp\left(\frac{1}{2\ln q}((p - 1/2)\ln q + \ln x)^2\right)$$

$$= \frac{1}{\sqrt{(-2\pi)\ln q}} q^{\frac{(2p-1)^2}{8}} x^{p-\frac{1}{2}} \exp\left(\frac{1}{2\ln q}\ln^2 x\right),$$

(3.38)

relation (3.38) simplifies (3.37) in the form

$$\int_0^\infty x^{p-\frac{1}{2}} \exp\left(\left(\frac{1}{\ln q^2}\right) \ln^2 x\right) V_n(x;p,q) V_m(x;p,q)\, dx$$

$$= \left(\sqrt{-2\pi \ln q}\; q^{-n-\frac{1}{8}(2p+1)^2}(q;q)_n\right) \delta_{n,m}. \qquad (3.39)$$

We can now compute the Fourier transform of $V_n(x;p,q)$ and substitute it in Parseval's identity to generate a new sequence of biorthogonal exponential functions. For this purpose, we first define the specific functions

$$g(x) = \exp(p_1 x) \exp(\tfrac{1}{\ln q_1} x^2) V_n(e^x; a_1, b_1),\; 0 < b_1, q_1 < 1,\quad a_1, p_1 \in \mathbb{R},$$

and

$$h(x) = \exp(p_2 x) \exp(\tfrac{1}{\ln q_2} x^2) V_m(e^x; a_2, b_2),\quad 0 < b_2, q_2 < 1,\quad a_2, p_2 \in \mathbb{R},$$
$$(3.40)$$

in terms of the polynomials (3.36).

Clearly, for both functions defined in (3.40), the Fourier transform exists. For example, for the function g we get

$$\mathbf{F}\, g(s) = \int_{-\infty}^\infty e^{-isx} \exp(p_1 x) \exp\left(\frac{1}{\ln q_1} x^2\right) V_n(e^x; a_1, b_1)\, dx$$

$$= \int_0^\infty t^{-is+p_1-1} \exp\left(\frac{1}{\ln q_1} \ln^2 t\right) \sum_{k=0}^n (-1)^k \begin{bmatrix} n \\ k \end{bmatrix}_{b_1} b_1^{k(k+a_1)} t^k\, dt$$

$$= \sum_{k=0}^n (-1)^k \begin{bmatrix} n \\ k \end{bmatrix}_{b_1} b_1^{k(k+a_1)} \int_0^\infty t^{-is+p_1-1+k} \exp\left(\frac{1}{\ln q_1} \ln^2 t\right) dt$$

$$= \sqrt{-\pi \ln q_1} \sum_{k=0}^n (-1)^k \begin{bmatrix} n \\ k \end{bmatrix}_{b_1} b_1^{k(k+a_1)} q_1^{\frac{-(-is+p_1+k)^2}{4}}, \qquad (3.41)$$

where we have used the general definite integral

$$\int_0^\infty x^\alpha \exp(\beta \ln^2 x)\, dx = \sqrt{-\frac{\pi}{\beta}} \exp(-\frac{(\alpha+1)^2}{4\beta}),\quad \beta < 0.$$

Now, by substituting (3.41) into Parseval's identity, we have

$$2\pi \int_{-\infty}^{\infty} \exp((p_1 + p_2)x) \exp\left(\left(\frac{1}{\ln q_1} + \frac{1}{\ln q_2}\right)x^2\right) V_n(e^x; a_1, b_1) V_m(e^x; a_2, b_2)\, dx$$

$$= 2\pi \int_0^{\infty} t^{-1+p_1+p_2} \exp\left(\left(\frac{1}{\ln q_1} + \frac{1}{\ln q_2}\right)\ln^2 t\right) V_n(t; a_1, b_1) V_m(t; a_2, b_2)\, dt$$

$$= \sqrt{\pi^2 \ln q_1 \ln q_2}$$

$$\times \int_{-\infty}^{\infty} q_1^{-(p_1-is)^2/4}\, \overline{q_2^{-(p_2-is)^2/4}}\, J_n(is; a_1, b_1, p_1, q_1)\, \overline{J_m(is; a_2, b_2, p_2, q_2)}\, ds\,,$$

$$(3.42)$$

in which $0 < q_1, q_2 < 1$; $p_1, p_2 \in \mathbb{R}$, and

$$J_n(x; a, b, c, d) = \sum_{k=0}^{n} (-1)^k \begin{bmatrix} n \\ k \end{bmatrix}_b b^{k(k+a)}\, d^{-\frac{1}{4}k(k+2c)}\, \left(d^{\frac{x}{2}}\right)^k.$$

On the other hand, if on the left-hand side of (3.42) we take

$$a_1 = a_2 = p_1 + p_2 - \frac{1}{2} \quad \text{and} \quad b_1 = b_2 = \exp\left(\frac{\ln q_1 \ln q_2}{2 \ln q_1 q_2}\right),$$

then according to orthogonality (3.39), relation (3.42) changes to

$$2\pi \int_0^{\infty} t^{-1+p_1+p_2} \exp\left(\left(\frac{1}{\ln q_1} + \frac{1}{\ln q_2}\right)\ln^2 t\right) V_n\left(t;\ p_1 + p_2 - \frac{1}{2}, \exp\left(\frac{\ln q_1 \ln q_2}{2 \ln q_1 q_2}\right)\right)$$

$$\times\ V_m\left(t;\ p_1 + p_2 - \frac{1}{2}, \exp\left(\frac{\ln q_1 \ln q_2}{2 \ln q_1 q_2}\right)\right)\, dt$$

$$= 2\pi \delta_{n,m} \times$$

$$\left(\sqrt{-\pi \frac{\ln q_1 \ln q_2}{\ln q_1 q_2}}\, \exp\left(-\left(n + \frac{1}{2}(p_1+p_2)^2\right)\frac{\ln q_1 \ln q_2}{2 \ln q_1 q_2}\right)\left(\exp\left(\frac{\ln q_1 \ln q_2}{2 \ln q_1 q_2}\right); \exp\left(\frac{\ln q_1 \ln q_2}{2 \ln q_1 q_2}\right)\right)_n\right)$$

$$= \sqrt{\pi^2 \ln q_1 \ln q_2} \int_{-\infty}^{\infty} q_1^{-(p_1-is)^2/4}\, \overline{q_2^{-(p_2-is)^2/4}}\, J_n\left(is;\ p_1 + p_2 - \frac{1}{2}, \exp\left(\frac{\ln q_1 \ln q_2}{2 \ln q_1 q_2}\right), p_1, q_1\right)$$

$$\times\ \overline{J_m\left(is;\ p_1 + p_2 - \frac{1}{2}, \exp\left(\frac{\ln q_1 \ln q_2}{2 \ln q_1 q_2}\right), p_2, q_2\right)}\, ds\,.$$

$$(3.43)$$

For convenience, we suppose in (3.43) that

$$q_1 \to \exp(-4q_1), \quad q_2 \to \exp(-4q_2), \quad \text{and} \quad p_1 \to -p_1,$$

to eventually obtain the following theorem.

Theorem 3.2 *The exponential functions*

$$E_n(x; q_1, p_1; q_2, p_2) =$$

$$\sum_{k=0}^{n} (-1)^k \begin{bmatrix} n \\ k \end{bmatrix}_{\exp(\frac{-2q_1 q_2}{q_1 + q_2})} \exp\left(\frac{q_1 k}{q_1 + q_2}((q_1 - q_2)k + q_2 - 2(q_1 p_1 + q_2 p_2)) \right) \exp(-2q_1 k\, x)$$

satisfy a biorthogonality relation

$$\int_{-\infty}^{\infty} \exp\left(q_1(ix + p_1)^2 + q_2(ix + p_2)^2 \right) E_n(ix; q_1, p_1; q_2, p_2) E_m(-ix; q_2, -p_2; q_1, -p_1)\, dx$$

$$= \left(\sqrt{\frac{\pi}{q_1 + q_2}} \, \exp\left((2n + (p_2 - p_1)^2) \frac{q_1 q_2}{q_1 + q_2} \right) \left(\exp\left(\frac{-2q_1\, q_2}{q_1 + q_2} \right); \exp\left(\frac{-2q_1\, q_2}{q_1 + q_2} \right) \right)_n \right) \delta_{n,m},$$

$$(3.44)$$

provided that

$$q_1, q_2 > 0 \quad \text{and} \quad p_1, p_2 \in \mathbb{R}.$$

Some Remarks on Theorem 3.2

1. The equality (3.44) is in fact a biorthogonality relation in complex weighted spaces, and $E_n(ix; q_1, p_1; q_2, p_2)$ in this relation are complex-valued functions that are biorthogonal with respect to $\exp\left(q_1(ix + p_1)^2 + q_2(ix + p_2)^2 \right)$ on $(-\infty, \infty)$.

   In general, complex orthogonal functions are defined on a rectifiable curve like C such that

$$\int_C W(z)\Phi_n(z)\, \overline{\Phi_m(z)}\, dz = \left(\int_C W(z)\, |\Phi_n(z)|^2\, dz \right) \delta_{n,m},$$

where

$$|\Phi_n(z)|^2 = \Phi_n(z)\, \overline{\Phi_n(z)},$$

and $W(z)$ is a positive function of the complex variable $z = x + iy$ defined on C.

2. According to Euler's identity $\exp(it) = \cos(t) + i \sin(t)$, the real and imaginary parts of the sequence $E_n(ix)$ are respectively defined as

$$\operatorname{Re} E_n(ix; q_1, p_1; q_2, p_2) = C_n(x; q_1, p_1; q_2, p_2)$$

$$= \sum_{k=0}^{n} (-1)^k \begin{bmatrix} n \\ k \end{bmatrix}_{\exp(\frac{-2q_1q_2}{q_1+q_2})} \exp\left(\frac{q_1 k}{q_1 + q_2}((q_1 - q_2)k + q_2 - 2(q_1 p_1 + q_2 p_2))\right)$$

$$\times \cos(2q_1 k x) \qquad (3.45)$$

and

$$\operatorname{Im} E_n(ix; q_1, p_1; q_2, p_2) = S_n(x; q_1, p_1; q_2, p_2)$$

$$= \sum_{k=0}^{n} (-1)^{k+1} \begin{bmatrix} n \\ k \end{bmatrix}_{\exp(\frac{-2q_1q_2}{q_1+q_2})} \exp\left(\frac{q_1 k}{q_1 + q_2}((q_1 - q_2)k + q_2 - 2(q_1 p_1 + q_2 p_2))\right)$$

$$\times \sin(2q_1 k x), \qquad (3.46)$$

where $q_1, q_2 > 0$ and $p_1, p_2 \in \mathbb{R}$.

3. The density of biorthogonality relation (3.44) is expandable as

$$W(z; q_1, p_1; q_2, p_2) = \exp(q_1(iz + p_1)^2 + q_2(iz + p_2)^2)$$

$$= \exp(q_1 p_1^2 + q_2 p_2^2) \exp(-(q_1 + q_2)z^2) \exp(2i(q_1 p_1 + q_2 p_2)z)$$

$$= K^* \exp(az^2 + ibz), \qquad (3.47)$$

where

$$a = -(q_1 + q_2) \quad \text{and} \quad b = 2(q_1 p_1 + q_2 p_2).$$

Hence, by noting that $|\exp(it)| = 1$, we have

$$|W(z)| = K^* \exp(-(q_1 + q_2)z^2), \quad z \in \mathbb{R},$$

in which $K^* = \exp(q_1 p_1^2 + q_2 p_2^2)$ is a real constant that has no effect on the weight function. This means that the complex weight function (3.47) is real in the weighted L^2-space if and only if $\frac{b}{2} = q_1 p_1 + q_2 p_2 = 0$.

By noting this important comment, let us reconsider the orthogonality relation (3.44) and suppose that

$$q_1 = q_2 = q/2, \quad q > 0, \quad \text{and} \quad p_1 = -p_2, \ p_1 \in \mathbb{R}.$$

So the special sequence

$$E_n^{(q)}(x) = E_n\left(-x; \frac{q}{2}, p_1; \frac{q}{2}, -p_1\right) = \sum_{k=0}^{n} (-1)^k \begin{bmatrix} n \\ k \end{bmatrix}_{e^{-\frac{q}{2}}} e^{k\frac{q}{4}} e^{kqx},$$

which is independent of p_1, satisfies

$$\int_{-\infty}^{\infty} e^{-qx^2} E_n^{(q)}(ix) E_m^{(q)}(-ix)\,dx = \int_{-\infty}^{\infty} e^{-qx^2} E_n^{(q)}(-ix) E_m^{(q)}(ix)\,dx$$

$$= \left(\sqrt{\frac{\pi}{q}}\ e^{n\frac{q}{2}} (e^{-\frac{q}{2}}; e^{-\frac{q}{2}})_n\right) \delta_{n,m}, \qquad (3.48)$$

for $q > 0$.

By noting the definitions (3.45) and (3.46), we can now define the following trigonometric sequences,

$$C_n^{(q)}(x) = C_n\left(-x; \frac{q}{2}, p_1; \frac{q}{2}, -p_1\right) = \sum_{k=0}^{n} (-1)^k \begin{bmatrix} n \\ k \end{bmatrix}_{e^{-\frac{q}{2}}} e^{k\frac{q}{4}} \cos(kqx),$$

and

$$S_n^{(q)}(x) = S_n\left(-x; \frac{q}{2}, p_1; \frac{q}{2}, -p_1\right) = \sum_{k=0}^{n} (-1)^k \begin{bmatrix} n \\ k \end{bmatrix}_{e^{-\frac{q}{2}}} e^{k\frac{q}{4}} \sin(kqx),$$

and verify that

$$E_n^{(q)}(ix) = C_n^{(q)}(x) + i\, S_n^{(q)}(x) \quad \text{and} \quad E_n^{(q)}(-ix) = C_n^{(q)}(x) - i\, S_n^{(q)}(x). \qquad (3.49)$$

If the relations in (3.49) are replaced in (3.48), then we get

$$\int_{-\infty}^{\infty} e^{-qx^2} \left(C_n^{(q)}(x) C_m^{(q)}(x) + S_n^{(q)}(x) S_m^{(q)}(x)\right) dx$$

$$+ i \int_{-\infty}^{\infty} e^{-qx^2} \left(S_n^{(q)}(x) C_m^{(q)}(x) - S_m^{(q)}(x) C_n^{(q)}(x)\right) dx$$

$$= \left(\sqrt{\pi/q}\ e^{nq/2} (e^{-q/2}; e^{-q/2})_n\right) \delta_{n,m}. \qquad (3.50)$$

But $S_n^{(q)}(x)$ is always an odd sequence, while $C_n^{(q)}(x)$ is even in (3.50), i.e.,

$$\int_{-\infty}^{\infty} e^{-q x^2} S_n^{(q)}(x) C_m^{(q)}(x)\, dx = \int_{-\infty}^{\infty} e^{-q x^2} S_m^{(q)}(x) C_n^{(q)}(x)\, dx = 0.$$

Corollary 3.1 *Two trigonometric sequences defined as*

$$M_{n,+}^{(q)}(x) = \frac{S_n^{(q)}(x) + C_n^{(q)}(x)}{\sqrt{2}} = \sum_{k=0}^{n} (-e^{q/4})^k \begin{bmatrix} n \\ k \end{bmatrix}_{e^{-\frac{q}{2}}} \sin\left(kqx + \frac{\pi}{4}\right)$$

and

$$M_{n,-}^{(q)}(x) = \frac{S_n^{(q)}(x) - C_n^{(q)}(x)}{\sqrt{2}} = \sum_{k=0}^{n} (-e^{q/4})^k \begin{bmatrix} n \\ k \end{bmatrix}_{e^{-\frac{q}{2}}} \sin\left(kqx - \frac{\pi}{4}\right)$$

respectively satisfy the orthogonality relations

$$\int_{-\infty}^{\infty} e^{-q x^2} M_{n,+}^{(q)}(x)\, M_{m,+}^{(q)}(x)\, dx = \left(\sqrt{\frac{\pi}{4q}}\; e^{n\frac{q}{2}} (e^{-\frac{q}{2}}; e^{-\frac{q}{2}})_n \right) \delta_{n,m} \tag{3.51}$$

and

$$\int_{-\infty}^{\infty} e^{-q x^2} M_{n,-}^{(q)}(x)\, M_{m,-}^{(q)}(x)\, dx = \left(\sqrt{\frac{\pi}{4q}}\; e^{n\frac{q}{2}} (e^{-\frac{q}{2}}; e^{-\frac{q}{2}})_n \right) \delta_{n,m}, \tag{3.52}$$

in which $q > 0$. *However, since* $M_{n,-}^{(q)}(-x) = -M_{n,+}^{(q)}(x)$ *if* $x = -t$ *in (3.52), only one orthogonality relation between (3.51) and (3.52) should be considered the main relation.*

3.3 Three Finite Classes of q-Orthogonal Polynomials

In this section, we introduce three ordinary q-Sturm–Liouville problems and prove that their polynomial solutions are finitely orthogonal with respect to three weight functions that correspond to the inverse gamma, Fisher, and Student-t distributions as $q \to 1$. We obtain the general properties of these polynomials and show that all results in three finite continuous cases are recovered again as $q \to 1$.

Let $\varphi(x)$ and $\psi(x)$ be two polynomials of degree at most 2 and 1 respectively as follows:

$$\varphi(x) = ax^2 + bx + c \quad \text{and} \quad \psi(x) = dx + e \quad (a, b, c, d, e \in \mathbb{C},\ d \neq 0).$$

If $\{y_n(x; q)\}_n$ is a sequence of polynomials that satisfies the q-difference equation

$$\varphi(x)D_q^2 y_n(x; q) + \psi(x)D_q y_n(x; q) + \lambda_{n,q} y_n(qx; q) = 0, \tag{3.53}$$

where D_q^2 is defined in (3.12), then the following orthogonality relation holds:

$$\int_a^b w(x; q)y_n(x; q)y_m(x; q)d_q x = \left(\int_a^b w(x; q)y_n^2(x; q)d_q x\right) \delta_{n,m},$$

in which $w(x; q) > 0$ is a solution of the Pearson q-difference equation

$$D_q\left(w(x; q)\varphi(q^{-1}x)\right) = w(qx; q)\psi(x),$$

and $w(q^{-1}x; q)\varphi(q^{-2}x)x^k$ for $k \in \mathbb{N}_0$ must vanish at $x = a, b$.

If $P_n(x) = x^n + \cdots$ is a monic polynomial solution of Eq. (3.53), the eigenvalue $\lambda_{n,q}$ can be directly obtained by equating the coefficients of x^n in (3.53) as

$$\lambda_{n,q} = -\frac{[n]_q}{q^n}(a[n-1]_q + d).$$

3.3.1   First Finite Sequence of q-Orthogonal Polynomials Corresponding to the Inverse Gamma Distribution

Consider the q-difference equation

$$x(q^2x + 1)D_q^2 y_n(x; q) - \left(\frac{q^2 - q^p}{1 - q}x + \frac{1}{1 - q}\right)D_q y_n(x; q) + \lambda_{n,q} y_n(qx; q) = 0, \tag{3.54}$$

with

$$\lambda_{n,q} = -\frac{[n]_q}{q^n}\left(q^2[n-1]_q - \frac{q^2 - q^p}{1 - q}\right),$$

for all $n = 0, 1, 2, \ldots$ and $q \in \mathbb{R} \setminus \{-1, 0, 1\}$, where

$$\lim_{q \to 1} \lambda_{n,q} = -n(n + 1 - p).$$

If Eq. (3.54) is expanded as

$$x(q^2x+1)\left(\frac{y_n(q^2x)-(1+q)y_n(qx)+qy_n(x)}{q(q-1)^2x^2}\right)$$

$$-\left(\frac{q^2-q^p}{1-q}x+\frac{1}{1-q}\right)\left(\frac{y_n(qx)-y_n(x)}{(q-1)x}\right)-\left(\frac{[n]_q}{q^n}\left(q^2[n-1]_q-\frac{q^2-q^p}{1-q}\right)\right)y_n(qx)=0,$$

then its symmetric form is derived as

$$(qx+1)y_n(qx)-\left((q^{n+1}+q^{p-n})x+1\right)y_n(x)+q^px\,y_n(q^{-1}x)=0.$$

The following theorem holds for the polynomial solution of Eq. (3.54).

Theorem 3.3 *Let $\{N_n^{(p)}(x;q)\}_n$ be a sequence of polynomials that satisfies Eq. (3.54).
For $n=0,1,\ldots,N$ we have*

$$\int_0^\infty w_1^{(p)}(x;q)N_n^{(p)}(x;q)N_m^{(p)}(x;q)d_qx=\left(\int_0^\infty w_1^{(p)}(x;q)\left(N_n^{(p)}(x;q)\right)^2d_qx\right)\delta_{n,m},$$

*where $q>1$, $N=\max\{m,n\}<\dfrac{1}{2}(p-1)$, $p\in\mathbb{R}$, and $w_1^{(p)}(x;q)$ is the solution of the
q-Pearson equation*

$$D_q\left(w_1^{(p)}(x;q)(x^2+q^{-1}x)\right)=w_1^{(p)}(qx;q)\left(\frac{q^p-q^2}{1-q}x-\frac{1}{1-q}\right),$$

which is equivalent to

$$\frac{w_1^{(p)}(x;q)}{w_1^{(p)}(qx;q)}=\frac{q^p}{1+q^{-1}x^{-1}}.\qquad\qquad(3.55)$$

Proof First, we can verify that a solution of the q-Pearson equation (3.55) is

$$w_1^{(p)}(x;q)=\frac{x^{-p}}{(-q^{-1}x^{-1};q^{-1})_\infty}\quad\left(p\in\mathbb{R}\quad\text{and}\quad 0<|q^{-1}|<1\right).\qquad(3.56)$$

In this sense, note that

$$\lim_{q\to1}(q-1)^{-p}\,w_1^{(p)}((q-1)^{-1}x;q)=x^{-p}e^{-1/x}.$$

Now we write Eq. (3.54) in the self-adjoint form

$$D_q\left(w_1^{(p)}(x;q)(x^2+q^{-1}x)D_q N_n^{(p)}(x;q)\right)+\lambda_{n,q} w_1^{(p)}(qx;q)N_n^{(p)}(qx;q)=0, \quad (3.57)$$

and for m as

$$D_q\left(w_1^{(p)}(x;q)(x^2+q^{-1}x)D_q N_m^{(p)}(x;q)\right)+\lambda_{m,q} w_1^{(p)}(qx;q)N_m^{(p)}(qx;q)=0.$$

$$(3.58)$$

By multiplying (3.57) by $N_m(qx;q)$ and (3.58) by $N_n(qx;q)$ and subtracting one from the other, we get

$$(\lambda_{m,q}-\lambda_{n,q})w_1^{(p)}(x;q)N_m^{(p)}(x;q)N_n^{(p)}(x;q)$$

$$= q^2 D_q\left(w_1^{(p)}(q^{-1}x;q)(q^{-2}x^2+q^{-2}x)D_q N_n^{(p)}(q^{-1}x;q)\right)N_m^{(p)}(x;q)$$

$$- q^2 D_q\left((w_1^{(p)}(q^{-1}x;q)(q^{-2}x^2+q^{-2}x)D_q N_m^{(p)}(q^{-1}x;q)\right)N_n^{(p)}(x;q). \quad (3.59)$$

Hence q-integration by parts on both sides of (3.59) over $[0,\infty)$ yields

$$(\lambda_{m,q}-\lambda_{n,q})\int_0^\infty w_1^{(p)}(x;q)N_m^{(p)}(x;q)N_n^{(p)}(x;q)d_q x$$

$$= \int_0^\infty q^2\Big\{D_q\left(w_1^{(p)}(q^{-1}x;q)(q^{-2}x^2+q^{-2}x)D_q N_n^{(p)}(q^{-1}x;q)\right)N_m^{(p)}(x;q)$$

$$- D_q\left(w_1^{(p)}(q^{-1}x;q)(q^{-2}x^2+q^{-2}x)D_q N_m^{(p)}(q^{-1}x;q)\right)N_n^{(p)}(x;q)\Big\}d_q x$$

$$= q^2\Big[w_1^{(p)}(q^{-1}x;q)(q^{-2}x^2+q^{-2}x)$$

$$\times \left(D_q N_n^{(p)}(q^{-1}x;q)N_m^{(p)}(x;q)-D_q N_m^{(p)}(q^{-1}x;q)N_n^{(p)}(x;q)\right)\Big]_0^\infty. \quad (3.60)$$

But since

$$\max\ \deg\left(D_q N_n^{(p)}(q^{-1}x;q)N_m^{(p)}(x;q)-D_q N_m^{(p)}(q^{-1}x;q)N_n^{(p)}(x;q)\right)=m+n-1,$$

if $N<\dfrac{1}{2}(p-1)$ for $N=\max\{m,n\}$, the following boundary conditions hold in (3.60):

$$\lim_{x\to 0} w_1^{(p)}(q^{-1}x;q)(q^{-2}x^2+q^{-2}x)x^{2N-1}=0$$

and

$$\lim_{x \to \infty} w_1^{(p)}(q^{-1}x; q)(q^{-2}x^2 + q^{-2}x)x^{2N-1} = 0.$$

Therefore, the right-hand side of (3.60) tends to zero, and

$$\int_0^\infty w_1^{(p)}(x; q)N_m^{(p)}(x; q)N_n^{(p)}(x; q)d_qx = 0$$

if and only if $m \neq n$ and $N = \max\{m, n\} < \frac{1}{2}(p - 1)$. $\square$

Corollary 3.2 *The finite set* $\{N_n^{(p)}(x; q)\}_{n=0}^{N < \frac{1}{2}(p-1)}$ *is orthogonal with respect to the weight function* $w_1^{(p)}(x; q) = \dfrac{x^{-p}}{(-q^{-1}x^{-1}; q^{-1})_\infty}$ *on* $[0, \infty)$.

The monic polynomial solution of Eq. (3.54) can be represented as

$$\bar{N}_n^{(p)}(x; q) = \frac{(-1)^n q^{n(n-p)}}{(q^{n-p+1}; q)_n} \, {}_2\phi_0\left(\begin{matrix} q^{-n}, q^{n-p+1} \\ - \end{matrix} \,\middle|\, q; -q^p x\right), \tag{3.61}$$

because it is enough to first expand the polynomial solution as

$$N_n^{(p)}(x; q) = \sum_{k=0}^n a_{n,k} \frac{x^k}{[k]_q!} \quad (a_{n,n} \neq 0, \, n = 0, 1, 2, \dots).$$

Then the coefficients $\{a_{n,k}\}_{k=0}^n$ satisfy the two-term recurrence relation

$$[n - k]_q(q^p - q^{n+k+1})a_{n,k} = -q^n a_{n,k+1},$$

which is directly determined up to the normalizing constant $a_{n,n} \neq 0$ as

$$a_{n,k} = \left(\prod_{i=1}^{n-k} \frac{q^i}{[i]_q} \frac{-q^{n-i}}{(q^p - q^{2n+1-i})}\right) a_{n,n}, \quad k = 0, 1, \dots, n - 1.$$

By choosing $a_{n,n} = (1 - q)^{-n}(q; q)_n$, the monic polynomial solution is finally derived as

$$\bar{N}_n^{(p)}(x; q) = q^{\frac{n(n+1)}{2}} \sum_{k=0}^n \frac{(q^{-n}; q)_k}{(q; q)_k} \left(\prod_{i=1}^{n-k} \frac{-q^{n-i}}{q^p - q^{2n+1-i}}\right)(-x)^k.$$

Remark 3.1 From (3.61), it can be concluded that

$$\lim_{q \to 1} \frac{(q^{n-p+1}; q)_n}{q^{n(n-p)}} \bar{N}_n^{(p)}((q-1)^{-1}x; q) = N_n^{(p)}(x),$$

in which

$$N_n^{(p)}(x) = n! \binom{p-1-n}{n} x^n \, {}_1F_1 \left(\begin{array}{c} -n \\ p-2n \end{array} \middle| \frac{1}{x} \right)$$

$$= (-1)^n \, {}_2F_0 \left(\begin{array}{c} -n, \, n-p+1 \\ - \end{array} \middle| -x \right)$$

is the same as the second finite sequence of hypergeometric orthogonal polynomials.

The following formula is known as the Rodrigues representation of monic polynomials:

$$\bar{N}_n^{(p)}(x; q) = \frac{q^{\frac{1}{2}(n^2-n-2np+2p)}(1-q)^n(-q^{-1}x^{-1}; q^{-1})_\infty}{(q^{2n-p}; q^{-1})_n \, x^{-p}}$$

$$\times D_q^n \left(x^{n-p}(-x; q^{-1})_n \, e_{q^{-1}} \left(\frac{q^n x^{-1}}{1-q} \right) \right), \qquad (3.62)$$

where the q-exponential function $e_q(x)$ is defined by (3.11). This formula can be extracted in a similar way as in [4] from the general formula

$$y(x; q) = \frac{K_n}{w(x; q)} D_q^n \left((\mathfrak{L}_q^{-n} w) \prod_{k=1}^n (\mathfrak{L}_q^{-k-1} \varphi) \right), \qquad (3.63)$$

where

$$K_n = \frac{q^{n(n+1)}}{\prod_{k=1}^n (a[2n-k-1]_q + e)}, \qquad n = 1, 2, 3 \ldots,$$

because in the case (3.62) we have

$$w(x; q) = \frac{x^{-p}}{(-q^{-1}x^{-1}; q^{-1})_\infty} \quad \text{and} \quad \varphi(x) = q^2 x^2 + x.$$

Note that (3.62) is a q-analogue of the relation

$$N_n^{(p)}(x) = (-1)^n x^p e^{1/x} \frac{d^n (x^{-p+2n} e^{-1/x})}{dx^n}.$$

Computation of the Norm Square Value By noting the explicit representation (3.61), it is not difficult to find that the monic polynomials satisfy the recurrence relation

$$\bar{N}_{n+1}^{(p)}(x; q) = (x - c_n)\, \bar{N}_n^{(p)}(x; q) - d_n \bar{N}_{n-1}^{(p)}(x; q),$$

where

$$c_n = -\frac{q^{n-p}(1 - q^n - q^{n+1} + q^{2n-p+1})}{(1 - q^{2n-p})(1 - q^{2n-p+2})}$$

and

$$d_n = -\frac{q^{n+1}(1 - q^n)(1 - q^{n-p})(q^{n-p-1})^2}{(1 - q^{2n-p-1})(1 - q^{2n-p})^2(1 - q^{2n-p+1})},$$

with the initial terms

$$\bar{N}_0^{(p)}(x; q) = 1 \quad \text{and} \quad \bar{N}_1^{(p)}(x; q) = x - \frac{q}{q^p - q^2}.$$

Since $d_n > 0$ for all $n = 1, 2, \ldots, N < \frac{1}{2}(p - 1)$, Favard's theorem can be applied, so that we have

$$\int_0^\infty w_1^{(p)}(x; q)\bar{N}_m^{(p)}(x; q)\bar{N}_n^{(p)}(x; q)d_qx = \left(\prod_{k=1}^n d_k \int_0^\infty w_1^{(p)}(x; q)d_qx\right)\delta_{n,m},$$

and $w_1^{(p)}(x; q)$ is given by (3.56).

In order to compute

$$\int_0^\infty w_1^{(p)}(x; q)d_qx = \int_0^\infty \frac{x^{-p}}{(-q^{-1}x^{-1}; q^{-1})_\infty}d_qx,$$

we use relation (3.18) for $u(x) = q^{-1}x^{-1}$ to obtain

$$\int_0^\infty \frac{x^{-p}}{(-q^{-1}x^{-1}; q^{-1})_\infty}d_qx = q^p \int_0^\infty \frac{x^{(p-2)}}{(-x; q^{-1})_\infty}d_{q^{-1}}x. \tag{3.64}$$

But the integral on the right-hand side of (3.64) can be computed by the Ramanujan identity (3.19) as follows:

$$q^p \int_0^\infty \frac{x^{(p-2)}}{(-x; q^{-1})_\infty} d_{q^{-1}} x = q^p (1 - q^{-1}) \sum_{n=-\infty}^\infty q^{-n(p-1)} \frac{(-1; q^{-1})_n}{(-1; q^{-1})_\infty (0; q^{-1})_n}$$

$$= \frac{q^p (1 - q^{-1})}{(-1; q^{-1})_\infty} \sum_{n=-\infty}^\infty q^{-n(p-1)} \frac{(-1; q^{-1})_n}{(0; q^{-1})_n} = \frac{q^p (1 - q^{-1})}{(-1; q^{-1})_\infty} \Psi(-1, 0; q^{-1}; q^{1-p})$$

$$= \frac{q^p (1 - q^{-1})}{(-1; q^{-1})_\infty} \prod_{n=0}^\infty \frac{(1 - q^{-n-1})(1 + q^{p-n-2})(1 + q^{1-n-p})}{(1 + q^{-n-1})(1 - q^{1-n-p})}$$

$$= \frac{q^{p-1}(q-1)(q^{-1}; q^{-1})_{2-p}}{(-1; q^{-1})_{1-p}(-q^{-1}; q^{-1})_{p-1}}. \tag{3.65}$$

Hence the norm square value is explicitly computed as

$$\int_0^\infty \frac{x^{-p}}{(-q^{-1} x^{-1}; q^{-1})_\infty} (\bar{N}_n^{(p)}(x; q))^2 d_q x$$

$$= \frac{(-1)^n (q-1)(q, q^{1-p}; q)_n (q^{-1}; q^{-1})_{2-p} q^{\frac{1}{2}(3n^2 - 4np + n + 2p - 2)}}{(q^{1-p}, q^{2-p}, q^{2-p}, q^{3-p}; q^2)_n (-1; q^{-1})_{1-p}(-q^{-1}; q^{-1})_{p-1}}$$

$$\Longleftrightarrow n = 0, 1, \ldots, N < \frac{p-1}{2}. \tag{3.66}$$

For instance, the set $\{\bar{N}_n^{(21)}(x; q)\}_{n=0}^{10}$ is a finite sequence of q-orthogonal polynomials with the weight function $\dfrac{x^{-21}}{(-q^{-1} x^{-1}; q^{-1})_\infty}$ that satisfies the orthogonality relation

$$\int_0^\infty \frac{x^{-21}}{(-q^{-1} x^{-1}; q^{-1})_\infty} \bar{N}_m^{(21)}(x; q) \bar{N}_n^{(21)}(x; q) d_q x$$

$$= \left(\frac{(-1)^n (q-1)(q, q^{-20}; q)_n (q^{-1}; q^{-1})_{-19} q^{\frac{1}{2}(3n^2 - 83n + 40)}}{(q^{-20}, q^{-19}, q^{-19}, q^{-18}; q^2)_n (-1; q^{-1})_{-20}(-q^{-1}; q^{-1})_{20}} \right) \delta_{m,n} \Longleftrightarrow m, n < 10.$$

Also, relation (3.65) helps us compute the moments corresponding to the weight function (3.56) as follows:

$$\mu_k = \int_0^\infty \frac{x^{k-p}}{(-q^{-1} x^{-1}; q^{-1})_\infty} d_q x = \frac{q^{p-k-1}(q-1)(q^{-1}; q^{-1})_{2-p+k}}{(-1; q^{-1})_{1-p+k}(-q^{-1}; q^{-1})_{p-k-1}}.$$

It is important to note that if $n > \frac{1}{2}(p - 1)$ in (3.66), the above moments will be divergent and consequently the norm square value will be divergent too.

3.3.2   Second Finite Sequence of q-Orthogonal Polynomials Corresponding to the Fisher Distribution

Consider the q-difference equation

$$x(qx+1)D_q^2 y_n(x;q) - (q[s-2]_q x + [-t-1]_q)D_q y_n(x;q) + \lambda_{n,q} y_n(qx;q) = 0,$$

$$(3.67)$$

with

$$\lambda_{n,q} = [n]_q [s-n-1]_q,$$

for all $n = 0, 1, 2, \ldots$ and $q \in \mathbb{R} \setminus \{-1, 0, 1\}$, where

$$\lim_{q \to 1} \lambda_{n,q} = n(s-n-1).$$

If Eq. (3.67) is expanded as

$$x(qx+1)\left(\frac{y_n(q^2 x) - (1+q)y_n(qx) + q y_n(x)}{q(q-1)^2 x^2}\right)$$

$$- (q[s-2]_q x + [-t-1]_q)\left(\frac{y_n(qx) - y_n(x)}{(q-1)x}\right) + [n]_q [s-n-1]_q y_n(qx) = 0,$$

then its symmetric form is derived as

$$(x+1)y_n(qx;q) - \left((q^{s-1-n} + q^n)x + (q^{-t} + 1)\right) y_n(x;q) + (q^{s-1}x + q^{-t}) y_n(q^{-1}x;q) = 0.$$

Theorem 3.4 *Let $\{M_n^{(s,t)}(x;q)\}_n$ be a sequence of polynomials that satisfies Eq. (3.67). For $n = 0, 1, \ldots, N$ we have*

$$\int_0^\infty w_2^{(s,t)}(x;q) M_n^{(s,t)}(x;q) M_m^{(s,t)}(x;q) d_q x = \left(\int_0^\infty w_2^{(s,t)}(x;q) \left(M_n^{(s,t)}(x;q)\right)^2 d_q x\right) \delta_{n,m},$$

where $0 < q < 1, t > -1, N = \max\{m,n\} < \frac{1}{2}(s-1)$, and $w_2^{(s,t)}(x;q)$ is the solution of the q-Pearson equation

$$D_q\left(w_2^{(s,t)}(x;q)(q^{-1}x^2 + q^{-1}x)\right) = -w_2^{(s,t)}(qx;q)\left(q[s-2]_q x + [-t-1]_q\right),$$

which is equivalent to

$$\frac{w_2^{(s,t)}(x;q)}{w_2^{(s,t)}(qx;q)} = \frac{q^{s+t}x+1}{x+1}q^{-t}.$$ (3.68)

Proof It can be verified that

$$w_2^{(s,t)}(x;q) = \frac{x^t}{(-x;q)_{s+t}} \qquad (s,t \in \mathbb{R} \text{ and } 0 < |q| < 1)$$ (3.69)

is a solution of the Pearson equation (3.68), where

$$\lim_{q \to 1} w_2^{(s,t)}(x;q) = \frac{x^t}{(1+x)^{s+t}}.$$

Now we write Eq. (3.67) in the self-adjoint form

$$D_q\left(w_2^{(s,t)}(x;q)(q^{-1}x^2+q^{-1}x)D_q M_n^{(s,t)}(x;q)\right) + \lambda_{n,q}w_2^{(s,t)}(qx;q)M_n^{(s,t)}(qx;q) = 0,$$ (3.70)

and for m as

$$D_q\left(w_2^{(s,t)}(x;q)(q^{-1}x^2+q^{-1}x)D_q M_m^{(s,t)}(x;q)\right) + \lambda_{m,q}w_2^{(s,t)}(qx;q)M_m^{(s,t)}(qx;q) = 0.$$ (3.71)

Multiply (3.70) by $M_m^{(s,t)}(qx;q)$ and (3.71) by $M_n^{(s,t)}(qx;q)$ and subtract one from the other to get

$$(\lambda_{m,q} - \lambda_{n,q})w_2^{(s,t)}(x;q)M_m^{(s,t)}(x;q)M_n^{(s,t)}(x;q)$$

$$= q^2 D_q\left(w_2^{(s,t)}(q^{-1}x;q)(q^{-3}x^2+q^{-2}x)D_q M_n^{(s,t)}(q^{-1}x;q)\right)M_m^{(s,t)}(x;q)$$

$$- q^2 D_q\left(w_2^{(s,t)}(q^{-1}x;q)(q^{-3}x^2+q^{-2}x)D_q M_m^{(s,t)}(q^{-1}x;q)\right)M_n^{(s,t)}(x;q).$$ (3.72)

Using q-integration by parts on both sides of (3.72) over $[0, \infty)$, we obtain

$$(\lambda_{m,q} - \lambda_{n,q})\int_0^\infty w_2^{(s,t)}(x;q)M_m^{(s,t)}(x;q)M_n^{(s,t)}(x;q)d_qx$$

$$= \int_0^\infty q^2\left\{D_q\left(w_2^{(s,t)}(q^{-1}x;q)(q^{-3}x^2+q^{-2}x)D_q M_n^{(s,t)}(q^{-1}x;q)\right)M_m^{(s,t)}(x;q)\right.$$

$$- D_q \left(w_2^{(s,t)}(q^{-1}x; q)(q^{-3}x^2 + q^{-2}x) D_q M_m^{(s,t)}(q^{-1}x; q) \right) M_n^{(s,t)}(x; q) \Big\} d_q x$$

$$= q^2 \Big[w_2^{(s,t)}(q^{-1}x; q)(q^{-3}x^2 + q^{-2}x) \left(D_q M_n^{(s,t)}(q^{-1}x; q) M_m^{(s,t)}(x; q) \right.$$

$$\left. - D_q M_m^{(s,t)}(q^{-1}x; q) M_n^{(s,t)}(x; q) \right) \Big]_0^\infty. \qquad (3.73)$$

Since

$$\max \ \deg \left(D_q M_n^{(s,t)}(q^{-1}x; q) M_m^{(s,t)}(x; q) - D_q M_m^{(s,t)}(q^{-1}x; q) M_n^{(s,t)}(x; q) \right) = m+n-1,$$

if $t > -1$ and $N < \frac{1}{2}(s-1)$ for $N = \max\{m, n\}$, the following boundary conditions hold in (3.73):

$$\lim_{x \to 0} w_2^{(s,t)}(q^{-1}x; q)(q^{-3}x^2 + q^{-2}x)x^{2N-1} = 0$$

and

$$\lim_{x \to \infty} w_2^{(s,t)}(q^{-1}x; q)(q^{-3}x^2 + q^{-2}x)x^{2N-1} = 0.$$

Under these conditions, the right-hand side of (3.73) tends to zero and

$$\int_0^\infty w_2^{(s,t)}(x; q) M_m^{(s,t)}(x; q) M_n^{(s,t)}(x; q) d_q x = 0$$

if and only if

$$m \ne n, \quad t > -1 \quad \text{and} \quad N = \max\{m, n\} < \frac{1}{2}(s-1).$$

□

Corollary 3.3 *The finite set* $\{M_n^{(s,t)}(x; q)\}_{n=0}^{N < \frac{1}{2}(s-1)}$ *for* $t > -1$ *is orthogonal with respect to the weight function* $w_2^{(s,t)}(x; q) = \dfrac{x^t}{(-x; q)_{s+t}}$ *on* $[0, \infty)$.

The monic polynomial solution of Eq. (3.67) can be represented as

$$\bar{M}_n^{(s,t)}(x; q) = \frac{q^{\frac{1}{2}(n^2 + (1-2s-2t)n)}(q^{t+1}; q)_n}{(q^{n-s+1}; q)_n} \ {}_2\phi_1 \left(\begin{matrix} q^{-n}, q^{n-s+1} \\ q^{t+1} \end{matrix} \ \middle| \ q; -q^{s+t}x \right),$$

$$(3.74)$$

because it is enough to first expand the polynomial solution as

$$M_n^{(s,t)}(x;q) = \sum_{k=0}^{n} a_{n,k} \frac{x^k}{[k]_q!} \qquad \left(a_{n,n} \neq 0, \ n = 0, 1, 2, \dots\right).$$

Then the coefficients $\{a_{n,k}\}_{k=0}^{n}$ satisfy the two-term recurrence relation

$$[n-k]_q(q^{s-1} - q^{n+k})a_{n,k} = (q^{-(t+1)} - q^k)q^{n-k}a_{n,k+1},$$

which is directly determined up to the normalizing constant $a_{n,n} \neq 0$ as

$$a_{n,k} = \left(\prod_{i=1}^{n-k} \frac{q^i}{[i]_q} \frac{(q^{-(t+1)} - q^{n-i})}{(q^{s-1} - q^{2n-i})}\right) a_{n,n}, \quad k = 0, 1, \dots, n-1.$$

Once again, by choosing $a_{n,n} = (1-q)^{-n}(q;q)_n$, the monic polynomial solution is finally derived as

$$\bar{M}_n^{(s,t)}(x;q) = q^{\frac{1}{2}n(n+1)} \sum_{k=0}^{n} \frac{(q^{-n};q)_k}{(q;q)_k} \left(\prod_{i=1}^{n-k} \frac{(q^{-(t+1)} - q^{n-i})}{(q^{s-1} - q^{2n-i})}\right)(-x)^k.$$

Remark 3.2 From (3.74), it can be concluded that

$$\lim_{q \to 1} \frac{q^{-\frac{1}{2}(n^2 + (1-2s-2t)n)}(q^{n-s+1};q)_n}{(q^{t+1};q)_n} \bar{M}_n^{(s,t)}(x;q) = M_n^{(s,t)}(x),$$

in which

$$M_n^{(s,t)}(x) = (-1)^n n! \binom{t+n}{n} {}_2F_1\left(\begin{array}{c} -n, n+1-s \\ t+1 \end{array} \middle| -x\right)$$

is the same as the first finite sequence of hypergeometric orthogonal polynomials.

The Rodrigues representation of the monic polynomials is

$$\bar{M}_n^{(s,t)}(x;q) = \frac{q^{\frac{1}{2}(n^2 + (1-2s-2t)n)}(1-q)^n(-x;q)_{s+t}}{x^t(q^{2n-s+2};q^{-1})_n} \times D_q^n\left(x^{t+n} \frac{(-q^{-1}x;q^{-1})_n}{(-q^{-n}x;q)_{s+t}}\right),$$

which can be directly derived by the formula (3.63) when

$$w(x;q) = \frac{x^t}{(-x;q)_{s+t}} \quad \text{and} \quad \varphi(x) = qx^2 + x.$$

Computation of the Norm Square Value By noting the explicit representation (3.74), we see that the monic polynomials satisfy the recurrence relation

$$\bar{M}_{n+1}^{(s,t)}(x;q) = (x - c_n)\,\bar{M}_n^{(s,t)}(x;q) - d_n\,\bar{M}_{n-1}^{(s,t)}(x;q),$$

where

$$c_n = -\frac{q^{n+s-1}(1 - q^n - q^{n+1} + q^{2n-s+1}) + q^{2n-t}(q^{s-n-1} - q^{-1} - 1 + q^n)}{q^{2s-2}(1 - q^{2n-s})(1 - q^{2n-s+2})}$$

and

$$d_n = q^{2n-2s-t+1}\frac{[n]_q\,[n-s]_q\,[n-t-s]_q\,[n+t]_q}{[2n-s-1]_q\,([2n-s]_q)^2\,[2n-s+1]_q}.$$

with the initial terms

$$\bar{M}_0^{(s,t)}(x;q) = 1 \quad \text{and} \quad \bar{M}_1^{(s,t)}(x;q) = x - c_0.$$

Since $d_n > 0$ for all $n = 1, 2, \ldots, N < \frac{1}{2}(s - 1)$, applying Favard's theorem yields

$$\int_0^\infty w_2^{(s,t)}(x;q)\bar{M}_m^{(s,t)}(x;q)\bar{M}_n^{(s,t)}(x;q)d_qx = \left(\prod_{k=1}^n d_k \int_0^\infty w_2^{(s,t)}(x;q)d_qx\right)\delta_{n,m}.$$

To compute

$$\int_0^\infty w_2^{(s,t)}(x;q)d_qx = \int_0^\infty \frac{x^t}{(-x;q)_{s+t}}d_qx,$$

we can directly use (3.20) to get

$$\int_0^\infty \frac{x^t}{(-x;q)_{s+t}}d_qx = \frac{2\,\Gamma_q(t+1)\Gamma_q(s-1)}{(-1;q)_{t+1}(-1;q)_{-t}\,\Gamma_q(s+t)}. \tag{3.75}$$

Hence the norm square value is computed as

$$\int_0^\infty \frac{x^t}{(-x;q)_{s+t}}(\bar{M}_n^{(s,t)}(x;q))^2 d_qx$$

$$= \frac{2q^{(n^2+(2-t-2s)n)}(q, q^{1-s}, q^{1-s-t}, q^{1+t}; q)_n\,\Gamma_q(t+1)\Gamma_q(s-1)}{(q^{1-s}, q^{2-s}, q^{2-s}, q^{3-s}; q^2)_n\,(-1;q)_{t+1}(-1;q)_{-t}\,\Gamma_q(s+t)}, \tag{3.76}$$

which is valid for $n = 0, 1, \ldots, N < \frac{s-1}{2}$.

For instance, the set $\{\bar{M}_n^{(101,1)}(x;q)\}_{n=0}^{50}$ is a finite sequence of q-orthogonal polynomials with the weight function $w(x;q) = \dfrac{x}{(-x;q)_{102}}$ that satisfies the relation

$$\int_0^\infty \frac{x}{(-x;q)_{102}} \bar{M}_m^{(101,1)}(x;q) \bar{M}_n^{(101,1)}(x;q) d_q x$$

$$= \left(\frac{q^{(n^2-201n)}(q,q^{-100},q^{-101},q^2;q)_n \, \Gamma_q(2)\Gamma_q(100)}{(q^{-100},q^{-99},q^{-99},q^{-98};q^2)_n \, (-1;q)_2(-1;q)_{-1} \, \Gamma_q(102)} \right) \delta_{m,n} \iff m,n < 100.$$

Also, relation (3.75) helps us compute the moments corresponding to the weight function (3.69) as

$$\mu_k = \int_0^\infty \frac{x^{t+k}}{(-x;q)_{s+t}} d_q x = \frac{2\,\Gamma_q(t+1+k)\Gamma_q(s-1-k)}{(-1;q)_{t+1+k}(-1;q)_{-t-k}\,\Gamma_q(s+t)}.$$

Once again, if $n > \frac{1}{2}(s-1)$ in (3.76), the above moments will be divergent, and the norm square value will be divergent too.

3.3.3 Third Finite Sequence of q-Orthogonal Polynomials Corresponding to Student's t-Distribution

As another special case of Eq. (3.53), let us consider the q-difference equation

$$(q^2x^2+1)D_q^2 y_n(x;q) - q^2[2p-3]_q \, x D_q y_n(x;q) + \lambda_{n,q} y_n(qx;q) = 0, \qquad (3.77)$$

with

$$\lambda_{n,q} = q[n]_q[2p-n-2]_q,$$

for all $n = 0, 1, 2, \ldots$ and $q \in \mathbb{R} \setminus \{-1, 0, 1\}$, where

$$\lim_{q \to 1} \lambda_{n,q} = -n(n+2-2p).$$

The symmetric form of Eq. (3.77) is denoted by

$$(x^2+1)y_n(qx;q) - \left((q^{2p-2-n}+q^n)x^2 + (q+1) \right) y_n(x;q) + (q^{2p-2}x^2+q)\, y_n(q^{-1}x;q) = 0.$$

Theorem 3.5 *Let $\{I_n^{(p)}(x;q)\}_n$ be a sequence of symmetric polynomials that satisfies Eq. (3.77). For $n = 0, 1, \ldots, N$ we have*

$$\int_{-\infty}^{\infty} w_3^{(p)}(x;q) I_n^{(p)}(x;q) I_m^{(p)}(x;q) d_q x = \left(\int_{-\infty}^{\infty} w_3^{(p)}(x;q) \left(I_n^{(p)}(x;q) \right)^2 d_q x \right) \delta_{n,m},$$

where $q > 1$, $N = \max\{m, n\} < p - 1$, $(-1)^{2p} = -1$, and the symmetric function $w_3^{(p)}(x;q)$ is the solution of the q-Pearson equation

$$D_q \left(w_3^{(p)}(x;q)(x^2 + 1) \right) = -q^2 [2p - 3]_q \, w_3^{(p)}(qx;q),$$

which is equivalent to

$$\frac{w_3^{(p)}(x;q)}{w_3^{(p)}(qx;q)} = \frac{q^{2p-1} x^2 + 1}{x^2 + 1}. \tag{3.78}$$

Proof First, it can be verified that

$$w_3^{(p)}(x;q) = \frac{x^{1-2p}}{(-x^{-2}; q^{-2})_{p-\frac{1}{2}}} \qquad \left(0 < |q^{-2}| < 1 \right) \tag{3.79}$$

is a solution of Eq. (3.78) with $(-1)^{2p} = -1$, where

$$\lim_{q \to 1} w_3^{(p)}(x;q) = (1 + x^2)^{-(p-\frac{1}{2})}.$$

Now write Eq. (3.77) in the self-adjoint form

$$D_q \left(w_3^{(p)}(x;q)(x^2 + 1) D_q I_n^{(p)}(x;q) \right) + \lambda_{n,q} w_3^{(p)}(qx;q) I_n^{(p)}(qx;q) = 0, \tag{3.80}$$

and for m as

$$D_q \left(w_3^{(p)}(x;q)(x^2 + 1) D_q I_m^{(p)}(x;q) \right) + \lambda_{m,q} w_3^{(p)}(qx;q) I_m^{(p)}(qx;q) = 0. \tag{3.81}$$

By multiplying (3.80) by $I_m^{(p)}(qx;q)$ and (3.81) by $I_n^{(p)}(qx;q)$ and subtracting one from the other, we get

$$(\lambda_{m,q} - \lambda_{n,q}) w_3^{(p)}(x;q) I_m^{(p)}(x;q) I_n^{(p)}(x;q)$$

$$= q^2 D_q \left(w_3^{(p)}(q^{-1}x;q)(q^{-2}x^2 + 1) D_q I_n^{(p)}(q^{-1}x;q) \right) I_m^{(p)}(x;q)$$

$$- q^2 D_q \left(w_3^{(p)}(q^{-1}x;q)(q^{-2}x^2 + 1) D_q I_m^{(p)}(q^{-1}x;q) \right) I_n^{(p)}(x;q). \tag{3.82}$$

Hence q-integration by parts on both sides of (3.82) over $(-\infty, \infty)$ yields

$$(\lambda_{m,q} - \lambda_{n,q}) \int_{-\infty}^{\infty} w_3^{(p)}(x; q) I_m^{(p)}(x; q) I_n^{(p)}(x; q) d_q x$$

$$= \int_{-\infty}^{\infty} q^2 \left\{ D_q \left(w_3^{(p)}(q^{-1}x; q)(q^{-2}x^2 + 1) D_q I_n^{(p)}(q^{-1}x; q) \right) I_m^{(p)}(x; q) \right.$$

$$\left. - D_q \left(w_3^{(p)}(q^{-1}x; q)(q^{-2}x^2 + 1) D_q I_m^{(p)}(q^{-1}x; q) \right) I_n^{(p)}(x; q) \right\} d_q x$$

$$= q^2 \left[w_3^{(p)}(q^{-1}x; q)(q^{-2}x^2 + 1) \left(D_q I_n^{(p)}(q^{-1}x; q) I_m^{(p)}(x; q) - D_q I_m^{(p)}(q^{-1}x; q) I_n^{(p)}(x; q) \right) \right]_{-\infty}^{\infty}.$$

$$(3.83)$$

Since

$$\max \ \deg \left(D_q I_n^{(p)}(q^{-1}x; q) I_m^{(p)}(x; q) - D_q I_m^{(p)}(q^{-1}x; q) I_n^{(p)}(x; q) \right) = m + n - 1,$$

if $N < p - 1$ for $N = \max\{m, n\}$, the following boundary condition holds in (3.83):

$$\lim_{x \to \infty} w_3^{(p)}(q^{-1}x; q)(q^{-2}x^2 + 1) x^{2N-1} = 0.$$

Therefore, the right-hand side of (3.83) tends to zero, and

$$\int_{-\infty}^{\infty} w_3^{(p)}(x; q) I_m^{(p)}(x; q) I_n^{(p)}(x; q) d_q x = 0$$

if and only if $m \neq n$, $(-1)^{2p} = -1$ and $N = \max\{m, n\} < p - 1$.   □

Corollary 3.4 *The finite polynomial set* $\{I_n^{(p)}(x; q)\}_{n=0}^{N<p-1}$ *is orthogonal with respect to the even weight function* $w_3^{(p)}(x; q) = \dfrac{x^{1-2p}}{(-x^{-2}; q^{-2})_{p-\frac{1}{2}}}$ *on* $(-\infty, \infty)$.

The monic polynomial solution of Eq. (3.77) can be represented as

$$\bar{I}_n^{(p)}(x; q) = \frac{i^n q^{\frac{1}{2}n(n+3) - n(p+\frac{1}{2})} (q^{\frac{3}{2}-P}, -q^{\frac{3}{2}-P}; q)_n}{(q^{n-2p+2}; q)_n}$$

$$\times {}_3\phi_2 \left(\begin{matrix} q^{-n}, iq^{\frac{3}{2}-P}x^{-1}, q^{n-2p+2} \\ q^{\frac{3}{2}-P}, -q^{\frac{3}{2}-P} \end{matrix} \ \middle| \ q; iq^{P-\frac{1}{2}}x \right), \qquad (3.84)$$

because it is enough to first expand the polynomial solution as

$$I_n^{(p)}(x;q) = \sum_{k=0}^{n} a_{n,k} \frac{(-cx^{-1}q;q)_k}{(q;q)_k}(1-q)^k x^k \quad (c \in \mathbb{C},\ a_{n,n} \neq 0,\ n = 0,1,2,\ldots).$$

If c satisfies the relation

$$q^{2p-1}c^2 + q^2 = 0,$$

then it can be verified that the coefficients $\{a_{n,k}\}_{k=0}^{n}$ satisfy the two-term recurrence relation

$$[n-k]_q(q^{2p-1} - q^{n+k+1})c\,a_{n,k} = (q^{2k}c^2+1)q^{n-k+1}a_{n,k+1},$$

which is directly determined up to the normalizing constant $a_{n,n} \neq 0$ as

$$a_{n,k} = c^{k-n}\left(\prod_{i=1}^{n-k} \frac{c^2q^{2n-i+1}+q^{i+1}}{[i]_q(q^{2p-1}-q^{2n-i+1})}\right)a_{n,n}, \quad k = 0,1,\ldots,n-1.$$

By choosing $a_{n,n} = (1-q)^{-n}(q;q)_n$, the monic polynomial solution is finally derived as

$$\bar{I}_n^{(p)}(x;q) = i^n q^{\left(\frac{n(n+3)}{2}+n(p-\frac{3}{2})\right)} \sum_{k=0}^{n} \frac{(q^{-n};q)_k(iq^{\frac{3}{2}-p}x^{-1};q)_k}{(q;q)_k}$$

$$\times \left(\prod_{i=1}^{n-k} \frac{(1-q^{n-i-p+\frac{1}{2}})(1+q^{n-i-p+\frac{1}{2}})}{(q^{2p-1}-q^{2n-i+1})}\right)(iq^{\frac{1}{2}-p}x)^k.$$

Remark 3.3 From (3.84), we can conclude that

$$\lim_{q \to 1} \frac{(q^{n-2p+2};q)_n}{i^n q^{\frac{1}{2}n(n+3)-n(p+\frac{1}{2})}(q^{\frac{3}{2}-p},-q^{\frac{3}{2}-p};q)_n} \bar{I}_n^{(p)}(x;q) = I_n^{(p)}(x),$$

in which

$$I_n^{(p)}(x) = \frac{(-4i)^n(p-n)_n(\frac{3}{2}-p)_n}{(n+2-2p)_n}\ {}_2F_1\left(\begin{array}{c} -n,\,n+2-2p \\ \frac{3}{2}-p \end{array}\middle|\ \frac{1-ix}{2}\right).$$

The Rodrigues representation of the monic polynomials

$$\bar{I}_n^{(p)}(x; q) = \frac{q^{(n^2+n)}(1-q)^n(-x^{-2}; q^{-2})_{p-\frac{1}{2}}}{x^{1-2p}(q^{2n-2p+1}; q^{-1})_n} D_q^n \left(\frac{x^{1-2p}(-q^{-2}x^2; q^{-2})_n}{(-q^{2n}x^{-2}; q^{-2})_{p-\frac{1}{2}}} \right)$$

(3.85)

can be directly derived by the formula (3.63) when

$$w(x; q) = \frac{x^{1-2p}}{(-x^{-2}; q^{-2})_{p-\frac{1}{2}}} \quad \text{and} \quad \varphi(x) = q^2 x^2 + 1.$$

Computation of the Norm Square Value By noting the explicit representation (3.84), we see that the monic polynomials (3.85) satisfy the recurrence relation

$$\bar{I}_{n+1}^{(p)}(x; q) = x \, \bar{I}_n^{(p)}(x; q) - d_n^* \bar{I}_{n-1}^{(p)}(x; q),$$

where

$$d_n^* = -q^{n-2p+2} \frac{[n]_q [n - 2p + 1]_q}{[2n - 2p]_q [2n - 2p + 2]_q},$$

with the initial terms

$$\bar{I}_0^{(p)}(x; q) = 1 \quad \text{and} \quad \bar{I}_1^{(p)}(x; q) = x.$$

Since $d_n^* > 0$ for all $n = 1, 2, \ldots, N < p - 1$, applying Favard's theorem yields

$$\int_{-\infty}^{\infty} w_3^{(p)}(x; q) \bar{I}_m^{(p)}(x; q) \bar{I}_n^{(p)}(x; q) d_q x = \left(\prod_{k=1}^{n} d_k^* \int_{-\infty}^{\infty} w_3^{(p)}(x; q) d_q x \right) \delta_{n,m}.$$

In order to compute

$$\int_{-\infty}^{\infty} w_3^{(p)}(x; q) d_q x = \int_{-\infty}^{\infty} \frac{x^{1-2p}}{(-x^{-2}; q^{-2})_{p-\frac{1}{2}}} d_q x,$$

we can use (3.18) for $u(x) = x^{-\frac{1}{2}}$ to obtain

$$\int_{-\infty}^{\infty} \frac{x^{1-2p}}{(-x^{-2}; q^{-2})_{p-\frac{1}{2}}} d_q x = \frac{2q^2}{(q+1)} \int_0^{\infty} \frac{x^{p-2}}{(-x; q^{-2})_{p-\frac{1}{2}}} d_{q^{-2}} x.$$

(3.86)

Since the integral on the right-hand side of (3.86) is directly computable as

$$
\int_{-\infty}^{\infty} \frac{x^{1-2p}}{(-x^{-2};q^{-2})_{p-\frac{1}{2}}} d_q x
$$

$$
= \frac{4q^2}{(q+1)(-1;q^{-2})_{p-1}(-1;q^{-2})_{2-p}} \frac{\Gamma_{q^{-2}}(p-1)\Gamma_{q^{-2}}(\frac{1}{2})}{\Gamma_{q^{-2}}(p-\frac{1}{2})}, \tag{3.87}
$$

the norm square value is eventually computed as

$$
\int_{-\infty}^{\infty} \frac{x^{1-2p}}{(-x^{-2};q^{-2})_{p-\frac{1}{2}}} \left(\bar{I}_n^{(p)}(x;q)\right)^2 d_q x
$$

$$
= \frac{4(-1)^n q^{\frac{1}{2}(n^2+5n-4np+4)}(q;q)_n (q^{2-2p};q)_n}{(q+1)(-1;q^{-2})_{p-1}(-1;q^{-2})_{2-p}(q^{2-2p};q^2)_n(q^{4-2p};q^2)_n} \frac{\Gamma_{q^{-2}}(p-1)\Gamma_{q^{-2}}(\frac{1}{2})}{\Gamma_{q^{-2}}(p-\frac{1}{2})}
$$

$$
\Longleftrightarrow n = 0, 1, \ldots, N < p-1. \tag{3.88}
$$

For instance, the set $\{\bar{I}_n^{(\frac{23}{2})}(x;q)\}_{n=0}^{10}$ is a finite sequence of q-orthogonal polynomials with the weight function $w(x;q) = \dfrac{x^{-22}}{(-x^{-2};q^{-2})_{11}}$ that satisfies the relation

$$
\int_{-\infty}^{\infty} \frac{x^{-22}}{(-x^{-2};q^{-2})_{11}} \bar{I}_m^{(\frac{23}{2})}(x;q) \bar{I}_n^{(\frac{23}{2})}(x;q) d_q x
$$

$$
= \left(\frac{4(-1)^n q^{\frac{1}{2}(n^2-41n+4)}(q;q)_n (q^{-21};q)_n \Gamma_{q^{-2}}(\frac{21}{2})\Gamma_{q^{-2}}(\frac{1}{2})}{(q+1)(-1;q^{-2})_{\frac{21}{2}}(-1;q^{-2})_{-\frac{19}{2}}(q^{-21};q^2)_n(q^{-19};q^2)_n \Gamma_{q^{-2}}(11)} \right) \delta_{m,n}
$$

$$
\Longleftrightarrow \quad m, n < 10.
$$

Also, relation (3.87) helps us compute the moments corresponding to the weight function (3.79) as

$$
\mu_k = \int_{-\infty}^{\infty} \frac{x^{k+1-2p}}{(-x^{-2};q^{-2})_{p-\frac{1}{2}}} d_q x
$$

$$
= \frac{4q^2 \,\Gamma_{q^{-2}}(k+p-1)\Gamma_{q^{-2}}(\frac{1}{2}-k)}{(q+1)(-1;q^{-2})_{k+p-1}(-1;q^{-2})_{2-p-k}\,\Gamma_{q^{-2}}(p-\frac{1}{2})}.
$$

Notice that if $n > p-1$ in (3.88), the above moments will be divergent, and the norm square value will be divergent too.

3.3.4 A Characterization of Three Introduced Finite Sequences

In 2001, the authors in [1] classified all orthogonal polynomial families of the q-Hahn tableau and compared their scheme with the q-Askey scheme and Nikiforov–Uvarov tableau. In fact, they considered a second-order q-difference equation of the form

$$\phi(x)D_q D_{q^{-1}} y(x; q) + \psi(x)D_{q^{-1}} y(x; q) + \lambda'_{n,q} y(x; q) = 0, \tag{3.89}$$

in which

$$\phi(x) = a_2 x^2 + a_1 x + a_0 \quad \text{and} \quad \psi(x) = b_1 x + b_0 \quad \text{with} \quad b_1 \neq 0.$$

Now consider the function

$$\phi^*(x) = q^{-1}\phi(x) + (q^{-1} - 1)x\psi(x), \tag{3.90}$$

which can be derived via the Pearson equation

$$D_{q^{-1}}(\phi w) = q\psi w$$

or its expanded form

$$D_{q^{-1}}\phi(x)w(q^{-1}x) + D_{q^{-1}}w(x)\phi(x) = q\psi w. \tag{3.91}$$

After some computations, Eq. (3.91) changes to

$$\phi(q^{-1}x)w(q^{-1}x) = \left(\phi(x) + q(q^{-1} - 1)x\psi(x)\right)w(x),$$

which is finally equivalent to

$$\frac{w(q^{-1}x)}{w(x)} = \frac{q^{-1}\phi(x) + (q^{-1} - 1)x\psi(x)}{q^{-1}\phi(q^{-1}x)} = \frac{\phi^*(x)}{q^{-1}\phi(q^{-1}x)}.$$

The purpose of the authors in [1] was to classify all classical q-orthogonal polynomials using the roots of ϕ and ϕ^*. In this regard, if $x = 0$ is a root of ϕ (i.e., $a_0 = 0$), then it is also a root of ϕ^* and conversely. Hence, all families of q-orthogonal polynomials can be devided to two main groups of 0-families and $\oslash$-families. Afterward, the authors classified the two above-mentioned families based on the degree of the polynomials ϕ and ϕ^* and multiplicity of the 0-families.

Our purpose in this section is to compare the three introduced finite q-orthogonal polynomials with the authors' characterization in [1].

Comparison with the First Finite Sequence

Reconsider the equation

$$x(q^2x + 1)D_q^2 y_n(x; q) - \left(\frac{q^2 - q^p}{1 - q}x + \frac{1}{1 - q}\right) D_q y_n(x; q) + \lambda_{n,q} y_n(qx; q) = 0,$$

which can be written as

$$q^{1-p}x(x + q^{-1})D_q D_{q^{-1}} y(x; q) - \left(\frac{q^{1-p} - q^{-1}}{1 - q}x + \frac{q^{-p}}{1 - q}\right) D_{q^{-1}} y(x; q)$$

$$+ q^{1-p}\lambda'_{n,q} y(x; q) = 0. \qquad (3.92)$$

By comparing Eq. (3.92) with (3.89) and (3.90), we obtain

$$\phi(x) = q^{1-p}x(x + q^{-1}) \quad \text{and} \quad \phi^*(x) = q^{-2}x^2.$$

This means that the polynomial solution of Eq. (3.92) is a particular case of the 0-Jacobi/Bessel polynomials $j_n(x; a, b)$ in [1] with $a = q^{1-p}$ and $b = -q^{-1}$.

Comparison with the Second Finite Sequence

The q-difference equation of the second sequence, i.e.,

$$x(qx + 1)D_q^2 y_n(x; q) - (q[s - 2]_q x + [-t - 1]_q)D_q y_n(x; q) + \lambda_{n,q} y_n(qx; q) = 0,$$

can be written as

$$(x^2 + x)(x)D_q D_{q^{-1}} y(x; q) - \left(\frac{q^{s-2} - 1}{q - 1}x + \frac{q^{-t-1} - 1}{q - 1}\right) D_{q^{-1}} y(x; q) + q\lambda'_{n,q} y(x; q) = 0.$$

$$(3.93)$$

By comparing Eq. (3.93) with (3.89) and (3.90), we obtain

$$\phi(x) = x(x + 1) \quad \text{and} \quad \phi^*(x) = q^{-2}x(q^{s-1}x + q^{-t}).$$

Although the polynomial solution of Eq. (3.93) can be considered a 0-Jacobi/Jacobi polynomial, it is not similar to any other polynomial sequence given in [1], i.e., a new type of 0-Jacobi/Jacobi polynomial is presented in Sect. 3.3.2.

Comparison with the Third Finite Sequence

The q-difference equation of the third sequence, i.e.,

$$(q^2x^2 + 1)D_q^2 y_n(x; q) - q^2[2p - 3]_q x D_q y_n(x; q) + \lambda_{n,q} y_n(qx; q) = 0,$$

is equivalent to

$$(x^2+1)(x)D_q D_{q-1} y(x;q) + \left(\frac{1-q^{2p-3}}{q-1}\right) x\, D_{q-1} y(x;q) + \lambda'_{n,q} y(x;q) = 0. \quad (3.94)$$

By comparing Eq. (3.94) with (3.89) and (3.90), we find that

$$\phi(x) = x^2 + 1 \quad \text{and} \quad \phi^*(x) = q^{2p-4}x^2 + q^{-1}.$$

Once again, although the polynomial solution of Eq. (3.94) can be considered a $\oslash$-Jacobi/Jacobi polynomial, it is not similar to any other polynomial sequence given in [1], i.e., a new type of $\oslash$-Jacobi/Jacobi polynomial is presented.

3.4   A Symmetric Generalization of Sturm–Liouville Problems in q-Difference Spaces

Theorem 3.6 *Let $\phi_n(x;q) = (-1)^n \phi_n(-x;q)$ be a sequence of symmetric functions that satisfies the q-difference equation*

$$A(x)D_q D_{q-1}\phi_n(x;q) + B(x)D_q\phi_n(x;q) + \Big(\lambda_{n,q} C(x) + D(x) + \sigma_n E(x)\Big)\phi_n(x;q) = 0,$$
$$(3.95)$$

where $A(x)$, $B(x)$, $C(x)$, $D(x)$, and $E(x)$ are real functions, $\sigma_n = \frac{1-(-1)^n}{2}$, and $\lambda_{n,q}$ is a sequence of constants. If $A(x)$, $(C(x) > 0)$, $D(x)$, and $E(x)$ are even functions and $B(x)$ is odd, then

$$\int_{-\alpha}^{\alpha} W^*(x;q)\phi_n(x;q)\phi_m(x;q)d_q x = \left(\int_{-\alpha}^{\alpha} W^*(x;q)\phi_n^2(x;q)d_q x\right)\delta_{n,m},$$

where

$$W^*(x;q) = C(x)W(x;q), \quad (3.96)$$

and $W(x;q)$ is a solution of the Pearson q-difference equation

$$D_q (A(x)W(x;q)) = B(x)W(x;q), \quad (3.97)$$

which is equivalent to

$$\frac{W(qx;q)}{W(x;q)} = \frac{(q-1)xB(x) + A(x)}{A(qx)}.$$

Of course, the weight function defined in (3.96) must be positive and even, and $A(x)W(x;q)$ must vanish at $x = \alpha$.

Proof If Eq. (3.95) is written in a self-adjoint form, then

$$D_q\Big(A(x)W(x;q)D_{q^{-1}}\phi_n(x;q)\Big) + \Big(\lambda_{n,q}C(x) + D(x) + \sigma_n E(x)\Big)W(x;q)\phi_n(x;q) = 0,$$
(3.98)

and for m we have

$$D_q\Big(A(x)W(x;q)D_{q^{-1}}\phi_m(x;q)\Big) + \Big(\lambda_{m,q}C(x) + D(x) + \sigma_m E(x)\Big)W(x;q)\phi_m(x;q) = 0.$$
(3.99)

By multiplying (3.98) by $\phi_m(x;q)$ and (3.99) by $\phi_n(x;q)$ and subtracting one from the other, we get

$$\phi_m(x;q)D_q\Big(A(x)W(x)D_{q^{-1}}\phi_n(x;q)\Big) - \phi_n(x;q)D_q\Big(A(x)W(x)D_{q^{-1}}\phi_m(x;q)\Big)$$

$$+ \Big(\lambda_{n,q} - \lambda_{m,q}\Big)C(x)W(x;q)\phi_n(x;q)\phi_m(x;q)$$

$$+ \frac{(-1)^m - (-1)^n}{2}E(x)W(x;q)\phi_n(x;q)\phi_m(x;q) = 0.$$
(3.100)

Since the q-integral of an odd integrand over a symmetric interval is equal to zero, q-integrating both sides of (3.100) over the symmetric interval $[-\alpha, \alpha]$ yields

$$\int_{-\alpha}^{\alpha}\phi_m(x;q)D_q\big(A(x)W(x;q)D_{q^{-1}}\phi_n(x;q)\big)\,d_q x$$

$$- \int_{-\alpha}^{\alpha}\phi_n(x;q)D_q\big(A(x)W(x;q)D_{q^{-1}}\phi_m(x;q)\big)\,d_q x$$

$$+ \Big(\lambda_{n,q} - \lambda_{m,q}\Big)\int_{-\alpha}^{\alpha}C(x)W(x;q)\phi_n(x;q)\phi_m(x;q)d_q x$$

$$+ \frac{(-1)^m - (-1)^n}{2}\int_{-\alpha}^{\alpha}E(x)W(x;q)\phi_n(x;q)\phi_m(x;q)d_q x = 0.$$
(3.101)

Now, using the rule of q-integration by parts, relation (3.101) is transformed to

$$\Big[A(x)W(x;q)\phi_m(x;q)D_{q^{-1}}\phi_n(x;q)\Big]_{-\alpha}^{\alpha}$$

$$- \int_{-\alpha}^{\alpha}A(qx)W(qx;q)D_{q^{-1}}\phi_n(qx;q)D_q\phi_m(x;q)d_q x$$

$$- \Big[A(x)W(x;q)\phi_n(x;q)D_{q^{-1}}\phi_m(x;q)\Big]_{-\alpha}^{\alpha}$$

$$+ \int_{-\alpha}^{\alpha} A(qx)W(qx;q)D_{q^{-1}}\phi_m(qx;q)D_q\phi_n(x;q)d_qx$$

$$+ \big(\lambda_{n,q} - \lambda_{m,q}\big)\int_{-\alpha}^{\alpha} C(x)W(x;q)\phi_n(x;q)\phi_m(x;q)d_qx$$

$$+ \frac{(-1)^m - (-1)^n}{2}\int_{-\alpha}^{\alpha} E(x)W(x;q)\phi_n(x;q)\phi_m(x;q)d_qx = 0. \qquad (3.102)$$

Since

$$D_{q^{-1}}f(qx) = D_q f(x),$$

relation (3.102) is simplified as

$$\Big[A(x)W(x;q)\big(\phi_m(x;q)D_{q^{-1}}\phi_n(x;q) - \phi_n(x;q)D_{q^{-1}}\phi_m(x;q)\big)\Big]_{-\alpha}^{\alpha}$$

$$+ \big(\lambda_{n,q} - \lambda_{m,q}\big)\int_{-\alpha}^{\alpha} C(x)W(x;q)\phi_n(x;q)\phi_m(x;q)d_qx$$

$$+ \frac{(-1)^m - (-1)^n}{2}\int_{-\alpha}^{\alpha} E(x)W(x;q)\phi_n(x;q)\phi_m(x;q)d_qx = 0. \qquad (3.103)$$

On the other hand, $W(x;q)$ is a symmetric solution of the Pearson q-difference equation (3.97). Hence if in (3.103) we take

$$A(-\alpha)W(-\alpha;q) = A(\alpha)W(\alpha;q) = 0,$$

then to prove the orthogonality property it remains to show that

$$F^*(m,n) = \frac{(-1)^m - (-1)^n}{2}\int_{-\alpha}^{\alpha} E(x)W(x;q)\phi_n(x;q)\phi_m(x;q)d_qx = 0.$$

For this purpose, four cases should be considered for values m, n as follows:

1. If both m and n are even (or odd), then $F^*(n,m) = 0$, because we have

$$F^*(2i, 2j) = F^*(2i+1, 2j+1) = 0.$$

2. If one of the two mentioned values is odd and the other one is even, then

$$F^*(2i, 2j+1) = \int_{-\alpha}^{\alpha} E(x)W(x;q)\phi_{2j+1}(x;q)\phi_{2i}(x;q)d_qx. \qquad (3.104)$$

Since $E(x)$, $W(x; q)$ and $\phi_{2i}(x; q)$ are assumed to be even functions and $\phi_{2j+1}(x; q)$ is odd in (3.104), its integrand will be an odd function, and therefore

$$F^*(2i, 2j + 1) = 0.$$

These results similarly hold for the case

$$m = 2i + 1 \quad \text{and} \quad n = 2j, \quad \text{i.e.,} \quad F^*(2i + 1, 2j) = 0.$$

$\square$

3.4.1 Some Illustrative Examples

Example 3.1 Consider the q-difference equation

$$x^2(1 - x^2)D_q D_{q^{-1}}\phi_n(x; q) + qx \left(-(q^2 + q + 1)x^2 + q + 1\right) D_q \phi_n(x; q)$$
$$+ \left([n]_q \left(q(q^2 + q + 1) - [1 - n]_q\right)x^2 - q(1 + q)\sigma_n\right) \phi_n(x; q) = 0$$

as a special case of (3.95). If we set

$$\phi_n(x; q) = \sum_{j=0}^{\infty} a_j(n)x^j,$$

then a solvable recurrence relation for the coefficients $\{a_j(n)\}_{j=0}^{\infty}$ is derived, giving rise eventually to the following representation

$$\phi_n(x; q) = x^{\sigma_n} \, {}_2\phi_1 \left(\begin{matrix} q^{\sigma_n - n}, q^{n + \sigma_n - 1} \left(q^4 - q + 1\right) \\ q^{2\sigma_n + 1} \left(q^3 - q + 1\right) \end{matrix} \middle| \, q^2; q^2 x^2 \right). \tag{3.105}$$

According to Theorem 3.6, the above sequence satisfies the orthogonality relation

$$\int_{-1}^{1} W_1^*(x; q)\phi_n(x; q)\phi_m(x; q)d_q x = \left(\int_{-1}^{1} W_1^*(x; q)\phi_n^2(x; q)d_q x\right) \delta_{n,m},$$

in which $W_1^*(x; q) = C(x)W_1(x; q)$ is the main weight function and $W_1(x; q)$ satisfies the equation

$$\frac{W_1(qx; q)}{W_1(x; q)} = \frac{(q^4 - q + 1)x^2 - q^3 + q - 1}{q^2 \left(q^2 x^2 - 1\right)}. \tag{3.106}$$

Up to a periodic function, a solution of Eq. (3.106) is of the form

$$W_1(x; q) = \frac{(q^2 x^2; q^2)_\infty (q^3 - q + 1)^{\frac{\log(x^2)}{2\log(q)}}}{x^2 (\frac{q^4 - q + 1}{q^3 - q + 1} x^2; q^2)_\infty} = W_1(-x; q).$$

In this sense, note that

$$\lim_{q \uparrow 1} W_1^*(x; q) = \lim_{q \uparrow 1} x^2 W_1(x; q) = \frac{x^2}{\sqrt{1 - x^2}},$$

which indeed gives the weight function of the Chebyshev polynomials of the fifth kind.

To compute the norm square value of the symmetric polynomials (3.105), we can use Favard's theorem [2] in q-spaces, which says that if $\{P_n(x; q)\}$ satisfies the recurrence relation

$$x P_n(x; q) = A_n P_{n+1}(x; q) + B_n P_n(x; q) + C_n P_{n-1}(x; q), \qquad n = 0, 1, 2, \ldots,$$

where $P_{-1}(x; q) = 0$, $P_0(x; q) = 1$, A_n, B_n, C_n real, and $A_n C_{n+1} > 0$ for $n = 0, 1, 2, \ldots$, then there exists a weight function $W^*(x; q)$ such that

$$\int_{-\alpha}^{\alpha} W^*(x; q) P_n(x; q) P_m(x; q) d_q x = \left(\prod_{i=0}^{n-1} \frac{C_{i+1}}{A_i} \int_{-\alpha}^{\alpha} W^*(x; q) d_q x \right) \delta_{n,m}.$$

It is clear that Favard's theorem also holds for the monic type of symmetric q-polynomials in which $A_n = 1$ and $B_n = 0$. So, if $\bar{\phi}_n(x; q)$ is considered the monic form of the symmetric q-polynomials (3.105), then after some calculations, they satisfy the following three-term recurrence relation:

$$\bar{\phi}_{n+1}(x; q) = x \bar{\phi}_n(x; q) - \gamma_n \bar{\phi}_{n-1}(x; q) \text{ with } \bar{\phi}_0(x; q) = 1 \text{ and } \bar{\phi}_1(x; q) = x,$$

$$(3.107)$$

in which

$$\gamma_{2m} = \frac{q^{2m+1} (q^{2m} - 1) ((q^4 - q + 1) q^{2m} + (-q^3 + q - 1) q^2)}{(q^4 - q + 1)^2 q^{8m} - (q^2 + 1) (q^4 - q + 1) q^{4m+1} + q^4}$$

and

$$\gamma_{2m+1} = \frac{q^{2m} ((q^3 - q + 1) q^{2m+1} - 1) ((q^4 - q + 1) q^{2m} - q)}{(q^4 - q + 1)^2 q^{8m+1} - (q^2 + 1) (q^4 - q + 1) q^{4m} + q}.$$

Therefore, the norm square value takes the form

$$\int_{-1}^{1} \bar{\phi}_n^2(x; q) W_1^*(x; q) d_q x = d_n^2 \int_{-1}^{1} \frac{(q^2 x^2; q^2)_\infty \left(q^3 - q + 1\right)^{\frac{\log(x^2)}{2\log(q)}}}{(\frac{q^4-q+1}{q^3-q+1} x^2; q^2)_\infty} d_q x,$$

where

$$d_{2m}^2 = \frac{(q-1)^2 \left(q^2+q+1\right) \left(q^3+q^2+q-1\right) q^{m(2m-1)-2} \left(q^3-q+1\right)^{m+1}}{\left(q \left(q^4-q+1\right); q^4\right)_m}$$

$$\times \frac{(q^2; q^2)_m \left(q^3-1+\frac{1}{q}; q^2\right)_m \left(q \left(q^3-q+1\right); q^2\right)_m \left(\frac{q^4-q+1}{q^5-q^3+q^2}; q^2\right)_{m+1}}{\left(q^4-q^2+1\right) \left(q-\frac{1}{q^2}+\frac{1}{q^3}; q^4\right)_{m+1} \left(q^3-1+\frac{1}{q}; q^4\right)_m \left(q^3-1+\frac{1}{q}; q^4\right)_{m+1}}$$

and

$$d_{2m+1}^2 = \frac{(q-1)^2 \left(q^2+q+1\right) \left(q^3+q^2+q-1\right) q^{2m^2+m-2} \left(q^3-q+1\right)^{m+1}}{\left(q \left(q^4-q+1\right); q^4\right)_{m+1}}$$

$$\times \frac{(q^2; q^2)_m \left(q^3-1+\frac{1}{q}; q^2\right)_{m+1} \left(q \left(q^3-q+1\right); q^2\right)_{m+1} \left(\frac{q^4-q+1}{q^5-q^3+q^2}; q^2\right)_{m+1}}{\left(q^4-q^2+1\right) \left(q-\frac{1}{q^2}+\frac{1}{q^3}; q^4\right)_{m+1} \left(\left(q^3-1+\frac{1}{q}; q^4\right)_{m+1}\right)^2}.$$

Example 3.2 Consider the q-difference equation

$$x^2(1-x^2) D_q D_{q^{-1}} \phi_n(x; q) + qx \left(-[5]_q x^2 + q + 1\right) D_q \phi_n(x; q)$$

$$+ \left([n]_q (q[5]_q - [1-n]_q) x^2 - q(q+1)\sigma_n\right) \phi_n(x; q) = 0 \qquad (3.108)$$

as a special case of (3.95).

Following the approach of Example 3.1, we can obtain the polynomial solution of Eq. (3.108) as

$$\phi_n(x; q) = x^{\sigma_n} \, {}_2\phi_1 \left(\begin{array}{c} q^{\sigma_n - n}, q^{n+\sigma_n-1} \left(q^6 - q + 1\right) \\ q^{2\sigma_n+1} \left(q^3 - q + 1\right) \end{array} \Big| q^2; q^2 x^2 \right).$$

Again, this sequence satisfies an orthogonality relation of the form

$$\int_{-1}^{1} W_2^*(x; q) \phi_n(x; q) \phi_m(x; q) d_q x = \left(\int_{-1}^{1} W_2^*(x; q) \phi_n^2(x; q) d_q x\right) \delta_{n,m},$$

where $W_2^*(x; q) = C(x)W_2(x; q)$ is the main weight function and $W_2(x; q)$ satisfies the equation

$$\frac{W_2(qx; q)}{W_2(x; q)} = \frac{(q^6 - q + 1)x^2 - q^3 + q - 1}{q^2 (q^2 x^2 - 1)}. \tag{3.109}$$

Up to a periodic function, a solution of Eq. (3.109) is denoted by

$$W_2(x; q) = \frac{(q^2 x^2; q^2)_\infty (q^3 - q + 1)^{\frac{\log(x^2)}{2\log(q)}}}{x^2 \left(\frac{q^6 - q + 1}{q^3 - q + 1} x^2; q^2\right)_\infty} = W_2(-x; q),$$

where

$$\lim_{q\uparrow 1} W_2^*(x; q) = \lim_{q\uparrow 1} x^2 W_2(x; q) = x^2 \sqrt{1 - x^2},$$

which indeed shows the weight function of the Chebyshev polynomials of the sixth kind.

The monic polynomial solution of Eq. (3.108) satisfies a three-term recurrence relation of type (3.107) with

$$\gamma_{2m} = \frac{q^{2m+1} (q^m - 1)(q^m + 1)((q^6 - q + 1)q^{2m} + (-q^3 + q - 1)q^2)}{(q^6 - q + 1)^2 q^{8m} - (q^2 + 1)(q^6 - q + 1)q^{4m+1} + q^4}$$

and

$$\gamma_{2m+1} = \frac{q^{2m} ((q^3 - q + 1)q^{2m+1} - 1)((q^6 - q + 1)q^{2m} - q)}{(q^6 - q + 1)^2 q^{8m+1} - (q^2 + 1)(q^6 - q + 1)q^{4m} + q}.$$

Thus, the norm square value is derived as

$$\int_{-1}^{1} \bar{\phi}_n^2(x; q) W_2^*(x; q) d_q x = d_n^2 \int_{-1}^{1} \frac{(q^2 x^2; q^2)_\infty (q^3 - q + 1)^{\frac{\log(x^2)}{2\log(q)}}}{\left(\frac{q^6 - q + 1}{q^3 - q + 1} x^2; q^2\right)_\infty} d_q x,$$

where

$$d_{2m}^2 = \frac{(1 - q)(q^5 + q^4 + q^3 - 1)([6]_q - 2) q^{m(2m-1)-2} (q^3 - q + 1)^{m+1}}{(q(q^6 - q + 1); q^4)_m}$$

$$\times \frac{(-1; q)_{m+1}(q; q)_m (q(q^3 - q + 1); q^2)_m (q^5 - 1 + \frac{1}{q}; q^2)_m (\frac{q^6 - q + 1}{q^5 - q^3 + q^2}; q^2)_{m+1}}{2(q^5 + q^2 - 1)(q^5 - 1 + \frac{1}{q}; q^4)_m (q^5 - 1 + \frac{1}{q}; q^4)_{m+1} (\frac{q^6 - q + 1}{q^3}; q^4)_{m+1}}$$

and

$$d_{2m+1}^2 = (1-q)\left(q^5 + q^4 + q^3 - 1\right)\left([6]_q - 2\right)q^{2m^2+m-2}\left(q^3 - q + 1\right)^{m+1}$$

$$\times \frac{(-1;q)_{m+1}(q;q)_m\left(q\left(q^3 - q + 1\right);q^2\right)_{m+1}\left(q^5 - 1 + \frac{1}{q};q^2\right)_{m+1}\left(\frac{q^6-q+1}{q^5-q^3+q^2};q^2\right)_{m+1}}{2\left(q^5 + q^2 - 1\right)\left(\left(q^5 - 1 + \frac{1}{q};q^4\right)_{m+1}\right)^2\left(\frac{q^6-q+1}{q^3};q^4\right)_{m+1}\left(q\left(q^6 - q + 1\right);q^4\right)_{m+1}}.$$

3.5   A Basic Class of Symmetric q-Orthogonal Polynomials with Four Free Parameters

In this section, using the generalized Sturm–Liouville Theorem 3.6 in q-difference spaces, we introduce a class of symmetric q-orthogonal polynomials with four free parameters and obtain its standard properties, such as a generic second-order q-difference equation, the explicit form of the polynomials in terms of q-hypergeometric series, a generic three-term recurrence relation, and a general orthogonality relation. We also present some particular examples in the sequel.

Motivated by Eq. (3.95), let us consider the q-difference equation

$$x^2\left(ax^2 + b\right)D_q D_{q^{-1}}\phi_n(x;q) + x\left(cx^2 + d\right)D_q\phi_n(x;q) + \left(\lambda_{n,q}x^2 - \sigma_n d\right)\phi_n(x;q) = 0.$$

$$(3.110)$$

To find a symmetric monic q-polynomial solution for Eq. (3.110), let

$$\bar{\phi}_n(x;q) = x^n + \delta_{n,q}x^{n-2} + \cdots \tag{3.111}$$

satisfy the well-known three-term recurrence relation

$$\bar{\phi}_{n+1}(x;q) = x\bar{\phi}_n(x;q) - C_{n,q}\bar{\phi}_{n-1}(x;q) \quad \text{with} \quad \bar{\phi}_0(x;q) = 1 \quad \text{and} \quad \bar{\phi}_1(x;q) = x. \tag{3.112}$$

From (3.110) and (3.111), equating the coefficient in x^{n+2} gives

$$\lambda_{n,q} = -[n]_q\left(c - [1 - n]_q a\right), \tag{3.113}$$

provided that $|a| + |c| \neq 0$.

Using the eigenvalue $\lambda_{n,q}$ given in (3.113) and equating the coefficient in x^n, we obtain in (3.111) that

$$\delta_{n,q} = \frac{q^2\left(-b(q-1)q[n-1]_q[n]_q - dq^{2n} + dq^n(q\sigma_n + \sigma_{n-1})\right)}{(q+1)\left(aq^3 - q^{2n}(a + c(q-1))\right)}.$$

Also from (3.111) and (3.112) we have

$$x^{n+1} + \delta_{n+1,q}x^{n-1} + \cdots = x\left(x^n + \delta_{n,q}x^{n-2} + \cdots\right) - C_{n,q}\left(x^{n-1} + \delta_{n-1,q}x^{n-3} + \cdots\right),$$

which implies

$$C_{n,q} = \delta_{n,q} - \delta_{n+1,q} = \frac{B^*_{n,q}}{a^2q^4 + q^{4n}(a + c(q-1))^2 - a\left(q^3 + q\right)q^{2n}(a + c(q-1))},$$

$$(3.114)$$

where

$$B^*_{n,q} = \left(q^{n+1}\left(q^{2n}(a + c(q-1))((d - dq)\sigma_n - b)\right.\right.$$

$$\left.\left. + q^n\left(a\left(b\left(q^2 + 1\right) + d(q-1)q^2\right) + bc(q-1)\right) - aq^2(b + d(q-1)\sigma_{n-1})\right)\right).$$

The limit case of (3.114) is computed as

$$\lim_{q\uparrow 1} C_{n,q} = \frac{n(a(b(2-n)+d) - bc) - d\sigma_n(2a(n-1)+c)}{(a(2n-3)+c)(a(2n-1)+c)},$$

and for the eigenvalue (3.113) we have

$$\lim_{q\uparrow 1} \lambda_{n,q} = -n(c - (1-n)a),$$

which gives exactly the same result as in the continuous case [5] by taking into account that the three-term recurrence relation (3.112) has a minus sign in the coefficients $C_{n,q}$. Since the polynomial solution of Eq. (3.110) is symmetric, we use the notation

$$\phi_n(x; q) = S_n\left(\begin{array}{cc} c & d \\ a & b \end{array}\middle|\, q; x\right)$$

for mathematical display formulas and $S_n(x; q; a, b, c, d)$ in the text. This means that we have to deal with just one characteristic vector (a, b, c, d) for any given subclass.

For $n = 2m$ and $n = 2m + 1$, the $C_{n,q}$ in (3.114) are simplified as

$$C_{2m,q} = -\frac{q^{2m+1}[2m]_q(q-1)\left(bq^{2m}(a + c(q-1)) - aq^2(b + d(q-1))\right)}{a^2q^4 + q^{8m}(a + c(q-1))^2 - a\left(q^3 + q\right)q^{4m}(a + c(q-1))}$$

and

$$C_{2m+1,q} = -\frac{q^{2m}\left(q^{2m}(a + c(q-1)) - aq\right)\left(q^{2m+1}(b + d(q-1)) - b\right)}{a^2q + q^{8m+1}(a + c(q-1))^2 - a\left(q^2 + 1\right)q^{4m}(a + c(q-1))}.$$

Theorem 3.7 *The explicit form of the polynomial $S_m(x; q; a, b, c, d)$ is*

$$S_m\left(\begin{matrix} c\ d \\ a\ b \end{matrix}\,\middle|\, q;\ x\right)$$

$$= \sum_{k=0}^{[\frac{m}{2}]} q^{(k-1)k} x^{m-2k} \begin{bmatrix}[\frac{m}{2}] \\ k\end{bmatrix}_{q^2} \prod_{j=0}^{[\frac{m}{2}]-k-1} \frac{a[2j + \sigma_m + m - 1]_q + cq^{2j+\sigma_m+m-1}}{b[(2j + (-1)^{m+1} + 2)]_q + dq^{2j+(-1)^{m+1}+2}},$$

where the q-number $[n]_q$ has been defined in (3.8) and the q-binomial coefficient is denoted by

$$\begin{bmatrix} n \\ m \end{bmatrix}_q = \frac{(q;\ q)_n}{(q;\ q)_m(q;\ q)_{n-m}}.$$

Moreover, if $a, b \neq 0$, then

$$S_m\left(\begin{matrix} c\ d \\ a\ b \end{matrix}\,\middle|\, q;\ x\right) = x^{\sigma_m}\ {}_2\phi_1\left(\begin{matrix} q^{-m+\sigma_m},\ \frac{(a+c(q-1))q^{m+\sigma_m-1}}{a} \\ \frac{(b+d(q-1))q^{2\sigma_m+1}}{b} \end{matrix}\,\middle|\, q^2;\ -\frac{aq^2x^2}{b}\right). \qquad (3.115)$$

Proof Despite the degrees of $A(x)$, $B(x)$, $C(x)$, and $E(x)$, the proof can be done in a similar way as in [4, Section 10.2] for classical q-orthogonal polynomials. □

It is easy to check that

$$\lim_{q \uparrow 1} S_m\left(\begin{matrix} c\ d \\ a\ b \end{matrix}\,\middle|\, q;\ x\right) = S_m\left(\begin{matrix} c\ d \\ a\ b \end{matrix}\,\middle|\, x\right),$$

where the right-hand-side polynomial was introduced in (1.101).

Also, the monic form of the polynomials (3.115) is represented as

$$\bar{S}_m \left(\begin{array}{cc} c & d \\ a & b \end{array} \middle| q; x \right) = q^{\sigma_m - m} \left(-\frac{b}{a} \right)^{\left[\frac{m}{2}\right]} \frac{(q^2; q^{-m\sigma_m}; \frac{(b+d(q-1))q^{2\sigma_m+1}}{b}; q^2)_{\left[\frac{m}{2}\right]}}{(q^{-m}; q^{-(m-1)\sigma_m}; \frac{(a+c(q-1))q^{m+\sigma_m-1}}{a}; q^2)_{\left[\frac{m}{2}\right]}}$$

$$\times x^{\sigma_m} \, {}_2\phi_1 \left(\begin{array}{c} q^{-m+\sigma_m}, \frac{(a+c(q-1))q^{m+\sigma_m-1}}{a} \\ \frac{(b+d(q-1))q^{2\sigma_m+1}}{b} \end{array} \middle| q^2; -\frac{aq^2x^2}{b} \right).$$

By noting Theorem 3.6, since

$$\frac{W(qx; q)}{W(x; q)} = \frac{A(x) + (q-1)x B(x)}{A(qx)} = \frac{x^2(a + c(q-1)) + b + d(q-1)}{q^2 \left(aq^2x^2 + b \right)}, \tag{3.116}$$

if $a, b \neq 0$, then the solution of Eq. (3.116) will be

$$W(x; q) = \left(\frac{d(q-1)}{b} + 1 \right)^{\frac{\log x^2}{2\log q}} \frac{(-\frac{aq^2x^2}{b}; q^2)_\infty}{x^2 (-\frac{(a+c(q-1))x^2}{b+d(q-1)}; q^2)_\infty} = W(-x; q),$$

in which some restrictions on the parameters must be considered in order to have convergence for the infinite products.

We are now in a position to analyze some particular cases of the q-difference equation (3.110) that provide q-analogues of different families of continuous orthogonal polynomials.

Case 1: A Generalization of q-Ultraspherical Polynomials
Consider the q-difference equation

$$x^2 \left(1 - x^2 \right) D_q D_{q^{-1}} \phi_n(x; q) + (q+1)qx \left(\alpha - x^2(\alpha + \beta + 1) \right) D_q \phi_n(x; q)$$

$$+ \left(-[n]_q(-(1+\alpha+\beta)q(1+q) + [1-n]_q)x^2 - \alpha q(1+q)\sigma_n \right) \phi_n(x; q) = 0 \tag{3.117}$$

as a special case of (3.110) with the polynomial solution

$$\phi_n(x; \alpha, \beta | q) = S_n \left(\begin{array}{cc} -q(q+1)(\alpha+\beta+1) & \alpha q(q+1) \\ -1 & 1 \end{array} \middle| q; x \right)$$

$$= x^{\sigma_n} \, {}_2\phi_1 \left(\begin{array}{c} q^{\sigma_n - n}, q^{n+\sigma_n - 1} \left((\alpha + \beta + 1)q \left(q^2 - 1 \right) + 1 \right) \\ q^{2\sigma_n + 1} \left(\alpha q \left(q^2 - 1 \right) + 1 \right) \end{array} \middle| q^2; q^2 x^2 \right). \tag{3.118}$$

The sequence (3.118) satisfies an orthogonality relation of the form

$$\int_{-1}^{1} W_1^*(x; \alpha, \beta|q)\phi_n(x; \alpha, \beta|q)\phi_m(x; \alpha, \beta|q)d_q x$$

$$= \left(\int_{-1}^{1} W_1^*(x; \alpha, \beta|q)\phi_n^2(x; \alpha, \beta|q)d_q x \right) \delta_{n,m},$$

in which

$$W_1^*(x; \alpha, \beta|q) = x^2 \, W_1(x; \alpha, \beta|q)$$

is the main weight function, and the function $W_1(x; \alpha, \beta|q)$ satisfies the equation

$$\frac{W_1(qx; \alpha, \beta|q)}{W_1(x; \alpha, \beta|q)} = \frac{q \left(q^2 - 1\right) \left(x^2(\alpha + \beta + 1) - \alpha\right) + x^2 - 1}{q^2 \left(q^2 x^2 - 1\right)}. \tag{3.119}$$

Up to a periodic function, a solution of Eq. (3.119) is of the form

$$W_1(x; \alpha, \beta|q) = B^{\frac{\log(x^2)}{\log(q)}} \frac{\left(q^2 x^2; q^2\right)_\infty}{x^2 \left(-\frac{A^2 x^2}{B^2}; q^2\right)_\infty} = W_1(-x; \alpha, \beta|q),$$

where

$$A = \sqrt{\left(q - q^3\right)(\alpha + \beta + 1) - 1} \qquad \text{and} \qquad B = \sqrt{\alpha \left(q^3 - q\right) + 1}.$$

Notice that

$$\lim_{q \uparrow 1} W_1^*(x; \alpha, \beta|q) = \lim_{q \uparrow 1} x^2 W_1(x; \alpha, \beta|q) = x^{2\alpha}(1 - x^2)^\beta$$

looks the same as the weight function of generalized ultraspherical polynomials.

The monic type of polynomials (3.118) satisfies a three-term recurrence relation of type (3.112) with

$$C_{2m,q} = \frac{q^{2m+2} \left(q^{2m} - 1\right) \left(q^{2m} \left(q \left(q^2 - 1\right)\vartheta + 1\right) + q^2 \left(\alpha \left(q - q^3\right) - 1\right)\right)}{- \left(q^2 + 1\right) q^{4m+2} \left(q \left(q^2 - 1\right)\vartheta + 1\right) + q^{8m+1} \left(q \left(q^2 - 1\right)\vartheta + 1\right)^2 + q^5}$$

and

$$C_{2m+1,q}$$

$$= \frac{q^{2m}\left(q^{2m}\left(-(\alpha q^5 + (\beta+1)q^3 + q^2 - q\vartheta + 1)\right) + (\alpha q\,(q^2-1)+1)\,q^{4m+1}\left(q\,(q^2-1)\,\vartheta + 1\right) + q\right)}{-(q^2+1)\,q^{4m}\left(q\,(q^2-1)\,\vartheta + 1\right) + q^{8m+1}\left(q\,(q^2-1)\,\vartheta + 1\right)^2 + q},$$

where $\vartheta = \alpha + \beta + 1$.

Hence the norm square value takes the form

$$\int_{-1}^{1} \bar{\phi}_n^2(x; \alpha, \beta|q)\,W_1^*(x; \alpha, \beta|q)\,d_q x = d_{n,\alpha,\beta}^2 \int_{-1}^{1} W_1^*(x; \alpha, \beta|q)\,d_q x,$$

where

$$d_{2m,\alpha,\beta}^2 = \frac{\left(q^2; q^2\right)_m \left(q\left(\alpha q\left(q^2-1\right)+1\right); q^2\right)_m}{\left(\frac{(\alpha+\beta+1)q\,(q^2-1)+1}{q^3}; q^4\right)_{m+1}\left(\frac{(\alpha+\beta+1)q\,(q^2-1)+1}{q}; q^4\right)_m}$$

$$\times \frac{(q-1)q^{m(2m-1)-2}(q(q+1)(\alpha+\beta)-1)(q(q+1)(\alpha+\beta+1)-1)\left(\alpha q\left(q^2-1\right)+1\right)^{m+1}}{(q+1)\left(-q(\alpha+\beta+1)+\alpha q^3 + 1\right)}$$

$$\times \frac{\left(\frac{(\alpha+\beta+1)q\,(q^2-1)+1}{q}; q^2\right)_m \left(\frac{(\alpha+\beta+1)q\,(q^2-1)+1}{q^2(\alpha q\,(q^2-1)+1)}; q^2\right)_{m+1}}{\left(\frac{(\alpha+\beta+1)q\,(q^2-1)+1}{q}; q^4\right)_{m+1}\left(q\left((\alpha+\beta+1)q\left(q^2-1\right)+1\right); q^4\right)_m}$$

and

$$d_{2m+1,\alpha,\beta}^2 = \frac{\left(q^2; q^2\right)_m \left(q\left(\alpha q\left(q^2-1\right)+1\right); q^2\right)_{m+1}}{\left(q\left((\alpha+\beta+1)q\left(q^2-1\right)+1\right); q^4\right)_{m+1}}$$

$$\times \frac{(q-1)q^{2m^2+m-2}(q(q+1)(\alpha+\beta)-1)(q(q+1)(\alpha+\beta+1)-1)\left(\alpha q\left(q^2-1\right)+1\right)^{m+1}}{(q+1)\left(-q(\alpha+\beta+1)+\alpha q^3 + 1\right)}$$

$$\times \frac{\left(\frac{(\alpha+\beta+1)q\,(q^2-1)+1}{q}; q^2\right)_{m+1}\left(\frac{(\alpha+\beta+1)q\,(q^2-1)+1}{q^2(\alpha q\,(q^2-1)+1)}; q^2\right)_{m+1}}{\left(\frac{(\alpha+\beta+1)q\,(q^2-1)+1}{q^3}; q^4\right)_{m+1}\left(\left(\frac{(\alpha+\beta+1)q\,(q^2-1)+1}{q}; q^4\right)_{m+1}\right)^2}.$$

A special case of Eq. (3.117) generating q-Chebyshev polynomials of the fifth kind is $\alpha = 1$ and $\beta = [3]_q/[2]_q - 2$, which gives the monic polynomial solution

$$\bar{\phi}_n\left(x; 1, \frac{[3]_q}{[2]_q} - 2\Big|q\right) = \bar{S}_n\left(\begin{array}{cc} -q\left(q^2 + q + 1\right) & q(q+1) \\ -1 & 1 \end{array}\Big|\, q;\, x\right),$$

satisfying the orthogonality relation

$$\int_{-1}^{1} W_1^* \left(x; 1, \frac{[3]_q}{[2]_q} - 2|q \right) \bar{\phi}_n \left(x; 1, \frac{[3]_q}{[2]_q} - 2|q \right) \bar{\phi}_m \left(x; 1, \frac{[3]_q}{[2]_q} - 2|q \right) d_q x$$

$$= d_{n,1,\frac{[3]_q}{[2]_q}-2}^{2} \left(\int_{-1}^{1} W_1^*(x; 1, \frac{[3]_q}{[2]_q} - 2|q) d_q x \right) \delta_{n,m},$$

in which

$$W_1^* \left(x; 1, \frac{[3]_q}{[2]_q} - 2|q \right) = \frac{(q^2 x^2; q^2)_\infty (q^3 - q + 1)^{\frac{\log(x^2)}{2\log(q)}}}{(\frac{q^4-q+1}{q^3-q+1}x^2; q^2)_\infty} = W_1^* \left(-x; 1, \frac{[3]_q}{[2]_q} - 2|q \right).$$

The second special case of Eq. (3.117) generating q-Chebyshev polynomials of the sixth kind is $\alpha = 1$ and $\beta = [5]_q/[2]_q - 2$, which gives the monic polynomial solution

$$\bar{\phi}_n \left(x; 1, \frac{[5]_q}{[2]_q} - 2|q \right) = \bar{S}_n \left(\begin{array}{cc} -q[5]_q \, q(q+1) \\ -1 \quad 1 \end{array} \middle| q; x \right),$$

satisfying the orthogonality relation

$$\int_{-1}^{1} W_1^* \left(x; 1, \frac{[5]_q}{[2]_q} - 2|q \right) \bar{\phi}_n \left(x; 1, \frac{[5]_q}{[2]_q} - 2|q \right) \bar{\phi}_m \left(x; 1, \frac{[5]_q}{[2]_q} - 2|q \right) d_q x$$

$$= d_{n,1,\frac{[5]_q}{[2]_q}-2}^{2} \left(\int_{-1}^{1} W_1^*(x; 1, \frac{[5]_q}{[2]_q} - 2|q) d_q x, \right) \delta_{n,m},$$

in which

$$W_1^* \left(x; 1, \frac{[5]_q}{[2]_q} - 2|q \right) = \frac{(q^2 x^2; q^2)_\infty (q^3 - q + 1)^{\frac{\log(x^2)}{2\log(q)}}}{\left(\frac{q^6-q+1}{q^3-q+1}x^2; q^2 \right)_\infty} = W_1^* \left(-x; 1, \frac{[5]_q}{[2]_q} - 2|q \right).$$

Case 2: A Generalization of q-Hermite Polynomials
Consider the q-difference equation

$$x^2 \left(\left(1 - q^2\right) x^2 - 1 \right) D_q D_{q^{-1}} \phi_n(x; q) + (q + 1)x \left(x^2 + p \right) D_q \phi_n(x; q)$$

$$+ \left(q[-n]_q x^2 - \sigma_n p \right) \phi_n(x; q) = 0$$

as a special case of (3.110) with the polynomial solution

$$\phi_n(x; p|q) = S_n \left(\begin{matrix} 1+q & p(1+q) \\ 1-q^2 & -1 \end{matrix} \Big| q; x \right)$$

$$= x^{\sigma_n} {}_2\phi_1 \left(\begin{matrix} q^{\sigma_n-n}, 0 \\ q^{2\sigma_n+1} & (-pq^2+p+1) \end{matrix} \Big| q^2; q^2 \left(1-q^2\right) x^2 \right). \qquad (3.120)$$

The sequence (3.120) satisfies an orthogonality relation

$$\int_{-\alpha}^{\alpha} W_2^*(x; p|q)\phi_n(x; p|q)\phi_m(x; p; q)d_qx = \left(\int_{-\alpha}^{\alpha} W_2^*(x; p|q)\phi_n^2(x; p|q)d_qx \right) \delta_{n,m},$$

in which $\alpha = 1/\sqrt{1-q^2}$ and

$$W_2^*(x; p|q) = x^2 W_2(x; p|q)$$

is the main weight function, where $W_2(x; p|q)$ satisfies the equation

$$\frac{W_2(qx; p|q)}{W_2(x; p|q)} = \frac{-pq^2+p+1}{q^2+\left(q^2-1\right)q^4x^2}. \qquad (3.121)$$

Up to a periodic function, a solution of Eq. (3.121) is of the form

$$W_2(x; p|q) = \frac{\left(p\left(1-q^2\right)+1\right)^{\frac{\log(x^2)}{2\log(q)}} \left(q^2\left(1-q^2\right)x^2; q^2\right)_\infty}{x^2} = W_2(-x; p|q),$$

where

$$\lim_{q\uparrow 1} W_2^*(x; p|q) = \lim_{q\uparrow 1} x^2 W_2(x; p|q) = x^{-2p} e^{-x^2},$$

which is the weight function of generalized Hermite polynomials.

The monic type of polynomials (3.120) satisfies a three-term recurrence relation of type (3.112) with

$$C_{2m,q} = -\frac{\left(p\left(q^2-1\right)-1\right)q^{2m-1}\left(q^{2m}-1\right)}{q^2-1}$$

and

$$C_{2m+1,q} = \frac{\left(-pq^2+p+1\right)q^{4m+1}-q^{2m}}{q^2-1}.$$

Consequently, the norm square value takes the form

$$\int_{-\alpha}^{\alpha} \bar{\phi}_n^2(x; p|q) W_2^*(x; p|q) d_q x = d_{n,p}^2 \int_{-\alpha}^{\alpha} W_2^*(x; p|q) d_q x,$$

where

$$d_{2m,p}^2 = \frac{1}{2} q^{m(2m-1)} \frac{(-pq^2 + p + 1)^m}{(q^2 - 1)^{2m}} (-1; q)_{m+1}(q; q)_m \left(-pq^3 + pq + q; q^2\right)_m$$

and

$$d_{2m+1,p}^2 = \frac{1}{2} q^{m(2m+1)} \frac{(-pq^2 + p + 1)^m}{(q^2 - 1)^{2m+1}} (-1; q)_{m+1}(q; q)_m \left(-pq^3 + pq + q; q^2\right)_{m+1}.$$

We point out that if $p = 0$, the weight function of discrete q-Hermite polynomials appears as

$$W_2^*(x; 0|q) = \frac{1}{\left((1 - q^2) x^2; q^2\right)_\infty},$$

and therefore we have

$$\phi_n(x; 0|q) = k_n h_n(x\sqrt{1 - q^2}|q),$$

in which $h_n(x|q)$ denotes the discrete q-Hermite polynomials and k_n is a normalizing constant.

3.5.1   Two Finite Sequences Based on Ramanujan's Identity

Case 3: First Finite Sequence   For $u, v \in \mathbb{R}$, consider a special case of Eq. (3.110),

$$x^2 \left(x^2 + 1\right) D_q D_{q^{-1}} \phi_n(x; q) - 2x \left((u + v - 1)x^2 + u\right) D_q \phi_n(x; q)$$

$$+ \left([n]_q \left(2u + 2v - 2 + [1 - n]_q\right) x^2 + 2u\sigma_n\right) \phi_n(x; q) = 0, \qquad (3.122)$$

whose monic polynomial solution can be represented as

$$\bar{\phi}_n(x; q, u, v)$$

$$= K_1 x^{\sigma_n} {}_2\phi_1 \left(\begin{matrix} q^{-n+\sigma_n}, (1 - 2(q - 1)(u + v - 1))q^{n+\sigma_n-1} \\ (1 - 2u(q - 1))q^{2\sigma_n+1} \end{matrix} \,\middle|\, q^2; -q^2 x^2\right), \qquad (3.123)$$

where

$$K_1 = \frac{q^{[n/2]([n/2]-1)}(q^{n+\sigma_n-1}(1-2(u+v-1)(q-1)); q^2)_{[n/2]}}{(q^{(-1)^n+2}(1-2u(q-1)); q^2)_{[n/2]}}.$$

In order to prove the orthogonality of the finite set $\{\bar{\phi}_n(x; q, u, v)\}_{n=0}^{N}$ on $(-\infty, \infty)$, it is necessary to impose a specific condition,

$$N < \frac{1 - \log_q(1 - 2(q-1)(u+v-1))}{2},$$

because if Eq. (3.122) is written in a self-adjoint form, then

$$D_q\left(x^2(x^2+1)\varrho_1(x; q, u, v)D_{q^{-1}}\phi_n(x; q)\right) + \left(\lambda_{n,q}x^2 + 2u\sigma_n\right)\varrho_1(x; q, u, v)\phi_n(x; q) = 0$$

$$(3.124)$$

and

$$D_q\left(x^2(x^2+1)\varrho_1(x; q, u, v)D_{q^{-1}}\phi_m(x; q)\right) + \left(\lambda_{m,q}x^2 + 2u\sigma_m\right)\varrho_1(x; q, u, v)\phi_m(x; q) = 0,$$

$$(3.125)$$

where

$$\varrho_1(x; q, u, v) = x^{\log_q\left(\frac{1-2(q-1)(u+v-1)}{q^4}\right)} \frac{\left(-\frac{1}{q^2x^2}; q^2\right)_\infty}{\left(-\frac{1-2u(q-1)}{(1-2(q-1)(u+v-1))x^2}; q^2\right)_\infty}.$$

Now, by multiplying (3.124) by $\phi_m(x; q)$ and (3.125) by $\phi_n(x; q)$ and subtracting one from the other, we get

$$\phi_m(x; q)D_q\left(x^2(x^2+1)\varrho_1(x; q, u, v)D_{q^{-1}}\phi_n(x; q)\right)$$

$$- \phi_n(x; q)D_q\left(x^2(x^2+1)\varrho_1(x; q, u, v)D_{q^{-1}}\phi_m(x; q)\right)$$

$$+ \left(\lambda_{n,q} - \lambda_{m,q}\right)x^2\varrho_1(x; q, u, v)\phi_n(x; q)\phi_m(x; q)$$

$$+ ((-1)^m - (-1)^n)u\varrho_1(x; q, u, v)\phi_n(x; q)\phi_m(x; q) = 0. \qquad (3.126)$$

Since $\varrho_1(x; q, u, v)$ is an even function, q-integrating on both sides of (3.126) over $\mathbb{R}$ yields

$$\int_{-\infty}^{\infty} \phi_m(x; q) D_q \left(x^2(x^2 + 1)\varrho_1(x; q, u, v) D_{q^{-1}} \phi_n(x; q) \right) d_q x$$

$$- \int_{-\infty}^{\infty} \phi_n(x; q) D_q \left(x^2(x^2 + 1)\varrho_1(x; q, u, v) D_{q^{-1}} \phi_m(x; q) \right) d_q x$$

$$+ \left(\lambda_{n,q} - \lambda_{m,q} \right) \int_{-\infty}^{\infty} x^2 \varrho_1(x; q, u, v) \phi_n(x; q) \phi_m(x; q) \, d_q x$$

$$+ u\big((-1)^m - (-1)^n\big) \int_{-\infty}^{\infty} \varrho_1(x; q, u, v) \phi_n(x; q) \phi_m(x; q) \, d_q x = 0,$$

which can be transformed, using the rule of q-integration by parts, to

$$\left[x^2(x^2 + 1)\varrho_1(x; q, u, v) \phi_m(x; q) D_{q^{-1}} \phi_n(x; q) \right]_{-\infty}^{\infty}$$

$$- \left[x^2(x^2 + 1)\varrho_1(x; q, u, v) \phi_n(x; q) D_{q^{-1}} \phi_m(x; q) \right]_{-\infty}^{\infty}$$

$$+ \left(\lambda_{n,q} - \lambda_{m,q} \right) \int_{-\infty}^{\infty} x^2 \varrho_1(x; q, u, v) \phi_n(x; q) \phi_m(x; q) d_q x$$

$$+ u\big((-1)^m - (-1)^n\big) \int_{-\infty}^{\infty} \varrho_1(x; q, u, v) \phi_n(x; q) \phi_m(x; q) d_q x = 0. \qquad (3.127)$$

In other words, (3.127) is simplified as

$$\left[(x^2 + 1)\varrho_1^*(x; q, u, v) \big(\phi_m(x; q) D_{q^{-1}} \phi_n(x; q) - \phi_n(x; q) D_{q^{-1}} \phi_m(x; q) \big) \right]_{-\infty}^{\infty}$$

$$= \left(\lambda_{m,q} - \lambda_{n,q} \right) \int_{-\infty}^{\infty} \varrho_1^*(x; q, u, v) \phi_n(x; q) \phi_m(x; q) d_q x, \qquad (3.128)$$

in which

$$\varrho_1^*(x; q, u, v) = x^2 \varrho_1(x; q, u, v) = \varrho_1^*(-x; q, u, v),$$

provided that $(-1)^{\log_q \left(\frac{1 - 2(q-1)(u+v-1)}{q^4} \right)} = 1$. Now, since

$$deg \big(\phi_m(x; q) D_{q^{-1}} \phi_n(x; q) - \phi_n(x; q) D_{q^{-1}} \phi_m(x; q) \big) = m + n - 1,$$

the left-hand side of (3.128) is zero if

$$\lim_{x \to \pm\infty} x^{m+n+1} \varrho_1^*(x; q, u, v) = 0. \qquad (3.129)$$

By taking $\max\{m, n\} = N$, relation (3.129) becomes equivalent to

$$\lim_{x \to \pm\infty} x^{2N+1+\log_q\left(\frac{1-2(q-1)(u+v-1)}{q^2}\right)} \frac{\left(-\frac{1}{q^2x^2}; q^2\right)_\infty}{\left(-\frac{1-2u(q-1)}{(1-2(q-1)(u+v-1))x^2}; q^2\right)_\infty} = 0. \qquad (3.130)$$

But relation (3.130) is valid if and only if

$$2N - 1 + \log_q\left(1 - 2(q-1)(u+v-1)\right) < 0 \quad \text{or} \quad N < \frac{1 - \log_q\left(1 - 2(q-1)(u+v-1)\right)}{2}.$$

Hence the orthogonality relation of q-polynomials (3.123) takes the form

$$\int_{-\infty}^{\infty} \varrho_1^*(x; q, u, v)\bar{\phi}_n(x; q, u, v)\bar{\phi}_m(x; q, u, v)d_q x = \left(\prod_{j=1}^{n} C_{j,q}^{(u,v)} \int_{-\infty}^{\infty} \varrho_1^*(x; q, u, v)d_q x\right) \delta_{n,m},$$

where the $\{C_{j,q}^{(u,v)}\}$ are computed as

$$C_{j,q}^{(u,v)} =$$

$$[q^{j+1}(q^{2j}(1-2(u+v-1)(q-1))(2u(q-1)\sigma_j-1)+q^j((q^2+1)-2u(q-1)q^2-2(u+v-1)(q-1))$$

$$-q^2(1-2u(q-1)\sigma_{j-1}))]/[q^4+q^{4j}(1-2(u+v-1)(q-1))^2-(q^3+q)q^{2j}(1-2(u+v-1)(q-1))].$$

Therefore, in order to complete the orthogonality relation, it just remains to compute the q-integral

$$\int_{-\infty}^{\infty} \varrho_1^*(x; q, u, v)d_q x = \int_{-\infty}^{\infty} x^{\log_q\left(\frac{1-2(q-1)(u+v-1)}{q^2}\right)} \frac{\left(-\frac{1}{q^2x^2}; q^2\right)_\infty}{\left(-\frac{1-2u(q-1)}{(1-2(q-1)(u+v-1))x^2}; q^2\right)_\infty} d_q x. \qquad (3.131)$$

For this purpose, we can use Ramanujan's identity (3.19) directly for computing the q-integral (3.131) as follows:

$$\int_{-\infty}^{\infty} \varrho_1^*(x; q, u, v)d_q x$$

$$= 2(1-q) \sum_{n=-\infty}^{\infty} q^{n\left(\log_q\left(\frac{1-2(q-1)(u+v-1)}{q^2}\right)+1\right)} \frac{\left(-q^{-2}q^{-2n}; q^2\right)_\infty}{\left(-\frac{1-2u(q-1)}{1-2(q-1)(u+v-1)}q^{-2n}; q^2\right)_\infty}$$

$$= 2(1-q) \sum_{n=-\infty}^{\infty} q^{n\left(\log_q\left(\frac{1-2(q-1)(u+v-1)}{q^2}\right)+1\right)} \frac{\left(-\frac{1-2u(q-1)}{1-2(q-1)(u+v-1)}; q^2\right)_{-n}\left(-q^{-2}; q^2\right)_\infty}{\left(-\frac{1-2u(q-1)}{1-2(q-1)(u+v-1)}; q^2\right)_\infty\left(-q^{-2}; q^2\right)_{-n}}$$

$$= h_1 \sum_{n=-\infty}^{\infty} q^{n\left(\log_q\left(\frac{1-2(q-1)(u+v-1)}{q^2}\right)+1\right)} \frac{\left(-\frac{1-2u(q-1)}{1-2(q-1)(u+v-1)}; q^2\right)_{-n}}{(-q^{-2}; q^2)_{-n}}$$

$$= h_1 \sum_{n=-\infty}^{\infty} q^{n\left(-\log_q\left(\frac{1-2(q-1)(u+v-1)}{q^2}\right)-1\right)} \frac{\left(-\frac{1-2u(q-1)}{1-2(q-1)(u+v-1)}; q^2\right)_{n}}{(-q^{-2}; q^2)_{n}}$$

$$= h_1 \Psi \left(-\frac{1-2u(q-1)}{1-2(q-1)(u+v-1)}, -q^{-2}; q^2; q^{\left(-\log_q\left(\frac{1-2(q-1)(u+v-1)}{q^2}\right)-1\right)}\right),$$

where $h_1 = \dfrac{2(1-q)\left(-q^{-2}; q^2\right)_{\infty}}{\left(-\frac{1-2u(q-1)}{1-2(q-1)(u+v-1)}; q^2\right)_{\infty}}$.

Corollary 3.5 *The polynomial set $\{\bar{\phi}_k(x; q, u, v)\}_{k=0}^{N}$ is finitely orthogonal with respect to the weight function $\varrho_1^*(x; q, u, v)$ on $(-\infty, \infty)$ if and only if*

$$N < \frac{1 - \log_q(1 - 2(q-1)(u+v-1))}{2},$$

so that we have

$$\int_{-\infty}^{\infty} \varrho_1^*(x; q, u, v) \bar{\phi}_n(x; q, u, v) \bar{\phi}_m(x; q, u, v) d_q x = \left(\frac{2(1-q)\left(-q^{-2}; q^2\right)_{\infty}}{\left(-\frac{1-2u(q-1)}{1-2(q-1)(u+v-1)}; q^2\right)_{\infty}}\right.$$

$$\Psi\left(-\frac{1-2u(q-1)}{1-2(q-1)(u+v-1)}, -q^{-2}; q^2; q^{\left(-\log_q\left(\frac{1-2(q-1)(u+v-1)}{q^2}\right)-1\right)}\right)\left.\prod_{j=1}^{n} C_{j,q}^{(u,v)}\right) \delta_{n,m}.$$

For instance, the polynomial set $\{\bar{\phi}_k(x; 0.5, 128, 896)\}_{k=0}^{N=5}$ is finitely orthogonal with respect to the weight function $\dfrac{x^8(-4x^{-2}; \frac{1}{4})_{\infty}}{(-\frac{129}{1024}x^{-2}; \frac{1}{4})_{\infty}}$ on $(-\infty, \infty)$.

Case 4: Second Finite Sequence For $u \in \mathbb{R}$, consider a special case of Eq. (3.110),

$$x^4 D_q D_{q^{-1}} \phi_n(x; q) + 2x \left((1-u)x^2 + 1\right) D_q \phi_n(x; q)$$

$$+ \left([n]_q \left(2u - 2 + [1-n]_q\right) x^2 + 2\sigma_n\right) \phi_n(x; q) = 0, \qquad (3.132)$$

whose monic polynomial solution can be represented as

$$\bar{\phi}_n(x; q, u) = K_2 x^{\sigma_n} \, {}_2\phi_0 \left(\begin{matrix} q^{-n+\sigma_n}, (1 + (2 - 2u)(q - 1))q^{n+\sigma_n-1} \\ - \end{matrix} \middle| q^2; \frac{q^{1-2\sigma_n} x^2}{2(1-q)} \right)$$

$$= \dot{x}^n \, {}_2\phi_1 \left(\begin{matrix} q^{-n+\sigma_n}, 0 \\ q^{3-2n}(1 + (2 - 2u)(q - 1))^{-1} \end{matrix} \middle| q^2; \frac{2q^2(1-q)}{(1 + (2 - 2u)(q - 1))x^2} \right),$$

$$(3.133)$$

where

$$K_2 = \frac{(q^{n+\sigma_n-1}(1 - 2(u - 1)(q - 1)); q^2)_{[n/2]}}{(2 - 2q)^{[n/2]} \, q^{[n/2](2+(-1)^{n+1})}}.$$

Once again, it is necessary for orthogonality of the finite set $\{\bar{\phi}_n(x; q, u)\}_{n=0}^N$ to impose a specific condition,

$$N < \frac{1 - \log_q(1 + 2(q - 1)(1 - u))}{2},$$

because if we write Eq. (3.132) in a self-adjoint form, then

$$D_q \left(x^4 \varrho_2(x; q, u) D_{q^{-1}} \phi_n(x; q) \right) + \left(\lambda_{n,q} x^2 + 2\sigma_n \right) \varrho_2(x; q, u) \phi_n(x; q) = 0 \quad (3.134)$$

and

$$D_q \left(x^4 \varrho_2(x; q, u) D_{q^{-1}} \phi_m(x; q) \right) + \left(\lambda_{m,q} x^2 + 2\sigma_m \right) \varrho_2(x; q, u) \phi_m(x; q) = 0,$$

$$(3.135)$$

where

$$\varrho_2(x; q, u) = \frac{x^{\log_q \left(\frac{1+2(q-1)(1-u)}{q^4} \right)}}{(-\frac{2(q-1)}{(1+2(q-1)(1-u))x^2}; q^2)_\infty}.$$

By multiplying (3.134) by $\phi_m(x; q)$ and (3.135) by $\phi_n(x; q)$ and subtracting one from the other, we get

$$\phi_m(x; q) D_q \left(x^4 \varrho_2(x; q, u) D_{q^{-1}} \phi_n(x; q) \right)$$

$$- \phi_n(x; q) D_q \left(x^4 \varrho_2(x; q, u) D_{q^{-1}} \phi_m(x; q) \right)$$

$$+ \left(\lambda_{n,q} - \lambda_{m,q}\right) x^2 \varrho_2(x; q, u)\phi_n(x; q)\phi_m(x; q)$$

$$+ ((-1)^m - (-1)^n)\varrho_2(x; q, u)\phi_n(x; q)\phi_m(x; q) = 0. \qquad (3.136)$$

Since $\varrho_2(x; q, u)$ is an even function, q-integrating on both sides of (3.136) over $\mathbb{R}$ yields

$$\int_{-\infty}^{\infty} \phi_m(x; q) D_q \left(x^4 \varrho_2(x; q, u) D_{q^{-1}}\phi_n(x; q)\right) d_q x$$

$$- \int_{-\infty}^{\infty} \phi_n(x; q) D_q \left(x^4 \varrho_2(x; q, u) D_{q^{-1}}\phi_m(x; q)\right) d_q x$$

$$+ \left(\lambda_{n,q} - \lambda_{m,q}\right) \int_{-\infty}^{\infty} x^2 \varrho_2(x; q, u)\phi_n(x; q)\phi_m(x; q)\, d_q x$$

$$+ ((-1)^m - (-1)^n) \int_{-\infty}^{\infty} \varrho_2(x; q, u)\phi_n(x; q)\phi_m(x; q)\, d_q x = 0,$$

which is transformed to

$$\left[x^4 \varrho_2(x; q, u)\phi_m(x; q) D_{q^{-1}}\phi_n(x; q)\right]_{-\infty}^{\infty}$$

$$- \left[x^4 \varrho_2(x; q, u)\phi_n(x; q) D_{q^{-1}}\phi_m(x; q)\right]_{-\infty}^{\infty}$$

$$+ \left(\lambda_{n,q} - \lambda_{m,q}\right) \int_{-\infty}^{\infty} x^2 \varrho_2(x; q, u)\phi_n(x; q)\phi_m(x; q) d_q x$$

$$+ ((-1)^m - (-1)^n) \int_{-\infty}^{\infty} \varrho_2(x; q, u)\phi_n(x; q)\phi_m(x; q) d_q x = 0. \qquad (3.137)$$

On the other hand, (3.137) can be simplified as

$$\left[x^2 \varrho_2^*(x; q, u) \left(\phi_m(x; q) D_{q^{-1}}\phi_n(x; q) - \phi_n(x; q) D_{q^{-1}}\phi_m(x; q)\right)\right]_{-\infty}^{\infty}$$

$$= \left(\lambda_{m,q} - \lambda_{n,q}\right) \int_{-\infty}^{\infty} \varrho_2^*(x; q, u)\phi_n(x; q)\phi_m(x; q) d_q x, \qquad (3.138)$$

in which

$$\varrho_2^*(x; q, u) = x^2 \varrho_2(x; q, u) = \varrho_2^*(-x; q, u),$$

provided that $(-1)^{\log_q \left(\frac{1+2(q-1)(1-u)}{q^4}\right)} = 1$. Now, since

$$deg\left(\phi_m(x; q) D_{q^{-1}}\phi_n(x; q) - \phi_n(x; q) D_{q^{-1}}\phi_m(x; q)\right) = m + n - 1,$$

the left-hand side of (3.138) is equal to zero if and only if

$$\lim_{x\to\pm\infty} x^{m+n+1}\varrho_2^*(x;q,u) = 0. \tag{3.139}$$

Again if max$\{m, n\} = N$, relation (3.139) is equivalent to

$$\lim_{x\to\pm\infty} \frac{x^{2N+1+\log_q\left(\frac{1+2(q-1)(1-u)}{q^2}\right)}}{(-\frac{2(q-1)}{(1+2(q-1)(1-u))x^2};q^2)_\infty} = 0, \tag{3.140}$$

and in the sequel, (3.140) is valid if and only if

$$2N - 1 + \log_q(1 + 2(q - 1)(1 - u)) < 0, \quad \text{or} \quad N < \frac{1 - \log_q(1 + 2(q - 1)(1 - u))}{2}.$$

Hence the orthogonality relation of q-polynomials (3.133) takes the form

$$\int_{-\infty}^{\infty} \varrho_2^*(x;q,u)\bar\phi_n(x;q,u)\bar\phi_m(x;q,u)d_qx = \left(\prod_{j=1}^{n} C_{j,q}^{(u)}\int_{-\infty}^{\infty}\varrho_2^*(x;q,u)d_qx\right)\delta_{n,m},$$

where the $\{C_{j,q}^{(u)}\}$ are computed as

$$C_{j,q}^{(u)} = [q^{j+1}(q^{2j}(1+2(1-u)(q-1))(2-2q)\sigma_j + 2q^{j+2}(q-1) - 2q^2(q-1)\sigma_{j-1}))]$$

$$/[q^4 + q^{4j}(1+2(1-u)(q-1))^2 - (q^3+q)q^{2j}(1+2(1-u)(q-1))].$$

Therefore, to obtain the norm square value, it remains to compute the q-integral

$$\int_{-\infty}^{\infty}\varrho_2^*(x;q,u)d_qx = \int_{-\infty}^{\infty}\frac{x^{\log_q\left(\frac{1+2(q-1)(1-u)}{q^2}\right)}}{(-\frac{2(q-1)}{(1+2(q-1)(1-u))x^2};q^2)_\infty}d_qx. \tag{3.141}$$

Here we can again use the Ramanujan identity for computing (3.141) to directly obtain

$$\int_{-\infty}^{\infty}\varrho_2^*(x;q,u)d_qx = 2(1-q)\sum_{n=-\infty}^{\infty}\frac{q^{n\left(\log_q\left(\frac{1+2(q-1)(1-u)}{q^2}\right)+1\right)}}{\left(-\frac{2(q-1)}{1+2(q-1)(1-u)}q^{-2n};q^2\right)_\infty}$$

$$= 2(1-q)\sum_{n=-\infty}^{\infty}q^{n\left(\log_q\left(\frac{1+2(q-1)(1-u)}{q^2}\right)+1\right)}\frac{\left(-\frac{2(q-1)}{1+2(q-1)(1-u)};q^2\right)_{-n}}{\left(-\frac{2(q-1)}{1+2(q-1)(1-u)};q^2\right)_\infty}$$

$$= h_2 \sum_{n=-\infty}^{\infty} q^{n\left(\log_q\left(\frac{1+2(q-1)(1-u)}{q^2}\right)+1\right)} \left(-\frac{2(q-1)}{1+2(q-1)(1-u)}; q^2\right)_{-n}$$

$$= h_2 \sum_{n=-\infty}^{\infty} q^{n\left(-\log_q\left(\frac{1+2(q-1)(1-u)}{q^2}\right)-1\right)} \left(-\frac{2(q-1)}{1+2(q-1)(1-u)}; q^2\right)_{n}$$

$$= h_2 \Psi\left(-\frac{2(q-1)}{1+2(q-1)(1-u)}, 0; q^2; q^{\left(-\log_q\left(\frac{1+2(q-1)(1-u)}{q^2}\right)-1\right)}\right),$$

where $h_2 = \dfrac{2(1-q)}{\left(-\frac{2(q-1)}{1+2(q-1)(1-u)}; q^2\right)_{\infty}}$.

Corollary 3.6 *The polynomial set* $\{\bar{\phi}_k(x; q, u)\}_{k=0}^{N}$ *is finitely orthogonal with respect to the weight function* $\varrho_2^*(x; q, u)$ *on* $(-\infty, \infty)$ *if and only if* $N < \dfrac{1-\log_q(1+2(q-1)(1-u))}{2}$, *so that we have*

$$\int_{-\infty}^{\infty} \varrho_2^*(x; q, u)\bar{\phi}_n(x; q, u)\bar{\phi}_m(x; q, u)d_q x =$$

$$\left(\frac{2(1-q)}{\left(-\frac{2(q-1)}{1+2(q-1)(1-u)}; q^2\right)_{\infty}} \Psi\left(-\frac{2(q-1)}{1+2(q-1)(1-u)}, 0; q^2; q^{\left(-\log_q\left(\frac{1+2(q-1)(1-u)}{q^2}\right)-1\right)}\right) \prod_{j=1}^{n} C_{j,q}^{(u)}\right) \delta_{n,m}.$$

For instance, the polynomial set $\{\bar{\phi}_k(x; 0.5, 256)\}_{k=0}^{N=4}$ is finitely orthogonal with respect to the weight function $\dfrac{x^6}{\left(\frac{x^{-2}}{256}; \frac{1}{4}\right)_{\infty}}$ on $(-\infty, \infty)$.

References

1. R. Álvarez-Nodarse, J.C. Medem, q-classical polynomials and the q-Askey and Nikiforov–Uvarov tableaus. J. Comput. Appl. Math. **135**(2), 197–223 (2001)
2. T.S. Chihara, *An Introduction to Orthogonal Polynomials* (Courier Corporation, North Chelmsford, 2011)
3. J.S. Christiansen, The moment problem associated with the Stieltjes–Wigert polynomials. J. Math. Anal. Appl. **277**(1), 218–245 (2003)
4. R. Koekoek, P.A. Lesky, R.F. Swarttouw, *Hypergeometric Orthogonal Polynomials and Their q-Analogues*. Springer Monographs in Mathematics (Springer, Berlin, 2010)
5. M. Masjed-Jamei. A basic class of symmetric orthogonal functions using the extended Sturm-Liouville theorem for symmetric functions. J. Comput. Appl. Math. **216**(1), 128–143 (2008)
6. T.J. Stieltjes, *Recherches Sur les Fractions Continues*. Annales de la faculté des sciences de Toulouse, 8(1894), J1–122; 9(1895), A1–47; Qeuvres, vol. 2, pp. 398–566
7. S. Wigert, *Sur les Polynomes Orthogonaux et l'approximation des Functions Continues* (Almqvist & Wiksell, Stockholm, 1923)

Further Reading

- A.K. Agarwal, E.G. Kalnins, W. Miller, Jr., Canonical equations and symmetry techniques for q-series. SIAM J. Math. Anal. **18**(6), 1519–1538 (1987)
- R. Álvarez Nodarse, M.K. Atakishiyeva, N.M. Atakishiyev, A q-extension of the generalized Hermite polynomials with the continuous orthogonality property on $\mathbb{R}$. Int. J. Pure Appl. Math. **10**(3), 335–347 (2004)
- R. Álvarez Nodarse, N.M. Atakishiyev, R.S. Costas-Santos, Factorization of the hypergeometric-type difference equation on non-uniform lattices: dynamical algebra. J. Phys. A **38**(1), 153–174 (2005)
- R. Álvarez Nodarse, R. Sevinik Adigüzel, H. Taşeli, On the Orthogonality of q-Classical Polynomials of the Hahn Class. SIGMA **8**, 042 (2012)
- I. Area, M. Masjed-Jamei, A symmetric generalization of Sturm–Liouville problems in q-difference spaces. Bull. Sci. Math. **138**(6), 693–704 (2014)
- I. Area, M. Masjed-Jamei, A class of symmetric q-orthogonal polynomials with four free parameters. Bull. Sci. math. **141**, 785–801 (2017)
- I. Area, E. Godoy, A. Ronveaux, A. Zarzo, Inversion problems in the q-Hahn tableau. J. Symbolic Comput. **28**(6), 767–776 (1999)
- I. Area, E. Godoy, P. Woźny, S. Lewanowicz, A. Ronveaux, Formulae relating little q-Jacobi, q-Hahn and q-Bernstein polynomials: application to q-Bézier curve evaluation. Integr. Transf. Spec. Funct. **15**(5), 375–385 (2004)
- M. Arik, M. Mungan, q-oscillators and relativistic position operators. Phys. Lett. B **282**(1), 101–104 (1992)
- M. Arik, G. Übel, M. Mungan, q-oscillators, the q-epsilon tensor and quantum groups. Phys. Lett. B **321**(4), 385–389 (1994)
- R. Askey, S.K. Suslov, The q-harmonic oscillator and an analogue of the Charlier polynomials. J. Phys. A **26**(15), L693–L698 (1993)
- R. Askey, S.K. Suslov, The q-harmonic oscillator and the Al-Salam and Carlitz polynomials. Lett. Math. Phys. **29**(2), 123–132 (1993)
- R. Askey, N.M. Atakishiyev, S.K. Suslov, An analog of the Fourier transformation for a q-harmonic oscillator, in *Symmetries in Science, VI (Bregenz, 1992)* (Plenum, New York, 1993), pp. 57–63
- N.M. Atakishiev, S.K. Suslov, A realization of the q-harmonic oscillator. Teoret. Mat. Fiz. **87**(1), 154–156 (1991)
- N.M. Atakishiyev, A.U. Klimyk, K.B. Wolf, A discrete quantum model of the harmonic oscillator. J. Phys. A **41**(8), 085201 (2008)
- C. Berg, A. Ruffing, Generalized q-Hermite polynomials. Comm. Math. Phys. **223**(1), 29–46 (2001)
- J.S. Christiansen, E. Koelink, Self-adjoint difference operators and classical solutions to the Stieltjes–Wigert moment problem. J. Approx. Theory **140**(1), 1–26 (2006)
- A. De Sole, V.G. Kac, On integral representations of q-gamma and q-beta functions. Rend. Mat. Acc. Lincei **9**, 11–29 (2005)

- N.J. Fine, *Basic Hypergeometric Series and Applications*. Mathematical Surveys and Monographs, vol. 27 (American Mathematical Society, Providence, 1988)
- R. Floreanini, L. Vinet, $\mathcal{U}_q$ (sl(2)) and q-special functions, in *Lie Algebras, Cohomology, and New Applications to Quantum Mechanics*, Springfield, 1992. Contemporary Mathematics, vol. 160 (American Mathematical Society, Providence, 1994), pp. 85–100
- R. Floreanini, L. Lapointe, L. Vinet, A quantum algebra approach to basic multivariable special functions. J. Phys. A **27**(20), 6781–6797 (1994)
- G. Gasper, M. Rahman, Basic hypergeometric series, in *Encyclopedia of Mathematics and Its Applications*, 2nd edn., vol. 96 (Cambridge University Press, Cambridge, 2004)
- V. Gupta, P.N. Agrawal, D.K. Verma, A q-analogue of modified beta operators. Rocky Mt. J. Math. **43**(3), 931–947 (2013)
- W. Hahn, Über Orthogonalpolynome, die q-Differenzengleichungen genügen. Math. Nachr. **2**, 4–34 (1949)
- M.E.H. Ismail, Classical and quantum orthogonal polynomials in one variable, in *Encyclopedia of Mathematics and Its Applications*, vol. 98 (Cambridge University Press, Cambridge, 2005)
- M.E.H. Ismail, D.R. Masson, q-Hermite polynomials, biorthogonal rational functions, and q-beta integrals. Trans. Amer. Math. Soc. **346**(1), 63–116 (1994)
- F.H. Jackson, On q-definite integrals. Q. J. Pure Appl. Math. **41**, 193–203 (1910)
- A. Jirari, *Second-Order Sturm–Liouville Difference Equations and Orthogonal Polynomials*, vol. 542 (American Mathematical Society, Providenc, 1995)
- V. Kac, P. Cheung, *Quantum Calculus*, Universitext (Springer, New York, 2002)
- H.T. Koelink, On Jacobi and continuous Hahn polynomials. Proc. Amer. Math. Soc., **124**, 997–898 (1996)
- T.H. Koornwinder, Orthogonal polynomials in connection with quantum groups, in *Orthogonal Polynomials*, Columbus, OH, 1989. NATO Advanced Science Institutes Series C: Mathematical and Physical Sciences, vol. 294 (Kluwer, Dordrecht, 1990), pp. 257–292
- T.H. Koornwinder, Compact quantum groups and q-special functions, in *Representations of Lie Groups and Quantum Groups*, Trento, 1993. Pitman Research Notes in Mathematics Series, vol. 311 (Longman Scientific & Technical, Harlow, 1994), pp. 46–128
- A.J. Macfarlane, On q-analogues of the quantum harmonic oscillator and the quantum group SU(2)$_q$. J. Phys. A **22**(21), 4581–4588 (1989)
- F. Marcellán, J.C. Medem, q-classical orthogonal polynomials: a very classical approach. Electron. Trans. Numer. Anal. **9**, 112–127 (1999)
- M. Masjed-Jamei, F. Soleyman, I. Area, J.J. Nieto, Two finite q-Sturm-Liouville problems and their orthogonal polynomial solutions. Filomat **32**(1), 231–244 (2018)
- J.C. Medem, R. Álvarez-Nodarse, F. Marcellán, On the q-polynomials: a distributional study. J. Comput. Appl. Math. **135**(2), 157–196 (2001)

- S. Odake, R Sasaki, q-oscillator from the q-hermite polynomial. Phys. Lett. B **663**(1), 141–145 (2008)
- A. Ronveaux, A. Zarzo, I. Area, E. Godoy, Bernstein bases and Hahn–Eberlein orthogonal polynomials. Integral Transform. Spec. Funct. **7**(1–2), 87–96 (1998)
- R. Sevinik Adigüzel, On the q-analysis of q-hypergeometric difference equation, Doctoral Dissertation, Middle East Technical University, Ankara (2010)
- F. Soleyman, M. Masjed-Jamei, I. Area, A finite class of q-orthogonal polynomials corresponding to inverse gamma distribution. Anal. Math. Phys. **7**, 479–492 (2016)
- J. Thomae, Beiträge zur Theorie der durch die Heinesche Reihe: $1 + ((1 - q^{\alpha})(1 - q^{\beta})(1 - q)(1 - q^{\gamma}))x + \cdots$ darstellbaren functionen. J. Reine Angew. Math. **70**, 258–281 (1869)
- N.Ja. Vilenkin, A.U. Klimyk, *Representation of Lie Groups and Special Functions. Vol. 3*. Mathematics and Its Applications (Soviet Series), vol.75 (Kluwer, Dordrecht, 1992)

Index

Printed in the United States
By Bookmasters